Praise for *Death at SeaWorld*

"A chilling depiction . . . Kirby lays out a compelling scientific argument against killer-whale captivity."
—*New Scientist*

"As David Kirby so eloquently documents in this timely work, killer-whale captivity only benefits the captors. It is impossible to read *Death at SeaWorld* and come to any other conclusion." —Jane Goodall, Ph.D., D.B.E., founder, the Jane Goodall Institute, UN Messenger of Peace

"Entertaining, engaging, and enraging. The fairy-tale fantasy that the captive marine mammal industry has spun for the unwary public is expertly unraveled in this nonfiction crime thriller." —Louie Psihoyos, Academy Award–winning director of *The Cove*

"In this authoritative and superbly investigative page-turner, certain to ruffle feathers and fins, David Kirby . . . reports brilliantly on the escalating troubles and conflicts, the surprising and sordid underbelly of life—and death—at SeaWorld." —Erich Hoyt, author of the bestselling classic *Orca: The Whale Called Killer*

"*Death at SeaWorld* is one of the most important books, if not the most important book, ever written on the horrific plight of captive cetaceans. Kirby systematically dismantles the arguments used to justify keeping these incredibly intelligent and sentient beings in aquatic cages."
—Marc Bekoff, University of Colorado, author of *The Emotional Lives of Animals*

"This is a book everyone should read. . . . David Kirby's *Death at SeaWorld* outlines in grim detail just how bad captivity is for orcas and other marine mammals." —Richard O'Barry, director of Earth Island Institute's Dolphin Project and star of *The Cove*

"At last, both sides of the story behind the events at SeaWorld are being told and the truth is finally getting out there. Every budding orca trainer should consider this the must-read book of their career."
—Dr. Ingrid N. Visser, founder and principal scientist, Orca Research Trust

"One helluva book! David Kirby provides the most complete and accurate account of what I perceive as a transgression of morality toward the animal kingdom—the slavery of orcas, supreme beings in the aquatic world."
—Ken Balcomb, director, Center for Whale Research

"David Kirby's research is impeccable and his words unforgettable. You will never view dolphin and orca shows the same way again."

—Lori Marino, Ph.D., senior lecturer in neuroscience and behavioral biology, Emory University

"This remarkable book deserves to be acknowledged as the most significant and moving account of the often disastrous interaction of cetaceans and humans since *Moby-Dick*." —Richard Ellis, author of *Men and Whales, The Empty Ocean,* and *The Great Sperm Whale*

"Kirby shows that the reality (of orca captivity) is more akin to a circus, in which any benefits are outweighed by the cost to the whale—and sometimes to the keepers."

—*Financial Times*

"Thorough and disturbing . . . One of the great books of the summer."

—*The Columbus Dispatch*

"SeaWorld got a firm slap in the form of journalist David Kirby's fascinating and deeply disturbing book." —*The Christian Science Monitor*

"This is a warning-bell book on par with Upton Sinclair's *The Jungle* or Jessica Mitford's *The American Way of Death*. . . . After Kirby's brutal and groundbreaking work, we can't say we weren't warned."

—*The National* (UAE)

"An informed narrative that strongly suggests that despite their name, only when captured do the mammals become dangerous to humans. Free Willy, indeed." —New York *Daily News*

"Kirby has done his homework and does an excellent job of educating the public about orcas in the wild, as well as highlighting the dangers inherent in keeping these highly evolved, intelligent animals in captivity."

—Examiner.com

"Award-winning author David Kirby's book *Death at SeaWorld* meticulously chronicles the miserable lives and deaths of marine mammals in captivity."

—*The Sacramento Bee*

"Eye-opening poolside reading . . . Death isn't supposed to pop up in environments carefully choreographed for family fun."

—*San Francisco Bay Guardian*

"A new book examining the dark side of keeping killer whales in captivity has slammed SeaWorld for its treatment of the enormous beasts and for massive safety failings which still haunt the world-famous marine parks."

—*Daily Mail* (UK)

"A page-turning book . . . A disturbing account that will be hard for SeaWorld to transcend . . . Kirby makes it horrifyingly clear how serious [captivity] can be for human safety and orca well-being."

—*Wayne's Blog,* Wayne Pacelle,
CEO of The Humane Society of the United States

"Even if you're not an animal nut like me, David Kirby's *Death at Sea World* is a fascinating book."

—Sam Simon, cocreator of *The Simpsons*
and leading animal-rights activist

"The important and accurate information in this book is strong. . . . The hero of the book is Naomi Rose, whose doctoral research on wild orcas led to her current position as senior scientist at The Humane Society of the United States."

—*Charlotte Observer*

"Kirby's knockout format is articulate and mind-blowing. This riveting read is not one that will easily be dismissed."

—*Digital Journal*

"Lives are at stake here, and Kirby can be trusted to tell the story, having won a passel of awards for his investigative work."

—*Library Journal*

"Journalist Kirby offers another passionate industry exposé. . . . The narrative goes into high gear with its concluding confrontation."

—*Publishers Weekly*

ALSO BY DAVID KIRBY

Animal Factory

Evidence of Harm

DEATH
—AT—
SEAWORLD

SHAMU *and the* DARK SIDE *of* KILLER WHALES *in* CAPTIVITY

DAVID KIRBY

ST. MARTIN'S GRIFFIN
NEW YORK

TO CARLOS

and

THE WHALES

Printed in the United States of America. For information, address St. Martin's Press, 175 Fifth Avenue, New York, N.Y. 10010.

www.stmartins.com

The Library of Congress has cataloged the hardcover edition as follows:

Kirby, David, 1960–
Death at SeaWorld : Shamu and the dark side of killer whales in captivity / David Kirby. — 1st ed.
p. cm.
ISBN 978-1-250-00202-0 (hardcover)
ISBN 978-1-250-00831-2 (e-book)
1. Killer whale—Behavior. 2. Captive marine mammals—Behavior. 3. Killer whale—Florida—Orlando. 4. Animal attacks—Florida—Orlando. 5. Brancheau, Dawn, 1969–2010—Death and burial. 6. Sea World. 7. Aquatic animal welfare—Moral and ethical aspects—United States. 8. Rose, Naomi A. 9. Marine biologists—United States—Biography. 10. Humane Society of the United States. I. Title.
QL737.C432K577 2012
599.53'6—dc23

2012009433

ISBN 978-1-250-03125-9 (trade paperback)

10 9 8 7 6

Contents

Author's Note

Several attempts were made to reach out to display industry representatives and other experts who support killer whale captivity, including phone calls, e-mails, and queries posted on venues such as the online marine mammal list MARMAM. Few industry defenders were willing to speak on the record. SeaWorld turned down interview and fact-checking requests, citing the title of this book as well as *Huffington Post* commentaries and media appearances that I had done. I have tried to include pro-industry statements on a variety of controversies—when they were available in the public record. Former SeaWorld employees Mark Simmons, Thad Lacinak, and Don Goldsberry also declined interview requests, as did the family and friends of Dawn Brancheau. I have endeavored to treat her life and memory respectfully.

On another note, it is of course impossible to re-create verbatim the exact dialogue in unrecorded conversations that happened between characters years or decades ago. In some cases, what is written reflects the recollection of those present at the time.

Finally, the spelling of *SeaWorld* has changed over the years (it was once spelled *Sea World*); I have adopted the contemporary spelling throughout this book, for the sake of consistency. Likewise, though there are various spellings, I have chosen to use *Tilikum* and his nickname *Tilly.*

Introduction

There is something about killer whales.

Maybe it's their sheer size, some as big as school buses. It might be their force and power, their awesome ability to rocket into the air, or travel a hundred miles a day. It could be their masterful design, their hydrodynamic submarine-like forms, or those elegant black-and-white patterns, as if outfitted in tuxedos made of wet suits. Perhaps it's the fearsome name and ruthless, predatory reputation. Or maybe it's the whales' almost otherworldly intelligence, their sense of humor and play, their apparent love of sharing.

For many people, it is all of the above. They are, quite simply, mesmerized by *Orcinus orca.*

"Killer whales are the most amazing animals that currently live on this planet," wrote Robert Pitman, a leading US government marine ecologist not given to hyperbole, in the spring 2011 *Journal of the American Cetacean Society*. "They are probably the most universally recognizable animals that live in the sea, or perhaps anywhere on the planet. Add to that, they are predators nonpareil—the largest top carnivores on the earth today with killing power that probably hasn't been rivaled since dinosaurs quit the earth 65 million years ago."[1]

It's little wonder that orcas are popular, drawing millions of fans to marine-based theme parks each year, and a growing number of people to British Columbia, Washington State, Norway, New Zealand, and other places to see these majestic animals thrive in their natural habitat—the boundless ocean.

Few people realize that killer whales are members of the family Delphinidae, making them the planet's largest dolphins, giant cousins to the far more common white-sided, spotted, bottlenose (think TV's *Flipper*), and twenty-eight other species of seagoing dolphins. Orcas not only have the largest brain of any dolphin, at twelve pounds it is four times larger than the human brain (a great white shark's brain weighs 1.2 ounces). They are among the smartest animals in the world.

Killer whales have been prowling the earth's oceans for millions of years, and their large and complex brains have apparently continued to evolve over much of that time. They are the ocean's top predator and one of the most widely distributed animals on earth after humans. Unlike us, they fear nothing in the natural world.

Orcas have captivated the human imagination for millennia, and the first written reference to them dates back to AD 70 by Rome's Pliny the Elder, who thought them to be loathsome, "pig-eyed" assassins. Killer whales also figure prominently in indigenous people's myths and legends, including those of many Native American tribes of the Pacific Northwest, who revere the whales in their history, religion, and art. They call the orcas "blackfish," and the animal is a common motif in totems and other Native sculpture and drawings.

The Haida and Nootka tribes of British Columbia created legends about orcas living under the sea in homes and towns, taking human form beneath the waves. People who drown go to join them there. In much of the region's mythology, killer whales are believed to embody the souls of chiefs who have passed away. The Tulalip Tribes of Washington State have sagas about blackfish helping Tulalip people during famines, and they selected the orca as their tribal logo.

Scientifically, the animal is known by its genus *Orcinus*—from the Latin "kingdom of the dead," or belonging to Orcus, Roman god of the underworld—and its species *orca,* from the Latin and Greek for large whale or fish. *Killer whale* evolved from the term given to orcas by eighteenth-century sailors—*whale killer*—because some types of orcas feed upon other whales and dolphins. It is difficult to find a satisfactory explanation for why *whale killer* got reversed into *killer whale.*

In today's vernacular, the names *orca* and *killer whale* are interchangeable, though many in the media and animal activists seem to prefer the former, while scientists and the display industry tend to use the latter. Indeed, SeaWorld has chided reporters in the past for using the term *orca,* rightly pointing out that we do not call any other animal by its Latin species name alone. A human is sometimes referred to as *Homo sapiens* (our genus and species), but never just as "sapiens." SeaWorld's protestations aside, the word *orca* appears in dictionaries and stylebooks throughout the English-speaking world, where the species has also been known as grampus, sea wolf, and of course blackfish at various times.

The term *orca* began gaining popularity in the 1960s and 1970s, when people started realizing how intelligent the creatures are and how gentle they can be. Before orcas were first held captive, they were widely regarded as bloodthirsty monsters and brutal, shark-like killing machines.

"In whatever quarter of the world [killer whales] are found, they seem always intent upon seeking something to destroy or devour," wrote nineteenth-century whaler Charles Scammon. Swimmers feared them, fishermen hated them, many people fired weapons at them. Nearly one-quarter of all orcas captured for display during the late sixties and early seventies showed signs of bullet wounds. Royal Canadian fighter pilots used to bomb orcas during practice runs, and in 1960, private fishing lodges on Vancouver Island persuaded the Canadian government to install a machine gun at Campbell River to cull the orca population. In the end, it was never fired. Even as recently as 1973, US Navy diving manuals warned that these "extremely ferocious" predators "will attack human beings at every opportunity."

In the second half of the twentieth century, killer whales were targeted by the commercial whaling industry, which was running out of larger species to pursue. Between 1954 and 1997, Japanese whalers captured and slaughtered 1,178 orcas, while the Norwegians took 987. The former Soviet Union was responsible for taking more than 3,000 killer whales, many of them from antarctic waters.

Despite their name, teeth, and reputation, killer whales in nature are generally mild-mannered with people and with each other, aside from the occasional spat. Not until the captive marine mammal industry began to display orcas did we improve our understanding of their nature. The public display industry should be credited for changing our attitude toward killer whales from contempt to admiration and even affection.

There is little evidence of wild killer whales attacking people. An early expedition to Antarctica reported that orcas had unsuccessfully attempted to flip over an ice floe bearing a terrified group of men and their dog team, though the whales may have thought the barking dogs were some strange species of seal.

Decades later, in 1972, an orca bit down on the leg of a surfer in Big Sur, California. The animal probably confused the surfer's black wet suit for seal skin, until it got a taste and let go. Nonetheless, the victim required a hundred stitches. It was the only human injury ever recorded. In 2005, a twelve-year-old boy was bumped in the shoulder by a Transient orca while swimming in shallow water in Helm Bay, Alaska, home to many harbor seals. The whale presumably mistook the boy, who left the water uninjured, for prey. So wild killer whales rarely, if ever, cause deliberate harm to humans. Captive killer whales, however, are another animal altogether.

Killer whales are part of a large order of marine mammals known as cetaceans (*seh-TAY-shuns*) from the Greek word for whale, *kētos*. All cetaceans—which comprise whales, dolphins, and porpoises—have forelimbs modified into "flippers" (pectoral fins), a "tail" (fluke) that has been flattened horizontally, as opposed to the vertically arranged tails of fish, and one or two nostrils on the top of the head (blowhole), which provide the only air passage to the lungs. Cetaceans cannot breathe through their mouth.

There are three suborders of cetaceans: the Mysticeti, or baleen whales (blue, humpback, gray, etc.), which trap krill, other zooplankton, and small schooling fish in the long strands of brushlike baleen material that line their mouths; the Odontoceti, Latin for "toothed whales," which includes dolphins, porpoises, belugas, and others; and the Archaeoceti, or ancient whales, which are extinct.

Among the toothed whales, the largest is the sperm whale (*Physeter macrocephalus*), the leviathan immortalized in Herman Melville's *Moby-Dick*. The male of the species can grow up to sixty feet in length and weigh ninety thousand pounds (females are much smaller). One-third of his body is taken up by his head and its seventeen-pound brain, the largest on earth.

Other toothed whales include beaked whales, beluga whales, narwhals, and six species of porpoises. Many people think that porpoises and dolphins are the same animals, but they are not even in the same family. Porpoises tend to be shorter and stouter, with rounded rather than beaked or bottlenose rostrums. They are actually more closely related to belugas than dolphins.

The largest family of toothed whales, the Delphinidae, or oceanic dolphins, has some thirty-six species, including such popular theme park draws as the bottlenose dolphin, Pacific white-sided dolphin, short-finned pilot whale, false killer whale, and of course the killer whale, the second-largest toothed whale after the sperm whale.

The terminology gets a bit confused here. The word *whale* is an English common term, not a scientific one. It basically means "large cetacean." So even though killer whales, false killer whales, and pilot whales are all in the dolphin family, they are also relatively large cetaceans, and therefore called whales. And while the smaller dolphins (as well as porpoises) are cetaceans, they aren't large enough to generally be called whales.

The distinct black-and-white pattern of a killer whale helps camouflage the animal from prey swimming underwater, or perched above the sea on ice floes, beaches, and shoals. Like a military plane, the orca's white belly makes it harder to spot from below, and its dark dorsal side makes it more difficult to discern from above.

Orcas communicate through a complex and poorly understood system of vocalizations that are divided into two main types. They are, according to the Vancouver Aquarium web page:

Whistles: Killer whale whistles are used for communication and sound a lot like human whistles. They are continuous sounds and are referred to as pure tones. Their function isn't entirely known, but they seem to play a role in communicating the emotional state and location of individuals—as they do in humans.

Calls: Referred to technically as burst pulse calls. Killer whale calls are very rapid streams of sound pulses that sound continuous to our ears. Most sound somewhat like human cries or screams, some sound a bit like a squeaky door or creaking floorboard. Many of the calls used by killer whales are "stereotyped" or produced repeatedly by a given group of killer whales. Certain killer whales even use sound as a kind of family badge, and researchers have discovered much about their family relationships by simply listening to the sound of their calls.

The vast majority of data available on wild orcas was collected from the Northern and Southern Resident whale communities of British Columbia and Washington State. That's mostly because these animals swim close to shore, often near populated areas. Orcas can be spotted from the shores of Seattle, Tacoma, Port Angeles, Bellingham, and the popular San Juan Islands in Washington State; and Vancouver, Victoria, Nanaimo, Campbell River, and other cities in BC. These venues not only offer easy access to the whales, they are scenic and pleasant places to live: Researchers who study orcas tend to gravitate more toward this region than, say, Iceland.

As a result, we know a great deal about the Southern and Northern Resident communities, which today number some 353 in total,[2] and far less about other populations (though that is changing), including the Transient and Offshore orcas who share the same region of the Pacific Northwest as the Residents. While it is tempting to ascribe "Resident" and "Transient" attributes to killer whales in Norway, Iceland, Antarctica, and elsewhere, scientists say there are just not enough data to support such comparisons, with the exception of New Zealand and possibly Alaskan orcas.

For the past forty years, field scientists have exhaustively documented these Pacific Northwest animals, which they break down into three distinct *ecotypes:*

Residents: The most highly studied whales of all, these orcas are divided into two groups: the Northern Resident community, which ranges from mid–Vancouver Island north toward the Alaskan panhandle, and the Southern Resident community, which typically ranges from mid–Vancouver Island south to Puget Sound in summer and fall and, in the winter and spring, as far south as Monterey, California, and also north of Vancouver Island. The two generally do not mix. These whales live in extremely stable and large groups, or pods, marked by tightly knit family units dominated by females. They communicate at a highly sophisticated level and eat mostly fish.

Transients: These whales differ from Residents primarily by what they eat: other marine mammals, including dolphins, porpoises, seals, sea lions, and even larger whales (and sometimes sharks). Transients have been known to bat their prey around before finally consuming it. They travel in small groups (three to twelve whales), and their range is far greater than that of Resident orcas, though it does overlap. Transients do not mix with Residents, having split from their cousins, genetically speaking, tens of thousands of years ago.

Offshores: Little is known about this population, which tends to stay about thirty miles off the mainland coast, though they have been spotted in inland waterways on rare occasions. Offshores can travel in huge pods numbering up to seventy to a hundred orcas. Scientists believe they mostly forage on Pacific sleeper sharks and schooling fish.

In recent years, researchers have compiled new data on killer whale populations in other parts of the world, including Iceland, Alaska, Norway, the UK, Japan, and Russia in the northern hemisphere, and New Zealand, Argentina, Antarctica, and the Indian Ocean's Crozet Archipelago in the southern. There are several other ecotypes, and perhaps even separate species entirely, besides the Northwest American whales.

Orcas can be found in all oceans of the planet, even in the tropics. Various studies have estimated their total population at anywhere between fifty thousand and one hundred thousand, perhaps half of them around Antarctica.[3] Most populations seem to be stable (though data are limited), but the Northern Resident population of British Columbia has been listed as "threatened," and the Southern Residents have been added to the more serious "endangered" list. Part of the reason is a reduction in fish stocks, which environmentalists say is due to pollution, salmon farming, and the damming of rivers that wild salmon must navigate to spawn upstream. New Zealand orcas are also in peril.

Then there are the captive killer whales; minuscule in number when compared to their wild counterparts, each is a political and emotional lightning rod. As of this writing, forty-two captive orcas are at theme parks and aquariums in Canada, France, Spain, Japan, Argentina, and the United States, which has twenty-two whales—twenty of them at the SeaWorld chain of attractions in Orlando, San Antonio, and San Diego.[4]

This book illuminates the intensifying debate over keeping killer whales for "public display," and whether captivity is too stressful on some animals, leading to health problems such as impaired immunity, increased infections, and other serious issues, as well as behavioral problems such as aggression toward one another, and violence—at times deadly, as we shall see—against humans.

At play here are two vital questions:

1. Is captivity in an amusement park good for *orcas:* Is this the appropriate venue for killer whales to be held, and does it somehow benefit wild orcas and their ocean habitat, as the industry claims?
2. Is orca captivity good for *society:* Is it safe for trainers and truly educational for a public that pays to watch the whales perform what critics say are animal tricks akin to circus acts?

Not surprisingly, people who support SeaWorld and other marine-themed entertainment parks (*pro-caps* in the lingo of this particular argument) answer affirmatively to both questions, while *anti-caps* insist the answers are resounding noes.

People opposed to captivity include some scientists, academics, veterinarians, and environmentalists, nearly all animal activists, a handful of former orca trainers, and a worldwide network of people who say that killer whales are too big, smart, sentient, mobile, and close to their families to be kept in tanks and trained to perform for tourists. They assert that keeping killer whales in captivity is cruel and unusual, dangerous for animals *and* people, and should be phased out.

On the other side are aquarium and amusement park owners, managers, and investors, current and former trainers and staff, industry trade associations, some scientists and veterinarians, and most government officials, especially those whose constituents benefit from having a large oceanarium in the area. They argue that captive whales educate the public about wild whales, that the quality of life for captive orcas is superior to that in the ocean, and that whales in these collections receive world-class care, dine on "restaurant-quality fish," and are free from the worries of pollution and dwindling food supplies found in the wild. Captive orcas, they imply, are simply better off. They are supported by millions of fans who spend billions of dollars each year on ticket sales, food, beer, and merchandise at the parks.

One faction views SeaWorld as a Garden Hilton for killer whales, and the other views it as a Hanoi Hilton for killer whales.

Those divisions aside, people on both sides of this battle sincerely care about these animals, and many of them truly love orcas. One such person was orca-training supervisor Dawn Brancheau, who was living out her life's dream in Florida, working with nearly every "Shamu" (a stage name given to performing orcas) at SeaWorld Orlando. Dawn was killed during a "relationship session" with SeaWorld's twelve-thousand-pound bull, Tilikum, following the popular "Dine with Shamu" lunchtime show. The notorious whale had already been involved in two other deaths. Now he had claimed his third victim. "Tilly" brutally rammed, dunked, bruised, and lacerated his adoring trainer. Tilikum was not just "playing." This was a killing.

Four people have died in a pool with killer whales. Dozens more have been attacked, some left with lifelong injuries. SeaWorld calls these events rare accidents; critics call them preventable tragedies, the inevitable outcome of what they claim is the stress of captivity. Killer whale shows are not going to be closed down anytime soon, but opponents are pushing hard to convince the public that they are as outdated and inhumane as the circus dancing bears that still perform in parts of Russia and China.

What's more, the killing of Ms. Brancheau woke up a previously inattentive media to a gripping story and a bruising national debate—one that would soon drag the courts, Congress, and even the Obama administration into the roiling conflict.

Is captivity for orcas, on balance, a good thing? Readers must make up their own minds. But regardless of whether one is pro-cap, anti-cap, or somewhere in the ambivalent middle, one thing is abundantly clear: Dawn Brancheau's death at SeaWorld, on February 24, 2010, forever changed this emotionally charged debate.

Prologue

The young orca trainer, an attractive woman with a bright smile, enchanted the tourists who came to gawk at the killer whales on this cool and gloomy February day. They gasped in awe as the trainer, an athletic, hometown celebrity, sent the orcas flying into the air with a few discreet flicks of her hand.

By lunchtime, she would be dead.

For now, the obedient animals pumped their powerful flukes and hurtled themselves upward from the depths of their cold-water confinement, rocketing through the surface into elegant arabesques and water-pounding breaches. The killer whales leapt forward around the small pool in tight unified arcs—repetitive airborne maneuvers not typically seen in nature, but theatrically referred to as "bows" in marine-park parlance.

The whales swam with military-style precision. The two older, dominant females (who rule orca society), easily distinguished by their smaller size and more diminutive dorsal fins that curved rearward into a point, flanked the large adolescent male in the middle. His dorsal fin had once grown straight, on its way to a natural elevation of five or six feet above sea level. But captivity had caused the erect, triangular fin to topple over, the force of gravity having pulled the mighty appendage downward, folding it onto his back like a giant slab of black taffy cooling on the sill of a seaside candy factory.

The trainer went through her well-rehearsed paces, and so did the whales. They returned each time to the low-lying stage to collect fistfuls of thawed smelt scooped from a metal bucket—a reward for each properly executed "behavior," the industry name for an animal trick.

The audience cheered its approval, mesmerized by the black behemoths with the beguiling white patches next to those unknowable dark eyes, yards of glistening porcelain skin lining their enormous underbellies. Each time they surged from the water, people held their breath. There is nothing quite like seeing a live orca show. And today's had been a good one.

"Do you think it's over now, or do you think there's more?" Nadine Kallen, visiting from Calgary, asked her friend Corinne Cowell and sister Silvia, a student at a nearby university.[1]

"I don't know, but here she comes now," Silvia said of the trainer, who was ferrying a metal pail of fish toward the corner of the arena where the three women were watching the show.

"I am *so* jealous. I wish I had her job," said Nadine, a student at Alberta College of Art and Design. "Me, too," Silvia sighed.

The trainer was now offering some well-earned treats to the big male with the collapsed fin. This common postperformance ritual was another positive reinforcement telling the whale he'd done a good job at work—and to please keep it up. The "play session" was a trust-building gesture, an incentive after each show, like a tip given to a favorite waitress in anticipation of good service tomorrow.

The trainer held fifteen-inch-long herring over the water, and the hungry male popped up vertically through the surface to grab them. Nadine, Sylvia, and Corrine were six feet away; they could see the whale's conical teeth, pink, soft palate, and massive jaws, just inches from the trainer. But the women believed there was nothing dangerous about the move. They knew the term *killer whale* was an anachronistic misnomer from a less enlightened era of human misunderstanding about wildlife. After all, they had been to SeaWorld before; they had seen for themselves that people can swim with, surf on, and launch into the air from the heads of these gentle pandas of the sea. Orcas were docile as dalmatians.

Then it happened.

The pretty trainer began walking along a narrow ledge between the pool and a safety railing that kept the public from stumbling into the chilly salt water. The ledge, slippery from the show, was two feet above the surface. Suddenly, the trainer lost her balance and stumbled. One foot dipped into the brine as the opposite leg kept her body perched on the ledge. "Oh, no! She slipped!" Nadine cried.

The woman scrambled to get up. But one of the whales, the male, had another idea.

The instant he saw a foot break the surface, the male was riveted. He was not accustomed to seeing trainers in the water. This was an exciting development, and eight thousand pounds of curiosity got the best of him. Just as the trainer hauled herself up and pulled her foot from the brink, he grabbed it. The whale pulled her into the water. She cried out, more in surprise than pain. But it was too late. The orca had decided to deny access to the narrow ledge to safety. A new and amusing game had just presented itself, right there in his watery living room. He was determined to win it.

The trainer freed herself and swam toward the edge of the pool, but it was no use. There was no way to climb from the tank: no ladder, no foothold. Before she could cry out for help, the male grabbed her again and pulled her into the middle of the water.

Now the game had drawn the attention of the two females, who circled the skirmish with rascally delight, screeching in high-pitched bursts of clicks, crackles, and calls. But the big male had no interest in sharing this new toy with his bossy tankmates. He dragged the panicking trainer down to the bottom of the forty-eight-degree water and held her there.

"She's gone under!" Nadine cried. The three women held their breath as the trainer disappeared beneath the surface, which was dark and mottled under the gloomy winter sky.

Maybe it's okay, Nadine speculated. Maybe this is part of the performance: a short swim with the orcas, perhaps, at the end of the show.

Silvia had no idea. No one could see what was happening down there.

"Is she all right?" Corinne asked.

Soon they got their answer. The trainer pierced the dark surface and emitted a heart-rending scream. "Help me!" she cried to the other young trainers on duty. "Karen! Please! Help me. My god, help me!" she shouted to her colleague, twenty-five-year-old Karen McGee, who, along with other staff now on the scene, tried everything they could to distract the whales and rescue their imperiled friend.

Someone deployed a shepherd's hook attached to a long pole for the trainer to grab on to, but the whales, as though they knew if she reached the device, it would be game over for them, kept her away from the hook. They also kept her from a life ring tossed into the pool.

Some staff tried hand signals in a vain attempt to command the orcas to return to the stage. Others used paddles to slap the water's surface—an audio signal telling them to do the same thing—or banged metal fish buckets to distract the animals. On occasion, a whale would respond to a trainer's "stationing cue," specifically designed to pacify marine mammals that grow too aggressive and keep them immobile. But their obedience was only momentary.

The staff also tried luring the whales into their overnight holding pen (a dark metal tank inside an enclosure called the module) with a proffering of filleted salmon, but the orcas had clearly decided that they and they alone would dictate the endgame of this electrifying new sport.

The orca grabbed his trainer again and yanked her back under. Many seconds went by. He resurfaced, bouncing the woman on his rostrum (where an orca's nose would be, if it had one) like a giant seal with a beach ball. The victim continued to cry out. The two chittering females circled the water with intensifying interest. Now all three whales joined in, handing off their screeching plaything like a human rugby ball.

The rescue team members, who seemed unequipped to deal with such an emergency and were now shouting in panic, made a last-ditch and futile effort at deploying a large net across the pool to separate the rampaging orcas from the victim, who was losing strength by the minute.

Instead, the male grabbed the woman once again and dove to the bottom, chased by the two females. They stayed under a long time. Nadine, Sylvia, and Corrine craned their necks to see. A heavy quiet fell on the arena. Gone were the woman's screams and the shouts of the staff, the splashing and thrashing in the water, the god-awful screeching of orcas. Nothing was left but stunned silence, broken by heavy breathing and someone sobbing softly in the distance. The whole drama lasted maybe fifteen minutes, but it seemed like a lot longer.

Security staff finally escorted the dozen or so guests left away from the horror in the pool. Nadine, Corrine, and Sylvia refused to leave the pool area and instead huddled outside the gate, anxiously awaiting word on the trainer's fate.

It seemed to take forever to retrieve the corpse: The male had refused to relinquish his trophy. Finally, a weighted net dragged the trainer up from the depths. Her clothes had been ripped from her body, which was peppered with ten lacerations from the teeth of killer whales.

The three women peered through the gates as emergency technicians pulled the trainer's naked body up from the water. There was no attempt at resuscitation. Nadine had never seen anyone die before. She was only eighteen.

As the three women prepared to leave the park in shock and sadness, TV news crews had already arrived. One woman was trying to hawk her home video of the killing to local TV crews. Another approached grief-stricken staffers to ask if the gift shop would still be open.

The date was February 20, 1991, and the place was SeaLand of the Pacific, just east of Victoria, British Columbia. The victim was Keltie Lee Byrne, twenty-four, a champion swimmer and seasoned athlete. The female orcas were Haida II, named for the Haida Nation of the Pacific Northwest, and Nootka IV, Nootka being an old European name for the Nuu-chah-nulth people, whose territory encompasses Nootka Sound, a spectacular fjord carved into the west coast of Vancouver Island.

The leading perpetrator, the big male orca with the collapsed fin, was named Tilikum, a word from the Chinook language meaning "friend" or "friendly people."

It was the first time that anyone had been killed by a killer whale in captivity (or anywhere else), but it would certainly not be the last. Three more people would die over the next two decades; Tilikum would be responsible for two of them, twenty-six hundred miles away in SeaWorld Orlando.

Nineteen years later, almost to the date, he would savagely attack and dismember beloved orca trainer Dawn Brancheau. The killing would throw coals on the increasingly heated argument over whether it is appropriate, ethical, and safe to confine the ocean's top predators—highly intelligent, mobile, and family-bound animals—to what is, essentially, an Olympic-size swimming pool dug far away from their native waters and free-swimming kin.

The pleasant city of Victoria, British Columbia, clings like a cockle to the bottom of expansive Vancouver Island, North America's largest west-coast land mass stretching 290 miles over large tracts of alpine wilderness, despoiled by intermittent rectangles of deforested earth.

Back in 1991, the northern half of the island was relatively undeveloped, with an eerie end-of-the-earth feel deepened by the wisps of gray mist that swirl around the lonely granite peaks and filter through the thick stands of cedar and western fir that rise along the lower slopes.

At the time, the north island was a destination almost exclusively for those who wished to flee the world, fish for salmon, log timber, commune with Native people, or observe wildlife up close—especially killer whales. There was, quite literally, little else to do.

In February of 1991, marine biology graduate student Naomi Rose, twenty-eight, was holed up in a deserted north-island fishing lodge perched on low, wooded bluffs overlooking an isolated inlet 240 miles from Victoria's SeaLand. The appropriately named Hidden Cove Lodge is tucked into the island's northeastern coast, where the broad Queen Charlotte Strait funnels south into the river-like Johnstone Strait, which separates Vancouver Island from the Broughton Archipelago—a dense warren of forested islets—and the mainland coast.

Naomi was lodge-sitting. She had agreed to guard the place—a contemporary wooden-beam-and-glass inn with a cathedral-ceilinged great room that peers out over the chilly cove—against vandals and teenagers throughout the lonely winter. In exchange, she received weekly provisions, fuel to run the generator five hours per day, and an unperturbed place to analyze the whale data she'd collected. Naomi had isolated herself in this remote corner of Canada to complete the number crunching required for her dissertation, "The Social Dynamics of Male Killer Whales, *Orcinus orca,* in Johnstone Strait," which she was preparing for her PhD in biology from the University of California at Santa Cruz.

Naomi was utterly alone. At this time of year, two dozen people, maximum, might be staying within a twenty-minute boat ride of the lodge. A car ride was out of the question: There were no roads to Hidden Cove, no way in or out, except by vessel over the often-turbulent inland waterway.

Naomi didn't mind being alone. Every year since 1986 she had ventured to Johnstone Strait to observe killer whales up close in their native habitat—in particular the sixteen distinct pods that travel through these northern waters every summer in search of their favorite meal, chinook salmon. Each year, Naomi had camped in close quarters with a half dozen or so other students on the rocky shoreline of West Cracroft Island, about eleven miles south of Hidden Cove. Under those conditions, six easily became a crowd, and Naomi would wander off down the cliffs for some peaceful isolation, alone with her thoughts and perhaps a flying squirrel or two.

But those visits came in the summer months, when Johnstone Strait bustles with pleasure craft, fishing boats, and whale-watching tours. Now, however, Naomi was by herself on a silent cove in the bowels of winter. The low-hanging fog and freezing drizzle only magnified her isolation.

She had been warned. The two-story lodge had been built by Dan Kirby, a local entrepreneur and architect of no-nonsense Canadian-Irish stock who based his blueprints on what he calls a "two-beer" design. ("I sat down, drank two beers, and designed it.") Dan was skeptical about hiring this single young female to take care of his property over the winter. When Naomi met with him, he sized her up with an odd mix of curiosity and skepticism. Who *was* this short (five-feet-two-inch) Asian-American (Japanese-Korean mother, Jewish-American father) woman, from *California*? Was she really cut out, physically and mentally, for a protracted winter solo in Hidden Cove?

"Do you even know how to chop wood?" Dan politely queried. "We have a wood-burning stove. It's the only source of heat."

Naomi was unfazed. "I chopped plenty of wood on West Cracroft, Dan," she said flatly.

"So you think you know how to live in the woods?"

"I *have* lived in the woods."

"Yes," he said, "but that was in the summertime, when the strait is full of people and you don't freeze to death. You *do* realize there will be very few boats out on the water, right? You're on your own now. And when you go out on the water, take a radio with you. You need a life jacket and a bailing can, too. If you break down or hit a shoal, nobody's going to come by and rescue you." And, he added, not everyone could live in the big lodge all alone. Some had fled in a cold sweat, complaining of odd creaks, groans, and other ghost-like occurrences.

Naomi laughed. But just two months into her gig, she began showing signs of an unsettling syndrome that can afflict lone caretakers at secluded outposts: a restless cabin fever, aggravated by the deeply disturbing sensation that one is, perhaps, not quite as alone as one had imagined.

The tough young graduate student was secretly haunted by the feeling that someone might be out there, lurking in the black Canadian night. Some evenings, she preferred to turn off the generator and sit in the pitch dark, rather than run the risk of attracting an observer, staring at her every move through the picture windows.

One night in December, Dan Kirby and his wife, Sandra, found Naomi alone in the lodge without a single light on. Dan knew deep-woods heebie-jeebies when he saw them, and he staged an intervention of sorts, diagnosing his caretaker with "being bushed" (British Columbian for forest jitters) and boating her back to his well-lit house in Port McNeill for dinner and a mug of warm cider, followed by a few days of Christmastime relaxing with friends in town.

In late February, Naomi was getting ready to finish up her data analysis on whale behavior and head back down to Santa Cruz, which seemed like midtown Manhattan compared to this place. Her studies were finished, the five summers of observing orcas in Johnstone Strait were over, and—like most people who spend a decent amount of time close to killer whales—her life would never be the same.

Naomi was about to become a full-fledged scientist, but she was also a human being, with the entire range of emotions and reactions that goes with our complex species. We can't help but have a physical, even jolting reaction when encountering certain phenomena, and not just food, danger, or sex. We humans—even scientists—respond to beauty, majesty, power, size, and intelligence, and no animal possesses all those things more than *Orcinus orca*.

Naomi had seen killer whales before, at SeaWorld, Marineland, Vancouver Aquarium, and other venues, and she had taken in the orca shows with a mixture of scientific detachment and entertainment-park amusement. She was interested in, but hardly moved by, captive orcas. But SeaWorld was nothing like seeing killer whales in their natural habitat, surrounded by the unspeakable beauty of Johnstone Strait, with its countless rocky islands, dense forests, and snow-covered peaks that rise from the distant mist.

The mist: For a while it seemed to be permanent. When Naomi first arrived on West Cracroft Island, in June of 1986, the strait was socked in for several days. She couldn't even see the water, let alone any whales.

But she could certainly *hear* them, maybe fifty feet offshore, breaking the foggy silence with powerful bellows of air and vapor bursting from their blowholes. Listening to one of these colossal mammals come up to breathe is almost as electrifying as seeing it. The sense of size and power one gets simply from the *pah-WOOSH* of exhalation is exhilarating.

But Naomi had come to Canada to *see* wild whales, not just listen to them. She waited for the skies to clear.

"Whales today!" the first-year grad student wrote in her journal the night of June 28, 1986. "And very close up, too." She had been out with her fellow field assistants from UCSC, at the tiller. The mist had finally lifted, revealing a pale blue sky hovering over water as smooth, green, and glassy as a 7UP bottle. Naomi's heart raced as the little boat pulled tight alongside members of the A pod, one of the largest clans of killer whales in the Pacific Northwest.

There were several adult females with their calves, a few of them recently born infants. There were frisky adolescents and a number of full-grown males whose straight, tall dorsal fins towered like black sails high over the boat, which was dwarfed by the whopping multi-ton males.

On this morning, A pod was busy socializing as its members swam around in a small area, rubbing against each other, vocalizing wildly, and seeming to have the time of their lives. If killer whales throw summertime parties, this was surely one of them.

When orcas are in a frolicking mood, anything can happen. Every so often and without warning, a whale would career from the sea just yards from the boat—a massive explosion of black and white dripping in salty froth—before crashing back down from its breach, creating a thunderous clap that could be heard for miles around on such a calm day. Naomi's jaw was agape at so many whales amusing themselves around the boat when two of the infants, known as A49 and A45, leapt from the water not ten feet from her, dousing the boat's occupants with the icy splash of their combined downward thrust of a thousand pounds. The young calves, with two kicks of their flukes, playfully swam away, leaving behind a wake that rocked the boat like a leaf in a rapid.

"Quite a shocker!" Naomi marveled in her journal. "It was so amazing knowing who they were and following them so closely. It was quite startling to suddenly have A45 right next to us. I started moving away, but it turns out that running the engine, even a little, is the worst move. So I turned it to idle after the others shouted at me to stop. It was really something."

What struck Naomi that first day was the velocity of these animals. It was snail-paced compared to bottlenose dolphins, which dash about like frenetic teenagers on speed in a flash of quicksilver. Instead, these giant creatures seemed to move around deliberately, calm and unhurried. "It's all very majestic," she told her colleagues over a late-evening campfire under the stars. "It's like watching them in slow motion, comparatively speaking with dolphins. It's totally emotional. *Very* visceral."

One summer, when Naomi and some of her field assistants were following whales along the coast of West Cracroft, they came across a lone female whose rostrum was crammed up against a crevice in the wall of the shoreline. She was fishing. The whale had chased a large salmon into the crevice, which she had completely blocked with her head, depriving the fish of circulating seawater to replace the dwindling oxygen supply in what had become its rocky death chamber. Naomi had heard of this cunning practice before, which is apparently unique to Johnstone Strait orcas, but she had never dreamed she would actually see the inventive use of a craggy cliff as a tool. How on earth did this female learn to deprive fish of oxygen? Her mother must have taught her, Naomi surmised.

"Drop me off over by those rocks," she said. "I want to take a closer look."

Naomi climbed from the boat and tiptoed over to a ledge above where the whale was patiently awaiting her meal. Naomi had never been so close to a wild orca before, maybe five feet away, almost close enough to reach out and touch the animal—an understandable urge that she subdued.

Two things overwhelmed the grad student that day. One was the sheer mass of the creature—from this close, she seemed as big as an airplane, her enormous ebony back rising up over her bulbous head like the front of a 747. Naomi was in awe.

The other amazing thing was the whale's utter awareness. She had watched Naomi approach, but did not move: The whale was far too intent on wearing down her oxygen-deprived prey, waiting for the dazed fish to come bumbling from its crevice and into her mouth. Naomi figured that this ingenious form of foraging was not only a low-energy enterprise for the whale, it was probably entertaining as well.

Naomi crouched down, but the orca did not flinch. Instead, her big black eye, the size of a cow's, met Naomi's in an instant flash of interspecies recognition, perhaps even communion. The eye stared back at Naomi, who took the look to mean something like, "Don't think I don't notice you there. I do. But I have to concentrate on catching my lunch, so don't interrupt."

To the marine biologist, it was an extraordinary moment. This animal had more than mere intelligence. She had a consciousness. She had *opinions*. And, Naomi thought, the killer whale had no interest in harming her.

There's a *mind* down there in the water, Naomi marveled to herself later that day at camp. The way she felt when she saw these killer whales up close in the wild, well, it was hard to explain. She later described the emotion, but she couldn't express why she felt it.

"Why don't you feel that way when you see this really cool salmon leap from the water? Why don't you feel that way when you see a giant Steller's sea lion hanging out on a rock, or even a gray whale or a humpback breaching in the ocean? It's not the same. There's just not the same awe. With orcas, there's a lot of awe."

Naomi knew that, after humans, killer whales are the most socially and ecologically complex animals on earth, and certain types of orcas are the most socially stable animals of all. No other species but *Homo sapiens* is so diverse in its rules and traditions governing such things as diet, mating, family relations, group size, foraging, or communications. It's why renowned whale scientists Hal Whitehead and Luke Rendell concluded in their research, "The complex and stable vocal and behavioral cultures of . . . killer whales appear to have no parallel outside humans and represent an independent evolution of cultural faculties."[2] That's right: Orcas have their own *cultures*.

This was certainly true for the fish-eating and highly social Northern Resident orca community that Naomi chose to study. The Northern Residents inhabit the inland waters of northern Vancouver Island in the summer months. At the time, they numbered 185 animals.

Resident orca communities, dominated by females, are populated by intensely social whales that travel in large, stable groups centered on a matriarch, typically the oldest living female. Each pod has its own signature collection of clicks, whistles, creaks, and groans, though some vocalizations overlap among pods. They mainly eat fish and are particularly fond of chinook salmon, which are large and rich in energy: more bang for the whale's caloric buck. Thanks to their highly sophisticated echolocation (a form of sonar), orcas can distinguish a species of salmon by its size, or by echolocating inside the fish's body to determine the dimensions of its air bladder.

"Residents travel in matrifocal [centered on the mother] units called matrilineal groups," Naomi wrote in her dissertation. "A matrilineal group usually consists of a reproductive female (the matriarch), her dependent calves, her juvenile and adolescent offspring and her known or presumed adult son(s)," she said. "A matrilineal group can also consist of a post-reproductive matriarch and her presumed adult son(s)."

"The outstanding feature" of Resident orca society is that neither sex wanders from the natal family and its home range, something "rarely seen in birds or mammals," Naomi continued. "However, the degree to which both sexes associate with their mothers may be unique." As young Resident females begin to produce their own calves, they spend more time away from their mother, eventually establishing their own matrilines within their particular pod, from which they never disperse.

But male Residents are another story entirely. They spend most of their time by their mother's side, from infancy through old age. They may swim off for a few hours or days to mate with females from other matrilines or pods, but in the end they always come back to their mother.

Male Resident orcas, in other words, are the planet's ultimate mama's boys.

"A son stabilizes his association with his mother at about ten years of age at a relatively high level (40–75% of his time is spent within a body length of his mother) and appears to maintain this association throughout the rest of his life," Naomi wrote. "My study focuses on the social dynamics of male killer whales of the Northern Resident community."

On the evening of February 20, 1991, after the Pacific weather had offered up a fairly uncommon snowfall, Naomi was working on the computer at her favorite table in the dining area. She was engulfed in her data on the social behavior of Resident males, and how orca society could permit these testosterone-charged and disruptive males to stay with their mother for life.

Suddenly, there was a *bang!* at the picture window.

The loud noise startled Naomi and pierced her heart with terror. Was her imagined backwoods killer finally at her doorstep? She looked up. A large man was plastered against the window, his face contorted into an evil snarl, a string of drool descending from one corner of his mouth. Naomi gasped and jumped in her chair.

But it was only Dan Kirby, playing a practical joke on his young caretaker while delivering some supplies. She knew he would be coming, she just didn't expect him quite then. Recovered from her shock, she welcomed him inside and they began to chat.

Dan had closely followed the Gulf War against Saddam Hussein, which was now in its waning days. "Your President Bush just rejected the Soviet-Iraqi peace plan and issued a twenty-four-hour ultimatum," he told Naomi. Iraq would have to leave Kuwait in order to avoid a ground war. Dan said the whole conflict could be over by the weekend.

Naomi, a left-of-center progressive, was no fan of the war, but she was glad to hear it was wrapping up successfully. "That's good news, Dan. Thanks for telling me. Anything else going on in the world?"

"Oh, yeah, I almost forgot. There was some kind of incident down at Sea-Land, in Victoria. The killer whales were acting up. Apparently someone died."

"My god. When did that happen?"

"This afternoon. I didn't get all the details. But they're talking about it on the radio. Why don't you tune in?"

Naomi was exhausted, and she had used up her generator fuel for the day. She would look into it soon. "Good night, Dan," she said with a wry grin. "Thanks for scaring the crap out of me. That was nice."

The late winter weather was bright and cold. Naomi turned on the radio, but the news was almost exclusively from the Gulf. Nothing about a death at SeaLand. Eventually she bundled herself up, jumped into the lodge's little skiff, and steered it toward Telegraph Cove, a tiny wood-milling center located around the bend, to see what she could find out.

Naomi was off to see her good friend Jim Borrowman. He would know what had happened. Jim and his business partner, Bill MacKay, operated the Stubbs Island Whale Watching company aboard their custom-built metal-hulled vessel, the *Lukwa*. Jim was somewhat of a local expert on killer whales, having worked with author Erich Hoyt in researching his groundbreaking book *Orca: The Whale Called Killer,* a 1980 paean to the Northern Resident community.

Naomi pulled out of Hidden Cove and braved the choppy waters, heading into the strait and turning right, past Beaver Cove and its masses of logs—felled not by beavers but lumberjacks. From there she headed into the protected haven of Telegraph Cove. Jim was working on his boat.

"What happened at SeaLand?" she asked, curious for news.

"It was a trainer," Jim said slowly. "Young girl named Keltie Byrne, a championship swimmer from Victoria." She had slipped by the pool and the whales took her into the water. It took them two hours to recover her body.

Naomi was taken aback. It was so hard to believe those whales had killed someone. "My god, I was just there, little more than two years ago, when my mom came to visit," she replied. In the summer of 1988 the two of them had driven together from the north island to Victoria, where they had been wowed by the high tea at the sumptuous Empress Hotel before heading over to Oak Bay, where SeaLand was located, to see the orcas.

Naomi had watched as Haida, Nootka, and Tilikum performed their "behaviors," and she remembered thinking how rinky-dink the pool had been, how cheesy the little show had seemed. She had no recollection if any of the young trainers there that day were named Keltie.

"But that's strange," Naomi said. "Orcas don't attack *people*. Not that I know of."

"To tell you the truth, Naomi, I really don't think those whales killed her on purpose. You know as well as I do that, when you see them in the wild and you watch how they work and how they move and how they socialize and everything they do, you know that they won't hurt you. I've tried to photograph orcas underwater and they just turn and swim away."

Naomi had seen that happen herself. She had seen people get in the water with wild orcas, and the whales really weren't interested. There was a silence. "So what do you think could've happened in Victoria, Jim?"

"I think once Keltie was in there, she panicked a bit, and the whales freaked out and started getting rambunctious—maybe playing, maybe pissed off, who knows? But things happen, and a human dies. These are big whales."

"You think they were just playing?" Naomi asked.

"I don't know. I think orcas get neurotic in these swimming pools. When I watch them in a place like SeaLand, well, I just think that *neurotic* is a good word for what they become."

Did you hear about that trainer, up in Canada?" The question swirled around SeaWorld Florida like a bad virus on a wide-body jet. But it was difficult to get much information. Computer nerds were navigating something new, called the World Wide Web, on a thingamajig called America Online, but few animal trainers had access to such wizardry. Even if they did, downloading news from distant Victoria would have been difficult.

It was February 1991. Mobile phones still looked like bricks, and most people got their daily news the old-fashioned way: through newspapers, networks, or neighborly gossip.

That winter at SeaWorld, details on Tilikum's tantrum were managed from the top. Canadian papers were reporting that Keltie had been dragged screaming into the pool, was alive for many minutes during the incident, and that the whales repeatedly dunked her and prevented her from reaching the edge. But many trainers at SeaWorld, with no access to that news, came away with a different impression altogether.

Jeff Ventre brought up the tragedy with his friend and fellow trainer Samantha Berg one day behind the show pool at Whale and Dolphin Stadium, where they both worked in the popular "New Friends" show. He said he'd been told by management that Keltie died of hypothermia. "It sounded like she was an inexperienced trainer. The accident was probably her fault," he said. "Their tanks are filled with ocean water, and she didn't have a wet suit on."

"Holy shit," Sam said. "The water is really cold up there." Sam was right. Water temperature in Victoria was about forty-eight degrees.

"I heard she died quickly," Jeff said. "From what I can gather, the whales blocked her from getting out. Then Tilikum grabbed her body. He thought it was a toy, and he wouldn't let the females have it."

But, Sam wanted to know, why wouldn't the whales let Keltie get out of the pool? Jeff explained that the orcas there had never been trained to get used to people in the water with them. "There was no water work at SeaLand," he said. At SeaWorld, in contrast, there were regular safety and rescue drills. Trainers were taught to go limp if they were grabbed, so the whales might lose interest. Whales were trained to close their mouths while swimming and were desensitized to stay calm and circle the pool if someone unexpectedly fell in.

Jeff had already done a tour of duty over at Shamu Stadium, back in 1988. Even though he'd spent that year as a lowly fish-bucket scrubber, he did manage to get in the water for some practice training with a few of the orcas once in a while. They were undisturbed by his presence because they had been trained to be undisturbed.

Jeff, a former high school baseball player and fraternity president with a solid build, dark hair, and hazel eyes, had been at SeaWorld for more than three years now; Samantha, a wisecracking, five-foot-two New Yorker with brown eyes and hair, had been there just one year. They both had studied life science in college, Jeff at Florida State University and Sam at Cornell.

SeaWorld was not exactly a hotbed of scientific discourse. A few days earlier, staff at all three marine animal stadiums—Shamu, Whale and Dolphin, and Sea Lion and Otter—were summarily called to an informational briefing at a meeting room over at the water-ski stadium. These powwows happened every so often, Jeff explained to Sam as they walked to the meeting.

As they spoke, they strolled past swarms of tourists from all over the world, many in shorts and tank tops, pushing strollers and popping into souvenir shops crammed with whale and dolphin plush toys, or refreshing themselves from the heat at little stands selling Anheuser-Busch beer, ice-cold and on tap. SeaWorld really was a "world" unto itself, with 220 landscaped acres brimming with pavilions, stadiums, thrill rides, stores, restaurants, plazas, lakes, ponds, and a four-hundred-foot Sky Tower.

The park's training department was large and growing. Sam and Jeff were just two of nearly fifty people at SeaWorld who were qualified to work directly with the animals. Everyone began his or her career as an apprentice trainer, which typically lasted about a year, before moving up to associate trainer, for twelve to eighteen months, then trainer, and, a number of years after that, senior trainer. Some people remained at the senior trainer level, others went on to vie for the limited number of supervisor positions.

Trainer meetings were essentially pep talks aimed at getting the staff to present the park and its operations in the most flattering light possible to the public. A small dose of eighth-grade science was often peppered into the talks. At this gathering, some of the company's top brass showed up to lecture their underlings about nutrients, food, marine mammal energy requirements, and the "restaurant-quality fish" (a common SeaWorld mantra) that everyone was feeding to the animals.

"We measure food energy in units called *calories,*" one of the bosses said slowly, as if speaking to third-graders. Jeff and Sam looked at each other and rolled their eyes.

"One calorie is the amount of energy it takes to raise the temperature of a glass of water by one degree Celsius," the speaker continued.

Sam had to say something. She raised her hand. "Excuse me?" she said politely. "I think you just misspoke."

"I what?" the executive snapped.

"You said 'a glass.' You said one calorie raises a glass of water by one degree. It's not a *glass* of water; it's a *gram* of water."

"Well . . . you get my point," the boss said, flustered and red-faced. "What I was trying to say is that we meet the caloric needs of all animals in our collection."

When Jeff started at SeaWorld, in the fall of 1987, he thought he would be working at some kind of Jacques Cousteau–like oceanic research center, where marine biologists and other scientists would be huddled around dolphin tanks, clad in white lab coats, and making copious marks on clipboards every time an animal did something interesting.

Jeff knew SeaWorld had a show business angle. He grew up near Orlando, raised on heart-racing images of athletic trainers in colorful wet suits shooting from the rostrums of massive orcas and somersaulting into the clear turquoise water. It looked cool.

Jeff's new career, then, seemed an appealing marriage of physical and intellectual activity. He believed he would be applying his biology background to evidence-based science, in addition to flying through the air across the whale pool.

Jeff was wrong. Little scholarly endeavor seemed to be going on behind the swaying palms and dense tropical hedges that line SeaWorld's perimeter. But Jeff got over his disappointment. Working at the park *was* cool and *was* fun. Besides, Jeff was already there, on the inside—one of the lucky few to be selected from an overflowing pile of résumés. People would kill for this job, so why leave it?

And there were other benefits. Jeff was a daredevil, hotdogging waterwork star at SeaWorld. The job required a lot of physical training and athleticism, so being healthy, fit, and good-looking came with the work. A SeaWorld trainer was something of a local celebrity, and sometimes people would recognize Jeff in a bar and buy him a drink. Most people thought that being a trainer at SeaWorld must have been the most exciting job in the world. It didn't hurt when trying to meet women, either. Most of them didn't know he was only making about $7 an hour.

Now, in 1991, Jeff loved his life at Whale and Dolphin Stadium, where, instead of scrubbing fish buckets for orcas, he was working up close with dozens of extraordinary animals. He was swimming with, surfing atop, or launching from the gregarious bottlenose dolphins, white-sided dolphins, the mysterious and beguiling belugas, also known as white whales, and the incredibly smart and nimble false killer whales. Torpedo-like, all-black, and much sleeker than orcas, false killer whales are related to them through the family Delphinidae.

Jeff realized that the "New Friends" show at Whale and Dolphin Stadium was a bit corny and lacked a solid educational component that could teach visitors about how these animals lived in the wild. But it was a blast to perform. False killer whales are able to do nearly all the same behaviors as orcas, but they can do so within much tighter quarters. Hana and Yaki were his favorites. False killer whales are great for doing the kind of hotdogging work that Jeff excelled at, including the crowd-pleasing "stand-on," in which a whale would push him by the feet quickly through the water, then lunge up perpendicularly into the air and straight back down again, with Jeff balanced perfectly on its rostrum. Jeff was astonished by their power, speed, grace, and agility—like riding a finely tuned Ducati motorbike.

Jeff's friend and confidant Samantha Berg was not about to give up her job either. She loved working at SeaWorld, though she did feel out of place with her dark, curly hair and the only Jewish surname in the department (there was one African-American; the rest were white and decidedly gentile). Nearly all the other female trainers were tall, fair-haired, and willowy—except for one other dark-haired New Yorker, of Italian descent, who worked at Sea Lion and Otter Stadium. When Sam was transferred to the stadium, the two became great friends. They both thought it was pretty funny they had been hired at all.

Still, Sam had the credentials that SeaWorld valued. She was not only pretty, she had been a competitive swimmer, gymnast, and runner in high school and learned stagecraft by osmosis growing up in New York. There were after-school and summer drama programs and staged theater performances in high school. She also had an animal-science background. Sam was athletic, and smart, a former lifeguard with a sharp wit that could slice through pastrami.

Sam Berg was born in Queens and grew up on Long Island. As early as the first grade, she had the distinct idea of working with dolphins and whales, probably from a combination of watching *The Undersea World of Jacques Cousteau* along with *Mutual of Omaha's Wild Kingdom* on television.

In 1985, Sam went upstate to Ithaca, New York, to get her animal-science degree at Cornell University. She spent her junior year at the Shoals Marine Lab, a joint operation of Cornell and the University of New Hampshire, on Appledore Island, about six miles off the New England coast. Samantha also loved animals, especially her charges at the "New Friends" show, where she spent most of her time working with the bottlenose dolphins and the mysterious belugas—those midsize white whales with the gentle faces, bulbous heads, and playful personalities.

Sometimes the belugas grew a little too playful. One of them, a male named Shadow, whose stage name was Belinda, was part of the "Belinda and Bianca" underwater ballet segment of the show. (Stage names are used for most animals so, when an animal dies, the character can live on.) But during downtimes, when the whales were on display behind the Plexiglas walls of the main pool, the male would sometimes get excited, extrude his long, pink penis from his genital slit, and wave it about. Whenever this happened, some tourist would invariably cry out, "Quick! Get the trainers! Belinda's giving birth!"

Despite her science background, Sam couldn't help but sense a certain mystical kinship with the intelligent and empathetic dolphins at SeaWorld. She had felt that way since she'd walked through the front gates on her first day at work. A supervisor had met her near the main entrance and driven her over to Whale and Dolphin. He introduced Sam to the training staff, including Jeff Ventre, then excused himself to tend to a behavioral matter with one of the dolphins.

"Why don't you go have a look around outside, Sam?" he said. It was like asking a kid to wander around FAO Schwarz alone. Sam walked out to the backstage area, where several small pools were connected to each other and to the main pool up front. There, the magic of show business unfolded—up to eight times a day during peak periods—in the "New Friends" whale and dolphin spectacular. Sam walked past the medical pool, shallow and easily drained in order to dry-dock the animals to take urine and blood samples and perform other veterinary procedures.

In the next pool over, gated off from the rest, two bottlenose dolphins leapt from the blue water in elegant flips and tight, graceful bows. They kept trying to see out over their tank, to get a glimpse of this person who had newly entered their world. They were curious, dignified, and adorable. Sam fell in love. She stood there motionless and continued her communion with the dolphins for quite a while more, dumbstruck with admiration. Oh my god, Sam thought to herself, I can't believe how lucky I am! Thank you! The feeling of peace and happiness was almost overwhelming, just from standing that close to such sublime and joyful beings. "I can't believe I'm really here," Sam said, this time aloud and with a soft laugh.

Her rapture abruptly disappeared as the supervisor walked outside and stumbled upon Sam gazing at the dolphins with almost religious wonder. He had seen that look before among rookies. It never lasted long. The supervisor snuck up behind Sam and whispered:

"God? Are you there, God?"

Samantha was stunned and hurt. Clearly this guy was making fun of her dolphin worship. His underlying message, she would come to realize, was this: "Don't get so involved—these animals are here to do a *job*. There's no magic, there's nothing mystical. They're just animals."

PART ONE

BLACKFISH

I

Marine Biologist

Naomi Rose fell in love with dolphins at the age of thirteen. It happened in 1975 while she was watching *An Evening with John Denver,* a major television special that aired that year.

To Naomi, nobody was better than the Rocky Mountain songster with the boyish grin and dirty-blond mop. John Denver was the reason why she had purchased a cheap, used acoustic guitar and started strumming simple sounds from a chord chart. She had every John Denver album there was and soon taught herself to play many of his songs, belting them out with gusto.

Denver's 1973 smash hit, "Rocky Mountain High," had made Naomi a fan, but it also sparked her desire to work around wildlife, move to Colorado, and become a park ranger.

Another John Denver song, "Calypso" (1975), made her want to become a marine biologist. *Calypso* was the name of the retired minesweeper that Jacques-Yves Cousteau, a longtime friend of Denver's, converted into a floating marine research lab. John Denver wrote the song—one of his signature hits—in celebration of Cousteau, his crew, and the beloved white vessel they made world-famous.

Naomi had tuned in to see her pop-country idol extol the wonders of the mountains and free-roaming wildlife. She wasn't expecting a special appearance by the old marine biologist with the white hair, red cap, and cool French accent. But there he was on-screen with Denver, during a moving tribute to Cousteau's work—the two of them sailing together on the *Calypso* as a cluster of dolphins surfed in the bow wave.

Naomi was transfixed. She watched the music video, primitive as it was, her eyes pegged to the screen.

As Denver's song "Calypso" played over the images, Naomi stood and clapped along, bedazzled by the dolphins leaping through the white foam from the boat. She listened in amazement to the tune that changed her life:

Like the dolphin who guides you, you bring us beside you
To light up the darkness and show us the way.

The scene had a profound, lifelong effect on the young girl. Thanks to Denver and that seafaring Frenchman, Naomi was hooked on dolphins at a young age. ("John Denver was the gateway drug," she would joke years later. "Jacques Cousteau was the addiction.")

Naomi went into the living room to deliver the announcement to her folks. "I am going to study dolphins," the thirteen-year-old declared with a calm smile. Her parents smiled back. They told Naomi that they trusted her judgment, and they gave her a lot of credit for knowing what she wanted to do, even though she was only a teenager. Naomi realized they didn't believe her. After all, what thirteen-year-old kid knows what she wants to be?

But Naomi knew. She had never been so certain of anything in her life. There was something about those dolphins on the Denver special, just the sight of them playing at the bow of the boat. Naomi had watched *Flipper* as a kid, but that didn't make her want to work with dolphins. It was just another fictional wildlife show. Naomi also watched *Daktari,* but that didn't make her want to move to Africa and work with lions.

Someday, she promised herself, she would work on a boat and swim in the open sea, observing the dolphins, just like *Capitaine* Cousteau.

Naomi Anne Rose was born in Hastings, Michigan, a typical small town far from the ocean. But her family soon moved to the tidy suburbs of Milwaukee, where she spent her formative years. Her father was a chemist by training and worked as a medical technologist, testing blood, urine, and other samples in commercial labs. Her mother, who did not finish her college degree until she was fifty-three, worked with her husband in the medical-testing field. The couple moved frequently to take new jobs.

Naomi's mother, Reiko Kim, was born in Tokyo and lived there through the Pacific war. Her family moved to Okinawa soon after the fighting ended. There, Reiko learned to speak English and received her primary education at the local US Air Force base. Her Korean father was a translator for the US government, and all of her friends were American military brats.

The Kim family emigrated to Hawaii when Reiko was eighteen, and a few years later that's where she met Naomi's father, Raymond Rose, who was stationed there during his stint in the army. The two were married in 1958, and Naomi's oldest brother, Greg, was born in the territory of Hawaii, in 1959. Her other brother, Lawrence, was born in the state, in 1960.

Naomi's mother is, as Naomi has put it, "very Asian—inscrutable, quite reserved." But Reiko was a good mother, if not the warm, June Cleaver kind. She was a good cook and knew how to make terrific Halloween costumes and kept her sons busy with judo lessons and her daughter enrolled in dance class. Naomi's father, Raymond, never really understood Naomi, though he made it abundantly clear that he was proud of her. To a young Naomi, he was a distant dad, often away on business trips. Raymond moved his family around a lot because his ambitions sometimes got the better of him. It made for an unstable childhood.

Then there were the arguments between husband and wife. They weren't violent, but the conflict and bickering often made life at home uncomfortable. When Naomi was eleven, the precocious girl flatly suggested that her parents seek a divorce.

Naomi's brothers were fond of their kid sister, but often gave her a hard time. The bullying was typical sibling rivalry, but Naomi had no intention of putting up with it. The boys might win the physical fights, but Naomi got them back by finding ways to get them in trouble with their mom. Did that make her a tattletale? Perhaps, but it also kept Naomi from growing up as their personal doormat. Within a few years, they had worked out a suitable détente.

Naomi was always the good girl, and quite a little square: gifted in school, well behaved if a bit too opinionated for someone that young. Naomi had always been more confident than most people, even as a young girl telling her older friends what to do.

The Rose family moved several times as Naomi was growing up, living in Wisconsin, Illinois, New Jersey, and New York. When she was fifteen, they moved to Southern California. Though she was wary of yet another relocation, at least her new home offered access to two major marine entertainment parks. She could not wait to visit them: San Diego's SeaWorld, home of the original Shamu, and Marineland of the Pacific, on the Palos Verdes Peninsula south of Los Angeles. Marineland had two famous killer whales: Orky II, the male, and Corky II, the female. Naomi loved seeing all the shows at both places. Now that she knew she wanted to become a marine biologist, she wanted to experience cetaceans up close. At this young age, Naomi saw only the excitement and spectacle of Corky, Orky, and Shamu leaping from the water, without giving any thought to what might be going on behind the scenes of the marvelous display. Not until years later, when she saw orcas in the wild, did she begin to think about what life must be like for them in captivity.

That summer before her junior year, the short, scrappy Asian-American teenager with wavy, dark hair, brown eyes, and steely self-confidence went on a scientific field trip up the coast of California. It was part of a summer school course she took on intertidal organisms and marine biology offered by the LA

County Unified School District. After a few weeks in a classroom learning to identify tide-pool species, Naomi and several other students chaperoned by two adults drove a large RV up to Big Sur for a few days of seaside study. To her, it was the ultimate in student field trips.

The students were divided into small groups and assigned a tide pool to observe over time. They took measurements of salinity, dissolved oxygen, temperature, and pH. They created graphs and tables and did field drawings showing where all the organisms were located in each pool. They sketched individual organisms and conducted censuses by species. They did sediment analyses, took weather readings, and compiled other scientific measurements with an impressive arsenal of equipment. All the while, just offshore, Pacific sea otters played and foraged in the kelp, carefree as monkeys. Naomi loved every minute of it.

But Naomi wasn't like the other, wilder LA kids. They liked to procure illicit bottles of Boone's Farm white zinfandel and get rather buzzed and giggly while writing their field reports. Not so Miss Rose. When offered some wine from one of the boys, she politely declined. The boy thought that was pretty cool. "You can say no without being a buzz kill," he marveled.

At sixteen, Naomi asked if she could go away to study at the Colorado Rocky Mountain School—mostly because she wanted to stay in one place for the rest of high school. That the boarding school was near Aspen, John Denver's home, was an added benefit. Naomi was so square that she still liked the singer and admired his environmental work. She didn't think she'd run into the star, and she never did. But the secret hope remained.

School was easy for Naomi and she excelled in all her classes, earning straight A's without much effort. She loved science most, especially animal behavior and ecology. Mostly Naomi just liked knowing things. She possessed an extraordinary memory to store them in: a brimming internal database of assorted factoids, both weighty and trivial, that she could retrieve at will with unnerving alacrity.

In selecting a college, Naomi made a counterintuitive choice, given her desire to study marine creatures. She planned to attend school *away* from the coast and wanted to get a good, solid biology degree before she specialized, she explained to her friends.

She selected Mount Holyoke, the Massachusetts liberal arts college for women, and fled her warring California household. During spring break of her first year, she traveled to the outer elbow of Cape Cod to Woods Hole, the largest nonprofit oceanographic institution in the world and a mecca for aspiring marine biologists. She wanted to check the upcoming cruise schedule for student opportunities on research ships.

That summer, Woods Hole was going to run a ship from Cape Cod to Spain and then on to the Canary Islands before returning to Massachusetts.

Naomi contacted the scientists about coming along. "Sure, just show up," she was surprised to hear from the chief scientist on the first leg of the cruise. "We can keep you busy. There are always tasks for a college student to do."

On departure day, in early June, Naomi appeared on the dock, army-surplus duffel bag and guitar in hand. The team members hadn't expected her to show. "Well, you're here," one of them said. "C'mon, I'll show you your bunk." It was way down in the bilge, cramped, hot, and noisy.

Naomi ended up spending more time with the burly merchant marines than the scientists because she stayed on board for the whole three months, while the research team turned over at the end of each one-month leg. The ship was to study the physical oceanography (water temperature, salinity, etc.) and the biological oceanography (plant and animal specimens, travel patterns, etc.) along certain points of the route to profile a slice of the Atlantic Ocean. Naomi was assigned menial tasks—pulling filters out of the seawater, keeping track of depth recordings, washing flasks.

It was hard to say that she "liked" the cruise, though she would never forget the experience: A young female college student at sea with a crew of beer-swilling merchant marines, many of whom had signed up to escape their questionable pasts. It was rumored that one guy did time for second-degree murder.

Naomi did not yet drink, but she learned to tolerate people who do, watching her shipmates get blisteringly drunk and then pass out. From them, she learned how to swear, quite literally like a sailor. It didn't take long for her to win their respect. Impressed by her endurance for the hardships of being at sea, the crew rewarded Naomi with ever more comfortable quarters—from the bilge, to the second deck, and finally the top deck. Naomi also found herself in her first serious romantic relationship on the trip, with the ship's medic, who lived near Mount Holyoke. She stayed with him for almost two years.

The wayfarer returned to school even more hooked on the ocean and its inhabitants. The first semester of her junior year, Naomi journeyed to Hawaii to attend the University of Hawaii–Mānoa (the school where Barack Obama's parents met in 1959), near downtown Honolulu. She spent the waning days of the summer with her mother's family before renting a modest apartment with another student from Mount Holyoke and starting classes full-time at the university.

Naomi also did volunteer work at the school's Kewalo Basin Marine Mammal Laboratory, a two-acre research facility near Ala Moana Beach Park, walking distance from Waikiki. The lab housed two captive bottlenose dolphins, and its scientists were studying the animals' language acquisition abilities through hand and audio signals. When Naomi learned that the pair of female dolphins—Phoenix and Akeakemai, or Ake (pronounced *ah-KAY*) for short—knew some five hundred signals based on American Sign Language, she immediately signed up to work at Kewalo.

Naomi would ride a moped, on loan from her aunt, down to the lab about twice a week after school. Student volunteers were not given a lot of responsibility—or initially much access to the dolphins. It was grunt work mostly: pushing paper, mopping decks, and, most arduous of all, cleaning out the dolphin tank. The water level would be lowered to just a few feet, leaving Phoenix and Ake to skim around in the shallows at the bottom while students scrubbed algae from the sides of the concrete tank.

Not until the last few weeks of her semester was Naomi allowed to interact with the dolphins. She began by giving fish to Ake and also started learning some of the hand signals the researchers used to study the animal's aptitude for language acquisition. She also worked with Phoenix, who was learning computer-generated sounds instead of hand signals. Each time Phoenix performed as requested by the tones, Naomi would offer her fish or praise.

After weeks of work, the scum-scrubbing volunteers were finally rewarded with time in the water with the dolphins. Naomi was excited, but she only tried it once.

On the appointed day, she showed up clad in a two-piece bathing suit with a diving mask in her hand. She slipped into the warm Oahu seawater that filled the tank. Naomi felt slightly uneasy, but figured the dolphins would be kind to the hand that had fed them. She figured wrong.

Naomi got in the water and started swimming around the pool's perimeter. The dolphins were alongside her. The students were told not to stop or look at the animals or to appear in any way nervous. Naomi didn't even make it around one full circuit. She swam a few feet, but must have seemed timid. The dolphins turned on her.

Smash, bang, *boom.*

One of the 350-pound animals butted Naomi hard across the chest with her snout. The other slapped Naomi in the face with her fluke, sending Naomi's mask flying.

Naomi was dazed. She lost her bearings, blinded and unable to catch her breath. She felt helpless, but she had no chance to panic, though her ribs felt as if they had been crushed as the air was expelled from her lungs. Research staff rushed to her side and dragged her from the water. The dolphins turned and swam away, slinking around at the other end of the pool. The bridge of Naomi's nose throbbed and her ribs were seriously bruised.

Naomi looked across at the rogue dolphins. They seemed sheepish, as though they had no idea she would react so poorly to their roughhousing. But she couldn't be angry with them.

When Naomi got back to Mount Holyoke, she went to see her mentor, Dr. Susan Smith, an animal behavior professor who had been a big influence on Naomi as a biology student. Susan had taught her how to observe animals in the wild and how to take accurate field notes. Naomi was eager to fill the professor in on her trip to Hawaii, and especially her work at Kewalo Basin with the dolphin language acquisition study.

"I think the work going on out there is *very* cool," Naomi told her. "I was just amazed by the modified ASL gestures that Ake knew, and how Phoenix learned the computer-generated sound language. Their understanding of syntax alone is so remarkable. I just loved working with these animals!"

Naomi gushed about the project for quite some time, without noticing the skeptical look that had crossed her mentor's face. "Isn't it *fascinating*?" Naomi asked.

"Well," Susan began slowly, "teaching them an artificial language so we can communicate with them is all very interesting, but I would think it would be even more cool to learn what they are trying to say to *us*. What about research to try to decipher *their* language, as opposed to teaching them an artificial one that we created?"

Naomi stopped short at this unexpected perspective. It was an unusual feeling to have someone she admired be so unimpressed by what she was describing. She pondered her mentor's question. Naomi realized that, even though she had spent time with captive cetaceans, it had never occurred to her that they might have languages of their own, that their thoughts might work very differently from ours. For the first time, Naomi began thinking of what life in captivity must be like from a dolphin's perspective.

After graduating from Mount Holyoke in 1984, Naomi was accepted into the graduate program in biology at UC Santa Cruz and was bequeathed a modest merit scholarship of $5,000. But Naomi, who was skipping her master's degree and aiming headlong for a PhD, was undecided on a topic for her dissertation. She decided to take a year off, spending half of it traveling around Europe, following in the footsteps of her brother, who'd been a vagabond around Europe the year before. Though Naomi was able to defer graduate school by a year, she had to forgo the scholarship money.

It was the first time Naomi had ever done something so unstructured: six months without a fixed itinerary, traveling through eleven countries with a Eurail Pass and a youth-hostel card. The adventure taught her how to cope with the unexpected, handle emergencies, live on a shoestring, travel light, deal with cultural differences, and enjoy her own company. Her European tour tested her confidence, competence, and ingenuity to the utmost—all three would be critical for what was coming a bit later.

By the time Naomi returned to the States, in May 1985, she had been granted a prestigious National Science Foundation fellowship that would pay for three years of graduate school, with a small stipend for living costs.

Her mother was thrilled. Raymond Rose was also pleased for his daughter. But a heart attack he had suffered cast a dark pall of post-infarction depression upon him. His moroseness was too much for Naomi's mother, and she finally filed for divorce. For Naomi, it was a rough reentry from her carefree months in Europe. Not only did she have to prepare for a PhD program, she had to comfort a divorcing, depressed father at home.

Naomi left for Santa Cruz, six hours up the coast from LA, in August of 1985—partly because school was starting soon, and partly because, once again, she just needed to get away.

2

Trainer

Jeffrey Michael Ventre's all-American upbringing was quite a bit different from Naomi Rose's more cosmopolitan early life. But he was an equally avid student and eager learner. Jeff never wanted to miss school. When he got sick, he would fake being *well,* just to go to class.

Jeff's father, Jerry Ventre (the French surname's American pronunciation is *VEN-tree*), was an aerospace engineer, college professor, and solar energy guru; Jeff's mother, Nancy, was a stay-at-home mom deeply engaged in school fund-raising, PTA, and Little League.

Jeff and his younger sister, Kimberly, were born a year apart in the mid-1960s in a bedrock conservative neighborhood on Cincinnati's west side. While they were still quite young, their parents decided to relocate to central Florida. The notion of moving away to *anywhere*—let alone exotic tropical Florida—was a somewhat radical proposition in the traditional neighborhood. But Jerry and Nancy were up for something new.

In 1968, the family loaded up their Cutlass Supreme and made the nine-hundred-mile journey south. They rented an apartment for a few years in the Orlando suburb of Winter Park as Jeff's dad began his new job teaching engineering at the new Florida Technological University. The small school was designed to provide high-tech training for prospective employees at the Kennedy Space Center, thirty-five miles to the east. (Today, the school is the sprawling University of Central Florida, the nation's second largest.)

A few years later, the Ventres moved into a custom-built dream house outside the town of Oviedo, about twenty miles northeast of Orlando. The modern, L-shaped ranch-style home was on five acres of woods and groves on the shore of Lake Mills, its warm and clear water so pure at the time that it was drinkable. The house was surrounded by the exuberant foliage of central

Florida: southern cypress trees rising from the lake, southern red maple and laurel oak growing amid thick Florida saw palmetto and fragrant rows of oranges and tangerines.

With a rocket scientist for a dad, it's not surprising that the family spent a good amount of time at the Kennedy Space Center, taking tours and watching rocket launches. The kids grew up loving the moon, stars, and space, and the science needed to get us there. Science was a mainstay in the household as the kids conducted simple experiments or were taught to use slide rules by their father. Jeff also loved watching nature documentaries. Much like the young Naomi, he was hooked on anything featuring the French explorer Jacques Cousteau and his underwater world. Disney World had not yet opened, and the family would drive over to Lake Buena Vista to visit the information center and gaze longingly at an architectural model of the marvels to come. Nearby, a new SeaWorld park was also being built, though Jeff only went there once as a kid.

When Jeff moved on to Oviedo High School, he won a spot on the baseball team. By his junior year, Jeff found time to run for class president, and he won. He was not only a jock and a politico; he was also a thespian, having played a successful run as Ebenezer Scrooge in the eleventh-grade Drama Club's production of *A Christmas Carol*. All his activities were good preparation for his later years on the big stage at Shamu Stadium.

After school and during the summer, life for Jeff and Kim centered on Lake Mills. They knew every cove, shoal, and sandbar in the water, which they expertly navigated like pilots who bring ships into a harbor. With few other people around, the kids had their run of hundreds of acres—truly free-range children. On horseback, they could go even farther, their unleashed dogs running beside them. After a good long tour, they would bring the animals down to the lake and everyone would take a plunge.

Dangers lurked in the water and on land, but the kids never felt seriously threatened. They learned to steer clear of the occasional alligator or the more common and venomous coral snakes and water moccasins. Other wildlife roamed around the brushy landscape: armadillos, river otters, white-tailed deer, and Osceola wild turkeys.

Waterskiing at Lake Mills was practically a religion. Back then, there wasn't much fancy equipment, so after progressing from two skis to a single slalom, kids had to invent their own "next step." Some went for barefoot skiing, but the Ventres' boat lacked the needed horsepower. So Jeff and Kim would grab, say, a paddle and try to ski on that. They would stand barefoot without straps on the paddle, using it like a single ski, handle facing forward, and go.

Jerry Ventre was often there with them, working with his kids on the fundamentals of sports. It was all about observation, coordination, and timing, he told them. Balance was also key. The two kids were constantly inventing new balancing tricks: Can you ski on a piece of lumber? Can you stand up bareback on a horse while it's trotting? Can you do it while the horse is swimming, then do a back dive into the lake?

These were lessons in physical acuity that would also serve Jeff well at SeaWorld.

In the fall of 1981, Jeff left home for Florida State University in Tallahassee, where he would study biology and eventually become president of the Theta Chi fraternity. After college, Jeff returned to central Florida, certain that he wanted to work in the biological sciences. But he also wanted a job he could actually enjoy. That's when he started thinking about SeaWorld. Jeff had seen ads of the hotdogger trainers rocket-hopping off the big killer whales, and it sure looked like fun. But how scientific would a job like that be? Jeff assumed that cutting-edge marine research was going on at the park, and he wanted in on the action.

"I can either try this thing at SeaWorld because it sounds cool, I'm trained in biology, and I'm comfortable around animals," Jeff told his parents, "or I can look for a job collecting pond scum off of a bayou somewhere."

Kim was a little surprised: Working at an amusement park seemed like an odd choice. But she knew that Jeff would enjoy the work, and he was almost uniquely qualified. After thinking about it for a while, Kim agreed it was a good move. "You have a science background, you can ski on a paddle, you like animals, you've got *great* balance, and you can act," she said to Jeff. "It makes sense."

Jeff's career timing turned out to be fortuitous. SeaWorld was opening its fourth park, in San Antonio, and some of the Orlando staff were being transferred there. That left a few openings in Florida. Jeff read about it in the paper, picked up the phone, and called human resources.

An animal trainer position at SeaWorld was not easy to get, Jeff quickly learned. Hundreds of people had filled out applications and only a handful of them, at most, would be selected. Jeff was told he needed to get certified in scuba before he would be considered.

Jeff was exultant when he got called in for an interview, which took place in a nondescript one-story building on the Orlando park's perimeter. About a dozen applicants were there, young men and women, athletic, good-looking, and eager. They were waiting to meet two of the most important people at SeaWorld Orlando: Shamu Stadium supervisor Chuck Tompkins, and Thad Lacinak, the assistant curator of animal training.

During his interview, Jeff talked about his biology background and love of animals. Whatever he said must have worked: He got a callback for a swim test at Whale and Dolphin Stadium in Orlando. On the appointed day, Jeff showed up with his bathing trunks, along with another dozen or so applicants, and was told to dive into the seventy-degree water without a wet suit.

First, he was instructed to swim underwater from the stage to the Plexiglas partition across the main Whale and Dolphin Stadium pool as many times as he could without coming up for air. He was then told to dive to the bottom and retrieve a scuba weight, to swim laps, to carry heavy buckets filled with water around the deck, and then to finish off with push-ups. It was exhausting, but he pulled it off.

Jeff was completely out of breath when someone handed him a microphone and a script. To work in a show at SeaWorld, you not only had to do stunts, you had to narrate other people's stunts as well, and you had to do it without panting over the loudspeakers. Jeff was told to read some lines as if the stadium were filled with fans. It was not easy. He drew a deep breath, centered himself as he learned in drama club, and began to read.

Jeff got the job. He was told to report for work at SeaWorld the week of November 23, 1987. Two days before that, there was an accident at SeaWorld San Diego.

John Sillick, an orca trainer with less than two years' experience, had been riding on the back of a female named Nootka. Suddenly, Orky II (whom Naomi had seen at Marineland when she was young) breached sideways from the water and landed with a bone-crushing thud directly on top of Nootka, sandwiching Sillick between their bodies. He left the pool with twelve broken vertebrae, a shattered femur, and a fractured pelvis.

Jeff saw coverage of the accident on the news and heard talk of it during his first week of work. He learned all trainers at Shamu Stadium had been ordered out of the water with the orcas, at least for the time being. The general impression was that the underexperienced Sillick was at fault; this was a rare, freak accident. Nevertheless, management imposed a system-wide ban on all water work with killer whales, pending further notice.

But Jeff had not been hired to work at Shamu Stadium. That would come later. Instead, he was immediately assigned to SeaWorld Theater, a covered stadium that offered shows featuring birds of prey, parrots, dogs, maybe an emu or a miniature bison. The training staff of the show tried to teach the audience how to train their own animals back at home. Jeff worked mostly with dogs and parrots.

It wasn't exactly what he had expected, but it was a start. Besides, working with the animals was a lot of fun. It was even somewhat educational. Jeff knew a little bit about training animals thanks to the dogs and horses he grew up with, plus some of the courses he had taken in college had included the concepts of operant conditioning. But he was eager to learn more. One day soon after being hired Jeff was handed an official SeaWorld publication on the fundamentals of animal training. "Here, read this," a supervisor instructed him. "And then pay very close attention to everything we do here with the animals." Once Jeff knew the basics, he could start learning various "behaviors" the animals knew, and the command signals needed to get them to complete their tasks properly.

Many fundamentals of modern animal training are based on the work of B. F. Skinner, the famous (and somewhat controversial) psychologist and behaviorist who pioneered much of the field of behavioral science and became a leading authority on operant conditioning. This is a psychological form of teaching in which the type and frequency of behaviors are modified through associating desired behaviors with a "reinforcer"—usually a reward such as food, praise, stroking, a kiss, or a good grade.

If behavior could be changed by reinforcing consequences in the environment, then any number or combinations of discrete behaviors could be elicited from an animal. Animals could be trained to perform long sequences strung together, spurred along by intermittent rewards given according to a variable "reinforcement schedule."

Getting an animal to do what you want, then, is basically a matter of reinforcing the desired behavior whenever it occurs.

Jeff remembered watching a few grainy colored films of Dr. Skinner from back in the 1960s, discussing his work with students. In one famous clip, Skinner shows how remarkably quick and simple it is to train a pigeon to turn around counterclockwise. As the famous psychologist narrates, a young female assistant places a white pigeon on a platform painted institutional green. A back panel on the platform has a food slot at the bottom of it, with a bare lightbulb at the top. Behind the panel is an unseen machine that makes mechanical clicks. The pigeon has already been conditioned to look for food that appears in the slot each time the light goes on and the machine clicks.

"I will try to pick out some particular pattern of behavior and make it a more frequent part of the repertoire of the bird," Skinner tells his students. "Now, we will just watch its behavior a bit." At first, the bird just stands there, pecking at a piece of grain that appears in the slot each time the light flashes and the machine clicks. "It's not doing anything particular, you see, but I am going to try to get it to do something," Skinner says. "Suppose I shape up the behavior of making a complete turn? What I do to make that turn is to simply wait for some part of that behavior."

And that's what he does. Within seconds the pigeon turns one-quarter of the way around, in the desired counterclockwise direction. The light goes on and the machine clicks and the bird turns back to collect its reward (the positive reinforcement). Now, almost instantly, the bird has learned that if he turns to his left, counterclockwise, he will be rewarded with a flashing light, a clicking noise, and a piece of food. "You can see the effect is instantaneous," Skinner says as the bird repeats the movement and collects the reward.

But now the doctor wants more from his pigeon. He is not going to be satisfied with a mere quarter spin, and the bird will not be rewarded for it. "I wait for a more pronounced movement than that," he says. Almost as if on cue, the bird gives a full 360-degree turn, counterclockwise, and collects his reward. The pigeon has been properly and positively "reinforced"—it has been trained to turn entirely around to the left, all in about sixty seconds.

Within a few weeks, Jeff became proficient enough with hand signals that he could work parts of the SeaWorld Theater show: commanding dogs to perform retrievals; signaling birds to conduct "perimeter flights" around the enclosed stadium and then return to base; getting talking birds to say witty things from a memorized script.

It wasn't all fun all the time. Usually when he showed up for work, at about 7:00 a.m., Jeff had to clean out the cages and the feeding dishes of all the animals in the theater.

"They sleep in sky kennels at night," he explained to his sister, Kim. "And of course when you show up in the morning, they all gotta pee, or they've already peed, or worse." Jeff put up with the drudge work cheerfully, though he really wanted to work with marine mammals—especially the killer whales—much more than collies and macaws. On the other hand, nobody was getting in the water with the orcas right now anyway, he thought, because of that trainer who got smashed up in San Diego.

Jeff had no problem making new friends at SeaWorld Theater, including with some of the animal performers. He quickly bonded with Spunky, a two-year-old, black-and-white boxer-type mutt who performed in the show. Spunky was special, and Jeff recognized it right away.

Spunky was highly athletic and a bit unpredictable. He'd often chase after expensive parrots and other exotic and pricey flying fowl. And Spunky liked to take off. He would somehow sneak out of the show and split, roaming around the grounds like a VIP guest. Jeff was usually dispatched to track him down.

Jeff fell in love with Spunky, but SeaWorld's management were ready to rid themselves of the troublesome pooch. They were terrified he would end up killing one of their $10,000 exotic birds. Spunky was fired and Jeff gladly took him off SeaWorld's hands.

Jeff's mom and dad fell in love with the dog, too. Nancy Ventre quickly went about deprogramming Spunky of all the behaviors that had been drilled into him at SeaWorld, teaching him instead how to live like a non-showbiz canine in the "outside world."

In those early days at SeaWorld, Jeff also grew close to two new trainers who had started work at roughly the same time he did—members of the lucky batch of hires who were brought on board to replace those heading off to San Antonio. One was a tall guy with dark hair and a square jaw, Mark Simmons, and the other was Carol Ray, smart, beautiful, and blond. The two of them were assigned to Whale and Dolphin Stadium. Jeff would remain close with both coworkers for many years, even after they all had left SeaWorld.

Eventually, Jeff, Sam, Mark, and Carol would be rotated into Shamu Stadium. A lot of people wanted to work at Shamu—going there was like being kicked upstairs, getting one's break in the big time. Working with captive killer whales was different from working with the dolphins, belugas, and false killers. They were many times larger in size and could injure a trainer purely by accident. They were not as quick and sleek as the false killer whales, but they were far more powerful. They had their own opinions about what they wanted to do at any time. They could be obstinate and manipulative. Some of them were fascinated by images on the JumboTron.

The other big difference about Shamu Stadium was the relatively small animal population. With a higher ratio of trainers to animals and fewer animals to train, most people who worked at Shamu were more performers than actual trainers. Only a handful of senior staff taught the killer whales new tricks; everyone else learned those hand signals to make the animals perform properly. Most people working at Shamu, especially those in the lower ranks, were there to "maintain" behaviors, scrub buckets, and recite show lines.

The orca population was small, and falling: By August of 1991, there would only be three killer whales at Orlando.

Chief among the Florida orcas was the undisputed matriarch, Katina. Just two or three years old when captured off the coast of Iceland in October 1978, Katina was separated from her family and pod.[1] Usually a young female will stay by her mother's side for many years, even after she's had her first calf, typically at about age fifteen. Only later, after establishing her own matriline, does she start spending more time away from her mother.[2]

SeaWorld had no intention of letting Katina reach fifteen before having her first calf. In March of 1984, when she was about nine and living at SeaWorld California, she was impregnated by a large whale named Winston, a Southern Resident orca rounded up in a notorious mass capture at Penn Cove, Washington, in 1970.[3] Winston had spent six years at a UK amusement park before SeaWorld California acquired him in October 1976.[4]

In late 1984, a pregnant Katina was flown to her new home in Orlando. There, on September 26, 1985, she gave birth to Winston's daughter, Kalina, the world's first orca successfully bred in captivity.[5] She became a media sensation and was dubbed Baby Shamu. Millions flocked to Orlando to see the highly promoted, first-of-its-kind newborn orca behind glass. Baby Shamu's father, Winston, would die seven months later in San Diego from chronic cardiovascular failure, at about nineteen years of age.[6] (The average life expectancy for a wild male orca is approximately thirty years, with an estimated maximum life span of about sixty.)[7]

The world fell in love with Kalina. SeaWorld jumped at the opportunity her arrival presented by launching a Baby Shamu Celebration tour. In February of 1990, at the age of four and a half, the calf was sent to SeaWorld in Ohio,[8] where for eight months she performed in a show with whales she barely knew, followed by a seven-month stint at SeaWorld San Diego. In May 1991, she was flown to SeaWorld San Antonio.[9]

Once in the heart of Texas, Kalina would become pregnant, at approximately just *six* years of age, by the male Kotar, who'd had his own share of traveling around North America. Kotar was one of the smallest orcas ever captured when he was taken from the waters off southeast Iceland in October 1978 at one or two years of age. Kotar spent two and a half years at SeaWorld San Diego before being moved to Orlando, where he worked another seven years.

But in January 1987, SeaWorld Florida acquired another male from Canada's Marineland Ontario—a large and moody male named Kanduke, the only Transient whale in the collection. He had been captured in 1975 in Pedder Bay, a narrow inlet just south of Victoria, British Columbia.[10]

The mammal-eating Pacific whale and the fish-eating Icelandic whale did not get along. One day they got into a fierce altercation. The two males repeatedly beached themselves on the slide-out and made loud crying noises. At the peak of the battle, Kotar bit Kanduke's penis, severely wounding it and leaving a four-inch scar.[11] That attack got Kotar banished to San Antonio, in 1988. There he mated with young Kalina. Kalina gave birth to a son, Keet, in 1993.[12]

As for Kalina, not until October 1994 did the traveling whale return home to Orlando, where she was reunited with her mother, Katina, after a four-and-a-half-year separation. Baby Shamu had to leave her own eighteen-month-old baby, Keet, in San Antonio. She was also now pregnant by Kotar again, just one year after Keet's birth. The gestation period for killer whales is seventeen to eighteen months, and wild females tend to space their calves at intervals of about five years.[13] Kalina's second calf, a male named Keto, was born in Orlando in 1995.[14] He would grow up to become the most notorious killer whale in the world, after Tilikum.

Katina was kept busily pregnant, too. Twenty months after having Kalina, she was impregnated once again, this time by the aggressive Transient Kanduke. Their hybrid offspring, a female named Katerina, was born at Orlando on November 4, 1988.[15] Katerina would be shipped off to Ohio in April of 1991, after spending just two and a half years with her mother.[16] In November 1994 she would be reassigned to San Antonio, where she would die four and a half years later of bacterial pneumonia and severe hemorrhage at ten years of age.[17]

The average life expectancy for female orcas in the wild has been estimated in published studies to be forty-five to fifty years, with a maximum life span of about ninety.[18]

Three other lower-caste whales were in Orlando. Kenau had been caught in Iceland in 1976 at a young age, along with another female named Gudrun. Kenau's life in captivity was difficult: She was moved on several occasions between California, Florida, Ohio, and Texas. In 1988, Kenau gave birth to her second calf (the first died after eleven days), named Kayla.[19] In early 1991, Kenau was sent to Orlando without Kayla, and Kayla was transferred to Ohio, at just two years of age. Kenau would only last seven months in Orlando. On August 6, 1991, she died of hemorrhagic bacterial pneumonia. She was pregnant again already—in her twelfth month of gestation. Kenau was about seventeen.[20]

By August of 1991, when Sam went to work at Shamu Stadium, only two orcas remained there beside the matriarch Katina: the subdominant Gudrun and her daughter, Taima. Gudrun had been caught with Kenau in Iceland and shipped to an aquarium in the Netherlands. In late 1987 she was flown to SeaWorld Orlando.[21] Gudrun had a rough time of things in Florida, though she did manage to produce a calf with Kanduke. In July 1989, their half-Transient, half-Icelandic daughter, Taima, was born.[22] Gudrun was a good mother to Taima, but not so with her second calf, Nyar, whom she would have in late 1993.[23]

With its orca population dwindling, SeaWorld Florida must have been eager to secure some replacements. But it was now much harder to capture wild killer whales: Canada, Iceland, and Washington State had banned the practice—taking orcas from other US waters would have sparked a public relations catastrophe. The most viable options for replenishing the killer whale population were captive breeding and importing animals from other parks.

On top of that, Orlando had a sperm problem. After Kanduke died, SeaWorld Florida would have no other males left: Kotar. The park needed new orcas, but it was critical to find at least one more healthy adult male with a strong libido.

SeaWorld began looking at the options. One possibility was a pathetic, lesion-covered whale owned by a Mexico City amusement park, Reino Aventura. His name was Keiko (*KAY-koh*), and he was about to star in some Warner Bros. picture called *Free Willy.* The other candidate was an eleven-year-old male in Victoria, Canada. He didn't have lesions, but he did come with a bad rap sheet—along with his tankmates, Haida II and Nootka IV.

His name, of course, was Tilikum, and SeaLand of the Pacific was ready to unload him.

3

Capture

The East Fjords of Iceland are sliced from some of the most rugged and breathtaking chunks of earth found on the planet. Here, along this supremely isolated stretch of coast, as far across the island from the capital, Reykjavík, as one can journey, a sawtooth pattern of bays and deep-water fjords were carved into the volcanic stone by retreating glaciers during the last ice age.

The eastern landscape is a visual phantasmagoria of moss green, sea blue, ice white, and lava red-black, spilled across the tangled cliffs and undulating ridges of wind-carved pumice.

Life doesn't come easy in this part of the world, and the east coast of Iceland is not without its sagas of human suffering—both historical and apocryphal. One site of particular sorrow is the twelve-mile-long Berufjördur (or Bera's Fjord), one of the most hauntingly lovely corners of the island nation.

The area is named for an early settler, a hearty Nordic woman called Bera, who operated a farm on the fjord along with her husband, Soti. According to legend, one winter's afternoon, Bera, Soti, and the entire household mounted their horses and rode over the craggy pass to attend a neighbor's party in the Fljot Valley, near the present-day town of Egilsstadir.

On the trip home, a ferocious winter storm blew in over the glaciated peaks to the west. The entire group froze to death, except for Bera, whose horse found its way back to the farm. When they finally came upon the cozy homestead, Bera's mount grew so excited it galloped straight for the barn, with Bera still in the saddle. Bera hit the doorway, broke her neck, and died a forlorn death.

In the seventeenth century, marauding buccaneers from faraway North Africa arrived by sea, unleashing a bloody hell on Icelanders settled along the east coast. The pirates reportedly looted, killed, and burned down Berufjördur farms before moving down the shoreline, reportedly capturing another 240 people in the Westman Islands for their lucrative slave market back home.

Berufjördur might be tough on humans, but it's quite ideal if you're a killer whale. Each year, large numbers of Icelandic orcas descend on this end of the island and its sheltered fjords, the preferred overwintering grounds of Atlantic herring, their favorite meal.

Icelandic killer whales have not been studied extensively, but it's possible to envision a little bit of what their life might be like in the icy North Atlantic. Imagine for a moment that you are a young killer whale—a male, about two years of age, maybe nine feet long—and you are swimming alongside the safety and comfort of your mother through a cold, choppy sea, a generous stratum of blubber insulating you from the chill.

You began eating fish about a year ago, but you still try to coax your mom into squeezing a bit more fatty, white milk into the water near your probing rostrum. Once in a while she acquiesces. But she has made it clear that your nursing days are drawing to a close. If you want to eat, well, here you are at the mouth of a narrow fjord filled with silvery herring. Eat already.

As the youngest sibling, you get the traveling spot right next to your mother, which you won't have to abandon until she has another baby, probably in a couple more years (Icelandic males might stay near their mother for life, like Resident orcas in the American Northwest, but scientists are unsure). For now, you are the center of attention. Your brothers and sisters mill about and forage. Not too far away, your mom's sisters and their kids are also catching herring. Their mother—your grandmother, who sometimes keeps you in line—is within earshot.[1]

There are enough fish for everybody. But you are puzzled by those odd dark blobs bobbing on the surface. You don't know what boats are—yet—nor do you realize that the odd lacy curtains that drape from their decks can trap herring in large purses made of netting. You are dying to satisfy your curiosity and go investigate, but mom and grandma keep you in line.

Your world is primarily acoustic, filled with the sounds of the sea. It's a never-ending symphony of clicks, whistles, squeals, and yelps from your own family, backed up by a chorus carried beneath the waves: the clicks of other orca pods; the horsing around of white-sided dolphins; the rolling of stones on the ocean floor; the eerie song of a humpback whale, a hundred miles out at sea.

When not engrossed in the ocean's music, you splash about in the chop of a blustery afternoon or play with your siblings, with kelp, and with your food. When you "spy-hop" out of the water (pop up vertically to have a look around), your excellent eyesight can see towns and sheep and volcanic formations like nothing you've seen underwater. The sunsets are spectacular.

In other words, you are a happy little whale. (Science tells us that animal contentment cannot be measured, though SeaWorld employees have publicly stated that they "know" their captive orcas are content, an illogical inconsistency, critics point out.)

Then one gray morning in November, the world as you know it comes to an end.

You and your pod are foraging for herring near Berufjördur. A painfully loud *clackity-clack-clack* reverberates through the water as one of the blobs—a boat, as you are about to learn—comes toward you. The sea is dark and you can't see much, but you can tell from the sounds that the boat is moving in fast.

Your family senses that something is not right. They begin turning away from the boat, which has never before gotten so close. Your mother issues a shriek. *Danger! Flee! Now!* Confused and gripped with terror, you make a run for it, trying to catch up with your mom, swimming just ahead.

Then you are stopped. You kick your flukes furiously, trying to propel yourself through the water, but you cannot swim. You realize you have darted directly into the lacy curtain. So has your entire family.

Panic ensues. You and your family swim around the perimeter of the net, but there is no way out. You are kept in there for hours, flailing about with nowhere to go.

Suddenly, you are snagged in another, smaller net. You cry out in shock and fear, calling for your mother. *CREEEEEEEE-eeeeeee!* You feel the net being pulled through the water toward the boat. Your heart races and you surface to breathe, quickly and with difficulty. What is going on? Where is your *mom?*

Then you hear her. You have never heard this wretched wail before: mournful, ragged, spiked with rage and terror. Now your other relatives have joined the awful remonstration. You answer their panicked cries with your own chaotic vocalizations as you're hauled from the water on a canvas sling. You are now suspended in the air ten feet above the surface. The harsh wind on your wet skin feels alien and frightening. You can hear the cries of your frantic family, milling around in helpless grief. This morning you were happy. Now you are trapped in hell.

The sling is lowered onto the deck. You are placed in a large foam-lined tub half-filled with seawater. Men in parkas and wool hats yell to each other in vocalizations that are deep, fearsome, and indecipherable.

You miss your mother already. She has never been more than a few feet from your side. You can hear her, calling for you in despair. She is next to the hull, in the open sea. Then you hear a mechanical roar and you sense you are moving across the water. It's a terrifying feeling. Over the din of the engine you can just make out the screams of your family. They have been freed from the nets. They are following the boat. You wonder when this will end, so you can be reunited with your pod and go back to your herring and kelp.

That will never happen.

Instead, you chug along for a long time. Eventually your mother gets tired and stops following. You no longer hear your family. You are alone, sick with worry. You want your mom—and to go home.

Finally the racket of the engine ceases. The silence is overpowering. Then you hear men yelling again as your tub is put on a flatbed truck and driven for another hour or two. When the truck stops, you are lifted by the sling once again. This time, you are lowered into a small concrete tank built inside a shed. You cannot see the sky and there is little sound. You have never heard such silent water.

Two other animals like you are in the tank—one male, one female. They are about your age and size, but you do not know them. When they begin making sounds, you cannot recognize their vocalizations. They seem as confused and dejected as you.

A year goes by. Humans coax you into jumping out of the water and touching your rostrum to a big red ball suspended above the indoor tank. It gives you something to do, and they give you fish each time you do it.

More humans come to stare at you from the edge of the tank, pointing, laughing, and producing bright camera flashes that bother your eyes. Then your tankmates disappear one at a time. Finally, your day comes. You are lifted by the sling yet again, put into another Styrofoam tub, trucked to an airfield, and loaded onto a cargo plane.

Many hours later, you find yourself in a watery enclosure thousands of miles away. You don't know it, but you are now in a different ocean, called the Pacific. The small pool you are in is netted off from a larger bay and surrounded by floating docks, on which lots of people and their flashing lights have gathered to gawk at you. Beyond the nets, you can hear other killer whales swimming around out in the strait.

Two other orcas are in the tank with you—they are older, much larger, and female. Though they were both caught in Iceland, you do not recognize their vocalizations. You are now three years old. You miss your mother terribly. You think one of these older females will be kind to you, comfort you, protect you. But they are too busy fighting with each other for dominance. You hear the humans calling them names—one they call Nootka and the other Haida. Before long, you are being called something, too: Tilikum.

You are subdominant. The only time that Nootka and Haida pay any attention to you is when they harass you, rake your back with their teeth, and chase you around the tank. What you wanted was succor; now you just want escape. You are denied both.

On most nights, you and the older females are locked up in a small indoor tank adjacent to the pool. It is made of metal and barely large enough to accommodate all three of you (the owner is worried that one of you might chew through the net that separates you from the sea, or that some softhearted human might try to liberate you through sabotage).

The humans call your nocturnal quarters the module. You quickly learn to despise it. They often lock you up in there for fourteen hours on end—from 5:00 p.m., when most people go home, until 7:00 a.m., when they come back. There is barely room to turn around in the module, let alone escape from your cell mates. All three of you routinely cut and scratch yourselves on the metal sides of the module. When the two girls are feeling aggressive, your life can become hell. Your skin is perpetually covered in scars from Nootka and Haida.

One night when Nootka is particularly hostile, she swings her head at you, jaws agape, only to smash her rostrum into the metal wall. Her head starts to hemorrhage and she spouts blood from her blowhole. On other mornings, when the humans let you out of the tomblike structure, part of your fluke looks like hamburger.

But if you refuse to go into the module at night, which you sometimes do, the humans will significantly cut back your supply of dead fish (you sorely miss chasing live fish). In fact, there seems to be a food shortage here. You are kept perpetually peckish—always a bit underfed, never fully satiated.[2] But you learn that if you jump out of the water exactly as taught, you can tamp down your hunger quite a bit, though it never fully goes away. It gets to the point where you cannot wait to perform.

This goes on for seven years. You mature. You grow. Your dorsal fin sprouts and then flops over. You get horny. Even though Nootka and Haida are dominant over you, they still go into heat and want to breed. You oblige them. By the time you are ten, they are both pregnant.

Your life is so very different from what you can remember back in Iceland. You yearn for winters in Berufjördur, for live herring, for your mother. These humans are okay, you suppose, but why do they lock you up with such overbearing females each night?

You become unsettled, you get neurotic, you ache for change.

Then, one dreary February afternoon, just like that, one of your trainers accidentally dips her foot in the water. You have never quite seen this before.

You are bored. You grab her foot.

4

Santa Cruz

In the mid-1980s, the University of California at Santa Cruz held a kind of Shangri-la mystique for many prospective students trying to decide where to study. The sprawling campus, seventy-five miles south of San Francisco, is set amid the meadows and redwoods of the rugged mountains overlooking Santa Cruz and the broad sweep of Monterey Bay beyond.

The school, founded in 1965, retained much of its counterculture underpinnings: Transcendental Meditation and "aura adjustments" at sunset; mellow pot parties with Grateful Dead music spinning on the turntable; magic mushrooms sprouting in nearby cow pastures. It was a natural hotbed of antiwar liberalism and a mecca for organic farming and environmental activism long before *green* meant something other than envy or cash. More contemplative ashram than tailgating party school, Santa Cruz helped put the hip in *hippie*.

In the 1980s, undergraduates called UCSC "Uncle Charlie's Summer Camp," although the graduate programs were world-class and highly respected. The collegiate environment was progressive by design and decidedly laid-back. Most students were given written evaluations instead of grades (science majors could opt for both).

When Naomi arrived in fall of 1985, the school's chancellor was trying hard to make the official mascot the "Santa Cruz Sea Lion." It nearly caused a riot among students, most of whom preferred the banana slug. The following year they voted by a fifteen-to-one margin to retain the slimy, yellow land mollusk, *Ariolimax dolichophallus,* as their official symbol.

Naomi did not choose Santa Cruz for its groovy vibe and acid-wash eccentricity. The university was and still remains home to one of the country's finest PhD programs in marine mammalogy.

At the time, many marine biologists and ocean scientists considered the study of marine mammals to be an inferior stepchild of the "real" ocean science they were conducting. Naomi had instinctively sensed this condescending attitude. "They must think we're a bunch of whale-huggers," she joked. "They think, 'Well, everybody loves *Flipper,* for Christ's sake. This is just a cheap way of getting into marine biology.'"

Part of the reason for this bias, Naomi assumed, arose from the misguided belief that marine mammalogy was based largely on observation alone. Some detractors, she knew, believed that marine mammalogists did little more than putt-putt around on a little dory, do some afternoon whale watching, maybe record an underwater song or scribble down a few items about foraging and breaching, then return to dry land in time for dinner and zinfandel.

UCSC boasted some of the marquee names in the fields of marine mammalogy and animal behavior, including Burney Le Boeuf, an authority on elephant seal behavior, Robert Trivers, the respected evolutionary biologist and "sociobiology" theorist, and Kenneth S. Norris, father of American marine mammalogy and perhaps the world's foremost expert on cetaceans.

Norris's presence at UCSC was a major reason why Naomi selected the school, and she arrived there in the fall of 1985 hoping he would agree to become her PhD adviser. Norris had pioneered the field of modern science on the intelligence, social lives, and echolocation abilities of a variety of cetaceans—the scientific order that includes whales, dolphins, and porpoises. He founded the University of California's Natural Reserve System (which today protects 120,000 acres of habitat), launched a national campaign to reduce dolphin deaths in tuna nets, and helped draft the landmark Marine Mammal Protection Act of 1972.

Norris was also a leading figure in the development of the marine mammal "display" industry. He had been appointed the founding curator of LA's Marineland of the Pacific in 1953, where he worked intensively with bottlenose dolphins in the captive collection when they were not entertaining tourists. While at Marineland, Norris concluded that an intelligent animal was waiting to be studied behind those mysterious "smiles." (Bottlenose dolphins do not smile: Their mouths naturally curve upward, giving the illusion of perpetual gaiety, even in death.) He set out to learn how dolphins perceived the world, what they saw and heard.

If not for the early captive display industry and innovators such as Dr. Norris, much of the groundbreaking work on whale and dolphin science would not have been accomplished.

Norris left Marineland in 1959 but remained active in the public display industry and a key supporter of marine mammal theme parks. He helped design Sea Life Park in Honolulu and Ocean Park in Hong Kong. In 1965, he was one of four founders of a world-class theme park on the shallow mudflats of Mission Bay, just north of San Diego. It was called SeaWorld.

Norris had also published several groundbreaking books, including *The Porpoise Watcher* in 1974, and was widely acclaimed for instructing his students to pursue science with rigor and passion. He was enormously curious and loved discussing his work with anyone who was interested. In Naomi's world, Dr. Ken Norris was a very big wig.

Unfortunately, by the time she arrived in Santa Cruz, he was winding down his academic career. Norris was helping a few grad students complete their degree, but he was not taking on anyone new. Naomi was at a loss: This was not part of her plan, and she liked having a plan.

Her first year at Santa Cruz proved thorny. Naomi knew few people on campus or in town. Even worse, without a master's degree under her belt, she didn't know what she was supposed to *do*. Everyone assumed she had her master's already and knew what grad school was all about. But Naomi did not know. She found the solitude, uncertainty, and the loss of her hoped-for adviser depressing.

Nor did she receive the type of guidance she would have liked from the adviser who stepped forward to help her: Dr. Robert Trivers.

Bob Trivers, a raffishly handsome man with a dark, wavy mane and a thick beard and mustache, was a popular figure on campus and a legendary thinker when it came to evolution, biology, DNA, and what makes people and animals do the things they do. Trivers had written two seminal and highly acclaimed papers while at Harvard in the 1970s—one on reciprocal altruism and one on parent-offspring conflict. They had become central canons of a growing discipline created by his mentor, E. O. Wilson, called *sociobiology*—the study of the genetic and evolutionary causes of social behavior. Why, for example, do males act differently from females? Why are parents so hardwired to "invest" in their children, and why are children so apt to challenge them? Can cooperation be explained by evolution? Can jealousy? Sociobiology examines all of these patterns of behavior, not only within species, but across them.

Trivers had just published a major book on the subject, called *Social Evolution,* and he gave Naomi an autographed copy as a welcoming gift. It fascinated her. His theory of reciprocal altruism, for instance, opened her eyes. It explained so much and she often described it to others as: "Reciprocal altruism is just what it sounds like: You do for me, I do for you. Or the prisoner's dilemma: You screw me, I screw you. Tit for tat. It's the basis of friendship; it's the basis of conflict." Sociobiology lies behind many of our interactions, and the same is true for animals, she explained. "Now, I know people might get upset when you boil down their friendships to some biological mechanism, but it really does explain just about everything. I just love this stuff. I really do."

Naomi was equally fascinated by the evolutionary biology that drives gender-specific social behaviors—something that would ultimately help her decide on her dissertation topic. For example, she knew about studies on the evolutionary roots of male social behavior among primates, elephants, and wolves, but what was known about male dolphins? Trivers's theories would eventually become central to Naomi's approach to killer whales.

Trivers had few graduate students under his wing—he was too eccentric for some students. He became Naomi's adviser, gave her some office space, handed her some theoretical papers to read and analyze, and even agreed to bat around dissertation ideas with her. He was terrific at theorizing, thinking big, and opening the minds of young people. But Naomi did not find him a helpful mentor. He had little interest in marine mammals. To her, he didn't seem interested in animals at all (the opposite of Ken Norris), but rather the *theory* of animals.

The lack of guidance was stressful. The paperwork alone at grad school was daunting, the various filing deadlines confusing. Naomi decided to just concentrate on her courses. She ended up taking more classes than she needed because she didn't know what else to do: intertidal biology, marine mammalogy, invertebrate zoology, statistics, even Spanish. The heavy workload was a welcome distraction; Naomi felt better when she was busy and engaged. One day in early 1986, she made an appointment with the campus counselor to discuss her feelings of loneliness and lack of support. The counselor suggested she approach other students and introduce herself.

For someone such as Naomi, who, like her mother, was quite reserved around strangers, that was gutsy. It seemed rather aggressive, but she figured she had nothing to lose. Her charm offensive paid off. Naomi quickly met a young woman named Janice Waite. Janice and another student, David Bain, were among the last grad students under the wing of Ken Norris. Naomi was envious of that, but she was also intrigued by what Janice and Dave were studying: the Northern Resident killer whale community, which spends each summer in and around Johnstone Strait, British Columbia.

Janice, more petite than Naomi even, with light brown hair and a soft smile, was a second-year master's student from Seattle. Her dissertation concerned the "alloparenting" behaviors of Northern Resident orcas. Alloparenting is when individuals other than the actual parents look after offspring, be it temporarily or permanently.

Janice began as Dave's research assistant before devising her own study on orca babysitting. Janice told Naomi about Resident orca mothers and how they occasionally part from their young to forage, socialize, or even rest, leaving their calf's aunt, grandmother, or older sibling to "babysit" the youngster.

Naomi was engrossed by the idea. She wondered how Bob Trivers's theories of social evolution might be applied to orca alloparenting.

Dave Bain, meanwhile, was finishing up his PhD work on the acoustical world of the Northern Residents. The tall, lanky student with blue eyes and pale, thinning hair was in his final season at the field camp he and Janice had helped set up on a large, rocky island, West Cracroft, which faces Johnstone Strait across from Vancouver Island. Dave had spent the past years recording the different pods' vocalizations through a homemade system of underwater microphones, known as hydrophones, designed to isolate individual vocalizers. He was trying to break down their dialects into something of a language, trying to decode their whistling racket.

Naomi became friends with Dave and Janice. She was intrigued by their work and excited by the thought of camping out up north and conducting actual field science on wild animals.

"Your project is really cool," she told them. "I still don't know what I'm doing for my PhD yet, but maybe I could go up there this summer and work as your field assistant?"

Janice and Dave loved the idea—they were always looking for people to help with collecting data and working around the camp and on the boats. They needed at least three or four others to help them, and volunteers were not exactly lining up at the door: The work was hard, the pay was nil, the place was remote, the weather was damp, and the horseflies were hungry.

The idea morphed into something more than a one-summer plan for Naomi. With Dave finishing his PhD that year and Janice leaving the following summer, they talked of handing off the field camp to Naomi. Janice said they could cede the whole Northern Resident project to her, as long as she could come up with an orca-oriented subject for her dissertation.

Naomi jumped at the idea. It didn't matter that the subject was killer whales. Janice and Dave could have been studying sea otters and Naomi would have been interested. Just the idea of working in the field with marine mammals was alluring. This is what I want to do, she told herself. I want to be in Canada this summer with these guys.

5
Johnstone Strait

On a cool and foggy morning in June 1986, Naomi Rose left Santa Cruz for the long drive north to Vancouver Island.

She joined up with an Australian friend she had met while traveling in Europe and set off on a road trip with him. The two stopped at Mount Saint Helens, just starting to show some green after the 1980 eruption, and visited the World Expo in Vancouver. They separated after that and Naomi caught the ninety-minute ferry across Georgia Strait from Vancouver to Nanaimo, an attractive Canadian municipality built around a protected harbor on the east coast of Vancouver Island. The island is named for Royal Navy captain George Vancouver, who charted the area for Great Britain aboard the HMS *Discovery* in the 1790s.

That night Naomi rendezvoused with Dave and Janice and the other field assistants, all of them crashing at the home of legendary whale expert Mike Bigg. He and his colleagues had developed the modern system of photo-identification for killer whales, which notates such things as size, gender, travel companions, dorsal fin marks (nicks and notches), surface scars, and the gray "saddle patch" area directly behind an orca's dorsal fin, with no two patches exactly alike.

Not until Mike and others demonstrated how orcas can be positively identified could genuine research on whale populations get underway in earnest. Mike Bigg was loved and respected for his gentle nature, generous spirit, and willingness to share his data with anyone interested in killer whales. His work was revolutionary. Mike was widely regarded as the father of modern orca research. Naomi was electrified by meeting him.

The following morning, the group began their final leg: the five-hour journey up Highway 19, an unforgettable stretch of roadway that heads northwest along the island's leeward coast, past Qualicum Beach, Fanny Bay, and Campbell River, where the Strait of Georgia narrows into Discovery Passage. The passage is a dazzling sight, with steep granite peaks rising from the cold waters of the channel. From there, Highway 19 winds westward over the island's mountainous spine, passing through sheer gorges veiled in daubs of mist. Finally they turned onto a winding lane that led downhill past Beaver Cove, with its families of black bears and nesting bald eagles, to minuscule Telegraph Cove.

The quaint settlement at Telegraph Cove could have been the setting for the TV series *Northern Exposure*. Colorfully painted clapboard cottages with white trim line the wooden boardwalk that curves around the cozy harbor. It got its name in 1912 when the village was selected as a lineman's station and terminus of the telegraph service from Campbell River.

At the cove, Dave and Janice introduced the new crop of field assistants to their old friends Jim Borrowman and Bill MacKay, who ran the Stubbs Island Whale Watching tours aboard their old sixty-foot workboat, the *Gikumi* (*GHEE-kah-mee*), which means "the chief" in the Kwakwala language. She had been custom-built for Mr. Fred Wastell back in 1954 and used to haul logs to the sawmill at Telegraph Cove and to deliver lumber and dry goods up and down the coast, before being converted to a tour boat in 1980. Bill and Jim always ferried the students out to West Cracroft at the beginning of the summer. They instinctively adopted a protective posture toward students who ventured into the area and checked in on them from time to time on the lonely island—out of concern for the young Americans and gratitude for their scientific interest in the Northern Resident whales.

The gang from UCSC loaded their gear—tents, tarps, and sleeping bags; stoves, pots, and pans; dried food, fresh food, and water; canteens, shovels, and axes; insect repellent, toilet paper, and bandages; maps, charts, and car batteries; binoculars, hydrophones, and cameras—into the *Gikumi* and another boat that would be their lifeline to the outside world for the next ten weeks: the *Tesseract* (named for the mysterious government project in Madeleine L'Engle's *A Wrinkle in Time*). She was a sixteen-foot C-Dory Classic Cruiser, white with a fiberglass hull. Everyone called her *Tessie*.

They launched their brimming vessels and set out into Johnstone Strait, turning right toward the southeast and plowing hard against the powerful and hazardous currents carrying the afternoon tide. The riverlike strait, a ribbon of seawater that separates Vancouver Island from the mainland, is named for James Johnstone, master of the armed supply ship *Chatham* that accompanied George Vancouver and the *Discovery* to the area. The strait's islets, forests, fjords, and granite spires could pass for Norway or parts of New Zealand. The passageway was sculpted by prehistoric glaciers that left it extremely deep—up to two thousand feet in some places—especially striking since the channel is only about a mile and a half wide in some places.

The strait's northern terminus, near Telegraph Cove, connects to Queen Charlotte Strait via a channel called Blackfish Sound. *Blackfish* is the word for "killer whale" used by the local Kwakwaka'wakw (*quah-QUAH-kah-wok*) people, as well as some white residents in the area. The First Nations people have lived here for thousands of years and control much of the region's land and water. Blackfish Sound is an apt name: In the summertime these waters pulse with two hundred Northern Resident orcas, plus a revolving number of Transients. There are also humpback, gray, and minke (*MINK-kee*) whales, Pacific white-sided dolphins, harbor porpoises, Dall's porpoises, harbor seals, Steller's sea lions, and other marine mammals.

Back then, a moving feast of salmon fed Resident orcas each year as several migrating species funneled through the strait en route to the freshwater streams of their birth. The first migrating salmon, the fatty chinooks, typically showed up in late May and kept passing through until August. Next came the masses of sockeye, usually from June to August, followed by the pinks in July and August, the coho in mid-July, the northern coho in September, and the chum from late August until October. In late December, winter chinooks made their own run through the passage. (Salmon stocks have been drastically reduced since then, presenting a significant threat to Resident orcas of the Pacific Northwest.)

It took the students close to an hour to reach their new wilderness quarters—twin sites called Cliff Camp and Boat Camp, etched from the rugged shoreline of West Cracroft, a large and lonely island with no permanent residents, buildings, hot water, sewage, or electricity but plenty of wild animals, some of them dangerous, including cougars and bears.

The two camps were directly across the channel from a sheltered Vancouver Island bay known as Robson Bight, where the Lower Tsitika River flows into Johnstone Strait. The site selection was no accident. In 1982 the government of British Columbia set aside the bight as a special ecological reserve and sanctuary for killer whales. The reserve includes beaches and tidal zones and extends one kilometer into the strait. No one is allowed to enter by land without a permit. Visitors on boats are asked to remain outside the one-kilometer perimeter.

For generations—if not centuries or more—Northern Resident killer whales have halted to spend an hour or so almost every summer day in the calm waters of the bight. The area's biggest attraction from an orca's perspective is a series of small beaches lined with smooth, round stones. Killer whales come in next to shore and rub their huge bodies on the ocean floor. They exhale to reduce buoyancy and spin around in apparent ecstasy on the stones, making repeated passes through the shallows. Often they vocalize wildly, like kids on a playground. At times several whales rub at once; other times they take turns, one by one.

The reason for this behavior remains unknown. The whales might be scraping parasites from their skin, they might crave the massaging stimulation of the smooth stones, they might find the whole thing fun and relaxing, or they might be motivated by all of those things. Whatever the reason, the rubbing beaches of Robson Bight almost guaranteed that the students on West Cracroft Island would see killer whales coming and going steadily: that is, Northern Resident whales. Transients and Southern Residents are not known to visit rubbing beaches or partake in this behavior.

Life on West Cracroft Island was at times exhilarating, in a survival-of-the-fittest kind of way. There were tents to erect, latrines to dig, fire rings to build, wood to chop, salmon to catch, and equipment to hook up to the car batteries.

Water was brought in by boat in big blue jugs. The island had a small source of water, but it was not considered safe to drink. A dribble of liquid oozed from a porous, mossy rock face several yards down the beach. Everyone called it the seep. A clever researcher with some engineering expertise had devised a makeshift wooden gutter lined with a tarp to catch the dripping water and collect it in a metal barrel. The water was warmed over a fire and used to fill a crude wooden hot tub by the sea.

Cliff Camp, where Naomi would spend many of her summer nights, was a small clearance on the narrow ledge atop a 150-foot bluff overrun with a thick forest of red cedars and Sitka spruce. Just getting yourself up there on the steep and often slippery deer trail was a bit of an ordeal, let alone while hauling gear, food, and field equipment. The main advantage of Cliff Camp, however, was its wonderful elevation and unobstructed views up and down Johnstone Strait, weather permitting. On a clear day, you could spot a pod of orcas from a mile away; even more with good binoculars or a spotting scope.

Cliff Camp's height was also ideal for deploying the theodolite, a surveyor's instrument that researchers used to track the movement and relative proximity of orcas. After carefully calibrating the instrument, one peered through the device to triangulate the animals and pinpoint them in the strait. It was a good way to notate their passage through one's arc of vision. It helped collect important data such as travel paths, speed, and distance between two whales.

Boat Camp, as its name would imply, was set inside a small cove just protected and deep enough to moor a boat and keep it safe during rough weather. That first summer, Naomi spent time in both camps, and each had advantages.

Up on the cliff, the view and the relaxed environment were sublime. Naomi found life slower there, more contemplative. She had a little bit of free time to read, to work on side projects, or just to sit and watch the eagles whoosh by—sometimes beneath her feet. "It is absolutely the most gorgeous place on earth," she told her mom on the phone from Alert Bay one afternoon while getting provisions and doing laundry.

Boat Camp, to Naomi, was a lot of hard work, but offered many chances for fun. She got to go out on the water and track alongside the orca pods, helping to gather audio recordings or compile written data on killer whale observations. And she didn't have to climb that cliff.

But keeping a boat secured was a serious challenge, Naomi soon discovered. Because the strait is so narrow, millions of cubic feet of water per second pass through during tidal shifts. The difference between low and high tide can be as exaggerated as seventeen vertical feet. This made mooring difficult. The length of line tethered to the anchor (an old tire filled with rocks) needed to be calibrated precisely. If the line was too short, the anchor might be lifted off the seafloor at high tide, causing the boat to drift loose and eventually bang up against the rocks. But if the line was too long, it might create excess slack at low tide, sending the boat onto the shore and leaving her grounded until the tide turned.

The lifestyle at camp could generally be described as damp and cramped—both sites were small, and your companions were most of your universe for ten weeks or more. As Naomi put it, "Either you get along, or you kill each other, and we had a little of both: getting along and killing each other." It was a serious challenge, to put it charitably, to share such close quarters, surrounded by wilderness, with only a few other people you didn't know well. Naomi found herself turning more and more to her personal journal to express anxieties and vent frustrations. Her patience was being rocked. On June 29, 1986, she wrote:

> *I was struck by a very frightening feeling this evening. I suddenly felt misplaced, or perhaps more accurately unplaced. I felt as if I had no real home, no identity. . . . And suddenly I even doubted my ambitions and future plans. What am I doing here in the middle of the wilderness, with a latrine in the woods, no outside world communications, limited resources. . . . What am I supposed to do here? How am I supposed to survive this summer?*

She was bored at times and terrified at times and had some sublime moments as well. On one of the first clear evenings at camp, a moonless night at about eleven o'clock, the clicks and wails of orcas came in over the hydrophone linked to the little RadioShack speaker tied to a tree. Everyone ran to the cliff's edge. Now they could hear the orcas' respirations, one after the other, as they emerged slowly from the water to breathe.

Then Naomi saw them, not twenty feet offshore swimming around in a bed of bull kelp. She didn't exactly see the whales themselves. It was too dark for that. What she saw was the *bioluminescence,* the eerily glowing blue-green phosphorescence created by swarms of plankton that flourish in these waters each summer. When agitated by motion, they light up like submerged fairy dust. Naomi watched in amazement as the glowing outlines of salmon pulsed through the black water, trying to flee from the giant green monsters that pursued them around the seaweed. Sometimes the light was so bright one could actually make out an orca eye patch or two.

On a typical day that summer, Naomi might either go out on the *Tesseract* with Janice and Dave to take "babysitting" data and record killer whale vocalizations, or head up the cliff to record theodolite tracks. The work was often quite rewarding even if the mood felt frantic and chaotic at times. Dave was determined to finish his data collection that summer, and Janice was equally determined to make this her second-to-last season in the boonies. It meant everyone had to adhere to a bruising and tight schedule.

To Naomi, Dave Bain could come across as a bit of a techno-geek, at times more interested in the machines that monitored living things than the living things they monitored. But he was clearly a technical genius, and Naomi was interested in his research on orca vocalizations, some of it beyond her understanding. Dave had invented an ingenious array of hydrophones that he attached to a homemade pyramid-shaped structure of PVC piping. With this setup, Dave could triangulate various calls as a way to discern which and how many whales were talking. Naomi's job was to help throw the pyramid contraption into the water or hold the wheel of *Tessie* until it was in, then help haul it back out again later.

Dave would bring the recordings back to camp and download them onto a crude computer program, with reel-to-reel tapes, that he had invented.

Naomi also loved listening to the killer whale sounds coming in over the camp hydrophone. She had heard humpback vocalizations before—as with humpback sounds, most of the orca sounds fell within the range of human hearing—but these were quite different, an unearthly array of whistles, creaks, and groans. "It really *is* a conversation, isn't it?" she marveled one day to Dave. He told her that many researchers now believed that orcas communicate through a complex combination of sounds, body language, and physical touch. It was not that orcas and other dolphins had "words"—that was not even a constructive question to ponder, to Naomi. While they can understand a little bit of our sign language and symbols (such as Phoenix and Ake did in Hawaii), we have no idea what they are saying, precisely because they lack the words.

Naomi couldn't forget the haunting clamor. "The whales sound eerie and mystical as they cry out underwater," she wrote in her journal. "Is it a language or just a signal of identification? And what do you hear underwater in the teeming ocean full of fish, whales and seals? Is it a symphony or a cacophony? But it is not silent, not silent at all."

She learned from Dave that Resident orcas do have whistles that signal to others in the pod, "Hey, this is me over here!" They make other sorts of sounds when they're excited, when they've found food, and so on. But what those sounds meant, exactly, scientists did not know. "We know the context of some of the sounds they make, but not what they mean," Dave explained.

"And there's quite a bit of crossover: Some sounds that they make in one context they will also make in other contexts."

Processing sound is essential for killer whales to eat, navigate, and socialize. Orcas have no sense of smell, and though their eyesight is excellent, it's not much help at night or in deep waters. That's where echolocation—emitting a series of clicks and listening for their echo—comes in. Killer whales use echolocation to avoid hazards such as rocks, shoals, and boats, to locate prey such as fish, seals, and other cetaceans. Orcas vocalize to keep track of each other when milling, foraging, or traveling.

Naomi knew that all toothed whales have a large sac that rests atop their skulls in their foreheads, called the melon, filled with a fine, thick waxy oil. They can manipulate nasal sacs located behind their melon to make clicking sounds, changing the shape of the melon and using it as an acoustic lens, which they can focus into a narrower or broader beam as needed and then send it in almost any direction.

The clicks, which sound like a finger running over a comb, last from one to five milliseconds. Sound travels through water about four and a half times faster than through air, around one mile per second. When each click pings off an object, part of the sound wave is sent back toward the dolphin, where it is received through fatty tissue located in the lower jaw. From there it is transmitted to the middle ear and the brain. Dolphins have pinholes where outer ears should be; their auditory canals are unused. They hear with their jaw.

Each click is exquisitely synchronized so that outgoing sounds do not interfere with incoming ones: Each echo is received before the next sound is dispatched. The amount of time that lapses between a sound and its echo tells the dolphin how far away an object is. Direction is determined by comparing the relative strength of the echo on each side of the animal's head. By sending and receiving a continuous string of clicks, all dolphins can follow moving objects (such as food) and home in on them.

Naomi also learned that the visual and auditory regions of the killer whale brain are set closely together and have become extraordinarily integrated over the ages, allowing the animal to construct a visual image based solely on the echoes being received through the lower jaw. The whales can even determine the *species* of fish they are tracking, not only by the echo of the prey's size, but also the dimensions of its air bladder. Orcas, and all dolphins, can differentiate between objects with less than 10 percent difference in size. They can do this in a noisy environment, even while vocalizing. And they can echolocate on near and distant targets simultaneously, something that boggles the imagination of human sonar experts. One thing dolphins cannot do is echolocate through thick beds of bull kelp (so named for its long, whiplike strands that can grow up to a foot a day). The salmon know this, and they try to hide themselves in the beds. That is why orcas can often be seen tearing up kelp—they are in search of food.

Even though Naomi was interested in Dave's acoustical experiments, Janice's work on orca babysitting behaviors was fascinating to her, as well as being more accessible. Alloparenting among Northern Resident whales was commonly observed: unrelated females babysitting an unrelated calf for an hour, adult males looking after younger siblings, a grandmother minding her grandkid. Naomi always looked forward to helping Janice collect observational data. As the summer wore on, the sun grew warmer in the sky and the mist parted for days. It could get quite sweaty on the strait when the boat was still, up to eighty-five degrees or more, even on the chilly water.

Naomi was surprised by how quickly she could recognize the various pods and subpods and the matrilines that made up their foundation. Before long, she was able to visually identify individual whales by their dorsal fin's nicks, notches, or waves, by various scars and scratches on their back, and by the pattern of their gray saddle patches. She also quickly learned which calves belonged to which mothers—a critical skill needed to gather observational data on orca alloparenting.

Once a group of whales was identified, the students followed them to begin a series of five-minute "observation sessions." Janice would call out the information as Naomi added notations to a data sheet. They recorded the time, the whales' number and letter identifications, distance between animals, travel speed and direction, and types of behavior observed. Naomi also put a check mark in a box for each respiration the whales took—usually about every ten seconds or so.

It wasn't easy, Naomi thought, but not as nerve-racking as trying to keep track of dolphins or porpoises. The orcas seemed so unhurried, so stately, in comparison to their smaller, frenetic cousins. "They just give you all the time in the world to figure out who they are, don't they?" Naomi said to Janice one day as they motored off Robson Bight.

When Resident orcas travel together, they often come up to breathe in unison—their black dorsal fins cutting the surface first, followed by a good portion of their entire shiny backs rolling forward, up and out of the water, punctuated by a whoosh of inhalation. Compared to dolphins, it does seem to be happening in slow motion. "It's like they're saying, 'Yeah, over here! See? It's me!' And then they slowly go down again," Naomi said, "deep under the water like phantoms." That first summer, Naomi would spend far more time staring at data sheets than actually observing orcas. Janice had that privilege, calling out names and behaviors to Naomi and her pen.

Naomi realized that time was going by—she had been in grad school for almost a year. She was certain she would be back the following summer, and she knew that then she would be the one calling out observations to a field assistant, and not the assistant who has to mark the sheet.

6

Hotdogging

Jeff Ventre used a running start, dove out into the middle of the pool, and waited for Katina the killer whale to make her steady approach for a foot push. Clad in the standard-issue black-and-red wet suit, he began treading water in a seated fashion with feet pointing out in the whale's direction, a signal for her to make contact with his feet and commence the maneuver. She made contact: like a giant mother ship gingerly docking with a small lunar space module. He could sense the firm roundness of her rostrum sink into the balls of his feet, with his toes helping to forge a proper fit. The killer whale had "hit her target."

Jeff blew his trainer's whistle—a signal to Katina, known as a bridge, to let her know it was a perfect docking.

Katina began pushing Jeff backward by his feet. He pivoted onto his belly, facing forward, arms at his side. Jeff had decided to open with a real crowd-pleaser, the jaw-dropping "hydro-hop"—or "hydro" for short—and Katina, or Tina, as most people called her, instantly understood. She knew the hotdogging part of the show, currently titled "Shamu: World Focus," was starting. And on this cloudless day, she and Jeff were playing the lead roles in A Pool—the main, glassed-in performance tank at Shamu Stadium. Tina gave a vertical heave to her mighty flukes, and off they went.

The fifty-five-degree water rushed past Jeff's ears and nose, the deep chill practically slamming him with an ice-cream headache. Salt, bromide, and other chemicals in the artificial ocean stung his eyes. But Jeff was used to it by now. Besides, any cold and discomfort instantly vanished when he was focused on the risky aerial ballet he was about to perform with a top predator nearly forty times his size.

Few things feel more exhilarating than bolting through water at ten to fifteen miles an hour with tons of thrust propelling you from behind by your feet. It's like having a jet pack fastened to your soles. With that much energy, all you need to do is steer.

Jeff loved steering, and he was good at it. He always wore one pair of white athletic socks under two pairs of black dress socks. They were better than rubber bootees, providing more flexibility for the toes and maximum traction against the whales' slick and rubbery skin.

To execute the hydro, he needed Katina to push him almost to the bottom of the pool, from where she could build up enough vertical speed to propel them both skyward and out of the water. Jeff lifted his head to gulp some air, then leaned both head and torso downward. Katina, the Cadillac of killer whales, effortlessly shifted course with her power steering and luxury ride. Now she knew what they were about to do: a hydro. She was definitely into it, Jeff could tell, and he smiled to himself underwater.

Tina nimbly drove her trainer downward by the balls of his feet. Jeff was getting over a cold and his sinuses were inflamed. He braced himself for the dive, knowing what was about to come. Katina continued the downward plunge, almost to the bottom of the thirty-six-foot-deep tank. Jeff's sinuses began squeaking and squealing. They felt as if they were about to pop. Water was still inside the cavities from the last show, and the pain was intensifying as bone and tissue tried to equalize the mounting pressure. Pressure at the bottom of A Pool was equal to two atmospheres, and Jeff was not exactly descending in a leisurely fashion. He was performing a hydro in Orlando, not snorkeling off Key West.

Jeff used his head and torso to steer Katina along the bottom contour of the pool, hugging its shape while allowing enough space for her to get underneath him. It was time to launch. Jeff steered toward the sky. Tina issued three strong pulses from her tail, and up they went. He could see the bright Florida sun, shattered by ripples on the surface. He felt the power of Katina at his feet as the brawny mammal contracted the powerful muscles of her peduncle, a whale's lower trunk extending from the dorsal fin to the fluke. The force of her thrust hurled them both upward at remarkable speed.

Hydros are tricky. If not performed properly, they can be disastrous. Jeff had to steer the whale just so, keeping her far away from the hazardous edges of the pool. He also had to ride Katina at a perfectly calibrated angle—about sixty degrees. If he shot from the water at a lesser degree, the whale might fling him entirely from A Pool and into the audience in a blood-splattered disaster. If they came up too vertically, Jeff might land on top of Katina on the way back down, seriously injuring both of them.

Katina knew what to do. She raced up through the water at the optimum angle, surfacing from A Pool at precisely the right spot. After several moments in the cold and quiet blue, it was something of a jolt to reemerge into the warm, muggy sunshine and the thunder of several thousand roaring fans. The cacophony was mixed with the blaring and rather bad jazzy sound track that sounded like theme music from a canceled eighties cop show, jarringly mixed with African-village spirituals.

But Jeff was focused on how he and Katina were going to finish the feat, not the racket and flashing cameras of Shamu Stadium. He wanted to execute a beautiful swan dive: It was extremely important to look good upon entering the water, much like an Olympic high dive.

Just as they reached their apex, about fifteen feet above the surface, Katina flicked her huge head skyward, lending Jeff an extra punch of energy that flung him even farther into the air. It was all projected on the huge JumboTron screens behind the stage, like two magnifying mirrors. The crowd bellowed its approval. A whistle blew above the din—recognition of a proper execution. It was coming from the "control trainer" onstage—another bridge to let Katina know she'd done the jump perfectly, and that her positive reinforcement (a reward) would soon follow.

Katina continued her forward trajectory, almost like an exaggerated bow, but instead of reentering the water on her rostrum, she landed with a big frontal belly flop. As she was descending, Jeff completed his swan dive and soared in a great arc before gliding back down into the frothing pool with a minimal splash: a perfect 10. He emerged from the water and pumped a fist wildly in the air, a gesture of pure jubilance projected onto the JumboTrons for all to see.

Hydros can be exhausting, and Jeff fought to catch his breath. But the hot-dogging portion of "Shamu: World Focus" was just getting started. Jeff swam over to the stage and hauled himself out. He grabbed a metal pail full of herring and carried it over to Katina, then shoveled handfuls of thawed silvery fish into her bright pink mouth and directly down her gullet. Jeff knelt beside Katina and rubbed her head and eye patch. He smiled, cooed gently, and kissed the whale on her rostrum. The crowd let out a collective "Ahhhh." Something was inherently endearing about an athletic young guy showing true affection for a fifty-five-hundred-pound sea creature—especially one with the power to shred him if she wanted to.

But Katina bobbed softly in the water, her shiny black head resting amicably on the rounded concrete ledge. She did not break her gaze with Jeff. He was in full control of the animal; or at least it seemed that way.

Now that Tina had been positively reinforced, it was time for the next stunt. Jeff got up and took three long strides back from the water to get a running start. Katina watched his every move, waiting to see which side of the pool he would dive into *this* time—it was always a mystery. Jeff took a step toward stage right and sprinted. Long before his feet left the concrete, Katina had already done a barrel roll and headed out.

Jeff swam out to one side. Katina saw him and banked a turn to come around from behind. Jeff parted his legs in a split. Katina knew that was the signal for a "water pickup." She swam up under him so that Jeff was straddling her back, about a foot in front of her dorsal fin. Katina surfaced, lifting Jeff's body out of the water, with just his feet and ankles dangling in the pool. It felt like riding a horse in the water back in Lake Mills.

They moved slowly, about walking speed. Jeff sprang to his feet. The fans were wowed. Jeff hunkered into a crouching position for balance and bent one knee to reach down and give Tina a quick tap-tap-tap of the hand.

That meant "Go!" Katina stepped on it, pumping her flukes with a steady, forceful rhythm. You need to gain a certain speed to stay up on a surfboard, just like riding a bicycle. Balancing on an orca involves the same principle, so Jeff surfed around on the whale's back at a fast clip, about 10 mph. He smiled and waved as they circled the perimeter. Jeff loved this behavior, as long as he didn't get washed off by Katina's own wake. It took a lot of practice for that not to happen.

They usually went around the entire pool once. Jeff blew his own whistle, which was hanging around his neck on a string, to indicate that the surf ride was ending. He reached down and tapped Katina again: a "tactile bridge" until he could reward her with fish. As they approached the stage, Jeff adroitly leapt from the orca surfboard and onto solid concrete. The crowd roared its approval.

The music kept up its loud, frenetic pace. Jeff took a running dive back into the water, and Katina rolled over once again to follow. She was such a great animal, Jeff thought, a true performer, the ultimate show woman. Katina the Cadillac was *so* much better to work with than some of the other Shamus—the stage name given to all performing orcas at SeaWorld. Katina's daughter Kalina, better known as Baby Shamu, had much of her mother's talents, but wasn't quite on par. Gudrun, the Icelandic whale, was generally sweet and had her good days, but she never became consistent with hotdog behaviors such as hydros and rocket hops. She spent much of her time in the back, in E and G Pools, with the males. Gudrun's daughter Taima, the half-Transient offspring of the late, great Kanduke, was exceedingly intelligent and, as a result, much less predictable than the other females. Eventually, she had to be banned from water work because of concerns for trainer safety.

Katina waited to see what trick they would execute next. She knew hundreds of behaviors, and she was ready to perform any of them. Jeff opted for the ever-popular "stand-on."

A stand-on was done very much like a hydro: You waited in the water on your back with legs outstretched; Katina pushed your feet and you flipped over and steered down toward the bottom; you both swooped back up toward the surface. The difference was that this time, instead of launching from the water at one end of the pool at an angle, you started in the middle, and the whale pumped her way perpendicularly to the surface. She sprang straight up from the water, like some massive black-and-white jack-in-the-box. And instead of facing stage left or right, you faced directly into the wide-eyed audience, seated right in front of you.

Jeff never moved his feet from Katina's rostrum. Standing straight and tall, he popped out of the water first, waving at the astonished crowd as he rose like a statue of Neptune balanced on the tip of a giant blackfish pedestal. The control trainer whistled a bridge. Jeff flashed a bright grin at the very zenith of the stand-on, his mirror image projected on the screen just behind him. Then the two of them dropped vertically together back into the pool and disappeared.

Jeff swam over to a small platform where he'd stashed some salmon, which Katina adored, behind a rock wall. He pulled himself from the water and grabbed some fish for the waiting orca.

The last behavior, the showstopping finale, was known as the rocket hop. Jeff climbed onto the stage and pointed to a far corner of the pool. Dolphins, unlike most animals, understand the abstract thinking behind a finger point. If you have a cat and you point your finger at its food, it will look at your finger, not the food. But a dolphin will quickly learn to look where you are pointing and can be trained to go there. Jeff made an exaggerated point to the corner of A Pool. This was a hand signal to Katina, telling her to execute a "sighting bow" when she got to her appointed spot. She complied. Jeff dove down beneath her about twenty feet and laced his fingers together, bringing his arms into a bread-basket shape.

Tina deftly slipped her rostrum into the interlaced fingers of Jeff's hands. She coasted briefly as Jeff caught his balance by firmly placing one foot onto each pectoral fin. It was critical to maintain physical contact at all three points (the head and the two flippers) or the rocket wouldn't hop.

They took off for the surface. This time, instead of doing the forward dive of a hydro, Katina knew she had to execute a high and precise back dive. At the apex, Jeff let go of Tina's head, pushed off from her pectoral fins, and, as she fell back to earth on her back like a booster rocket, decoupled from her and continued to soar through the air, finally diving into the water several yards ahead of the great whale.

Jeff caught his breath once more as he swam to the stage and got more fish for Katina. He had saved the best for last and lavished her with two plump salmon. He could rest up before the next show in a couple of hours. This round of "Shamu: World Focus" was almost over; only one segment was left, the big "splash" routine, when a large male came out and slapped his flukes on the surface, soaking the first few rows of squealing tourists, most of them children.

Everyone always waited until the last person was safely out of A Pool before opening the gates and ushering in the massive male for the big splash finale. Nobody was allowed to be in the water when Tilikum was swimming around out there.

7

Residents vs. Transients

Summers on West Cracroft Island were never easy, but Naomi still kept returning to the isolated field camp. She was hooked on the place, in love with the whales, and eager to get on with her dissertation. She knew her topic would be the Northern Residents. She wanted to apply the sociobiology theories she'd learned from Bob Trivers, her adviser, to her investigations and analysis.

At the start of her second summer, in 1987, Naomi, Janice, and a crew of field assistants left Santa Cruz in two cars packed full of camping gear, field equipment, granola, and dried beans, overnighting at the home of Janice's parents in Seattle. From there, it was a three-hour drive across the border to Vancouver, where Janice always stopped to spend a few hours at Vancouver Aquarium, set amid the firs and Sitka spruces of the city's waterfront jewel, Stanley Park.

That year, Janice had prearranged a VIP tour of the aquarium. The students watched the show starring three killer whales: Hyak II, Finna, and Bjossa (*bee-YO-sah*).

Hyak II was a huge male captured in 1968 off Pender Harbour, British Columbia, along with six members of his Southern Resident pod. Finna and Bjossa were caught in Iceland and brought to Canada. Hyak quickly befriended them, and in 1987 he impregnated Bjossa. Finna, originally thought to be a female, turned out on closer inspection to have male genitalia.

Naomi enjoyed the orca show—she harbored no qualms at that point about keeping killer whales in captivity and thought the performances were entertaining. After the show, the students went backstage where only trainers had access. They got to spend time getting close to the orcas, close enough to touch. Naomi admired the two charismatic adolescents and was overwhelmed by the sheer magnitude of the giant male, Hyak II.

They continued the journey over to Nanaimo, sleeping again at the home of Mike Bigg. By the end of the next day, Naomi was back in Telegraph Cove. Dave Bain and the *Tessie* had left Johnstone Strait, but Janice was still there, as was a used inflatable boat for their use, a fifteen-foot Avon borrowed from Ken Norris's lab, which they named the *Sparrowhawk*. "It is so good to be back here. Everything familiar. Everything feeling right," Naomi wrote in her journal at the opening of her second summer on the island, in 1987. "I already feel like this is home again. Rereading last year's entries for these days, I realize I have already made a lot of spiritual progress by now, and seen a lot of sunshine, slept out in the rain."

Naomi shared the responsibility with Janice of hiring the field assistants in 1987 and was solely in charge after that. Finding quality people was not easy. Some of the students she took on were terrific, but not all. Naomi found that managing complex relationships in tight quarters was taxing. She wasn't always at her most magnanimous when people messed up, which was often. But she was hardly running for Miss Congeniality. She had a PhD to earn. If people took umbrage that she occasionally bristled at their incompetence, so be it.

"Some people think I'm a bitch," she announced to her underlings one day. "I don't have a problem with that as long as it helps me to get done what I need to get done."

Mostly, however, Naomi confined the bitching to her journal. By her third summer, now fully in charge of the field camp, she felt the burden of leadership. On July 25, 1988, she wrote:

> *I am not happy. The stove isn't working. People didn't bring enough cooking gear. How did that happen? Nobody brought pots or pans or plates or utensils or pails. . . . Cooking is going to be a fairly dull and unpleasant business like this, and all the food buying got screwed up, and . . . AUGH. . . . I will end up spending so much more money than I needed to have, because of fuck-ups and malfunctions and just plain bad luck. I am so aggravated. . . . I don't know. If I can get the lantern to work and the stove to work (I can't believe all these brand-new things that don't work!) and the struts come on Wednesday and I buy a big pot . . . I don't know. Where's the granola?*

One night an ember from the campfire smoldered all night and, in the early-morning hours, set fire to part of the campsite while everyone was still sleeping. Naomi's tent was closest to the blaze—its crackling and popping woke her up. She scrambled to douse it before it spread. Other times bears foraged through the settlement in noisy nocturnal raids. The students kept their food locked up tight in a bear box. When that ran out of space, they hung groceries from tree branches. One day a woman camping on the island was jumped by a cougar. She got away, but was badly bitten and scratched. Jim and Bill raced over from Telegraph Cove to ferry her to a hospital.

Then there was the "swim with the killer whales" moment. It happened with one of Naomi's field assistants, a young New Yorker who'd quit his banking career and decided to spend a summer in British Columbia studying whales. He wanted to be a field biologist, and Naomi decided to take a chance on him. He was easygoing, had a great sense of humor, and knew how to cook—all valuable assets in a remote and sometimes tense campground.

The killer whales had been swimming past camp nearly daily, stopping to graze on salmon in the bull kelp anchored near the cliff. One day they heard a group of whales on the hydrophone. Radio calls started coming in from the *Gikumi* and other vessels: "Hey, you guys, they're gonna come right under your cliff!" It was Jim Borrowman, announcing the arrival of A5 pod.

The former banker smiled. "Cool!" he shouted, then scrambled down the bluff to see the whales up close. Naomi watched from above as he ran to the beach and plunged into the fifty-degree water, without a wet suit. She grimaced. If the approaching whales didn't finish him off, the icy water just might.

Six whales were steadily traveling down the island's edge, their backs and fins slowly rising above the surface then slipping beneath the water once again, like black submarines. Now the killer whales were thirty feet from the field assistant and extremely close to shore. He treaded water. Other students held their breath.

The orcas disappeared. The powerful whoosh from their blowholes ceased. Everyone waited in the silence. Things grew tense. A minute went by, then two minutes, then three. Finally, the A5s resurfaced, maybe a hundred yards away, farther down the strait. They had either swum under, or around, the human in the water.

Naomi breathed easier. She hadn't thought the animals would bother him, and besides, it was not her job to be his mom. He knew he was taking a stupid risk, Naomi thought. Everyone has the right to act stupidly. On the other hand, these were top predators, aggressive hunters without enemies, wild and unpredictable. Swimming with them was not advisable, in Naomi's view. Getting in the water with Resident orcas was risky. Swimming with Transients could be suicidal.

Naomi figured that the A5s had heard the loud splash when the man jumped in and then processed the weird echolocation pinging off his unfamiliar mammalian form. The whales may have mistaken him for an outsize seal. Perhaps they couldn't figure out *what* he was. Whatever their reasoning, they made the instant, collective decision to avoid the object altogether. They were too cautious and conservative to investigate the strange being.

It was one of the many fascinating lessons of life on Johnstone Strait. Another intriguing trait about killer whales in this part of the world was the clear distinction between the Transient and Resident communities. The ecotypes are distinct populations even though they overlap in territory. When Mike Bigg and his colleagues first began identifying killer whales, they believed the Transients were outcasts from the larger and stable Resident pods. Only later did they discover the two groups began dividing along evolutionary lines about 2 million years ago. Their DNA indicated they had not interbred for more than ten thousand years. (Some scientists now believe that the two ecotypes should officially be designated distinct species, but the idea remains controversial in marine mammal circles. More genetic research is needed.)

Naomi quickly learned to tell the two types apart, just by the pointed tips on the dorsal fins of the Transients, and by the way the mammal eaters moved in the water. Residents tended to travel in deliberate, forward-moving patterns. When they traveled together and surfaced to breathe and disappeared again, Naomi could predict their dive pattern. It was typically two or three short dives, ten to twelve seconds apart, followed by a "sounding," a dive that lasts two or three minutes. They would usually resurface traveling in the same trajectory. Transients didn't follow such predictable patterns. They were far more prone to roam about in erratic motions that made it hard to know where they would surface next, the better to sneak up on their intelligent prey.

Prey was the other main distinction between the two groups. Residents wanted fish and Transients craved mammal flesh. Residents also vocalized intensively as they foraged, Naomi observed, possibly to tell each other where the salmon were hiding, or perhaps simply to express glee at the challenge of the chase. But Transients stalked silently—otherwise their prey would hear them coming. After a good stealthy kill, though, they often vocalized up a triumphant storm, accompanied by exuberant breaches into the air.

Naomi learned that Transients were skilled and forceful killers. They typically hunted in packs of three to five, often patrolling the waters around shoals, rocks, and reefs where harbor seals and huge Steller's sea lions weighing up to twenty-five hundred pounds spent their days. Out in deeper water, Transients stalked Pacific white-sided dolphins, harbor porpoises, minke whales, and even young humpback and gray whales, which can grow several times their size. Some Transients even went after large sharks. A popular YouTube video shot near San Francisco shows a Transient orca ravaging a great white shark, then sharing its liver with another killer whale as the shark's carcass descends into the deep.

Watching a pack of Transients pursue and catch their quarry is one of the most terrifying spectacles in the animal kingdom—though Naomi herself never saw a Transient kill. The orcas combine speed, power, and cunning to surround and overcome their victims, oftentimes hurling themselves forward, up and out of the water, in a behavior called porpoising, to vanquish their terror-stricken prey. In addition to eating marine mammals, killer whales, almost certainly Transients, have been seen harassing or feeding upon deer and at least one moose as the land mammals swam across the straits and narrow passages of the archipelago.

Once the Transients nabbed a meal, however, the ordeal for their victim—a mammal far more intelligent than a salmon—was often just beginning. Many killer whales played with their food; violently, mercilessly, and for what seemed like cruel lengths of time. It was not uncommon to see Transients batting a seal around with their flukes for up to an hour or more, slinging the victim far into the air like a beach ball; a pathetic howling missile waiting to die.

Scientists gave a few possible explanations for such prolonged and energy-consuming battery of prey. Unlike fish, many mammals fought back. Sea lions in particular were strong and ferocious, and seals delivered painful bites as well—one reason why Transients usually bore more scars than Residents. Dolphins and porpoises could ram hard with their beaks. It made sense that these orcas would expend extra calories wearing down their lunch *before* finally dispatching it.

Another explanation was that orcas liked to process their food. Despite their sharp teeth, killer whales did not chew; they tended to tear and swallow chunks of flesh whole. Resident whales might gingerly remove a salmon's head before eating it; antarctic orcas used their tongues to surgically excise the breast muscle of penguins, leaving the rest of the bird intact. One advantage of beating up a seal might be that it made it easier to remove the skin and fur—a brutal form of tenderizing meat.

Killer whales in other parts of the world had equally fascinating cultures of hunting and eating. In the North Atlantic, some pods corralled large schools of herring into tight balls and then stunned the fish by slapping them with their flukes before taking turns eating from the dazed mass, Naomi had read. In the South Atlantic and Indian Oceans, some whales rode shallow waves onto remote beaches to snag unsuspecting seals or penguins. Around Antarctica, pods hunted seals cooperatively by lining up and charging an ice floe with a resting seal atop it, utterly unaware. Just before reaching the ice, the killers would dive, creating a wave that flushed the seal from the floe. The orcas often let the seal climb back up onto its icy refuge and then flushed it again, several times, before killing, skinning, and *finally* eating their meal.

Resident whales had also been seen chasing and harassing porpoises and seals. A number of Southern Residents were once observed killing harbor porpoises for no apparent reason off San Juan Island. But no Resident has ever been known to eat a mammal.

Transient killer whales were less vocal, less social, and less attached to their mother and matriline than Resident orcas. Unlike Residents, both female and male Transients were known to disperse from the pod of their birth; females left to join other Transient pods while some of the males remained solitary for much of their lives.

Naomi learned that Residents and Transients did not like to mix. "They really avoid each other," she explained to friends back home. One morning out on the *Sparrowhawk,* she and her assistant had spotted a male orca swimming alone, hugging the shore on the far side of the strait. Older Resident males often traveled a mile or so away from the females and calves. Naomi assumed that was the case here.

"Let's move in closer and try to ID him," she said. That's when she realized this was no Resident male. It was a lone male Transient. They followed him for several minutes before Naomi noticed a group of Residents traveling, in their leisurely manner, down the strait in the opposite direction. She had never seen Transients and Residents in the same waters simultaneously. Would there be blood?

There would not be. The Transient stayed on the far edge of *his* side of the strait, and the Residents kept to the far edge of *theirs*. "I think they were perfectly well aware of each other," Naomi speculated. "They were just staying as far away from each other as they could."

Naomi was coming to know these animals personally. Though scientific circles hugely frowned upon thinking of any animal in anthropomorphic terms, she couldn't resist a human analogy. The small groups of Transients were "kind of like the local trash family; people that nobody in town really gets along with," Naomi joked. "And every time they come through town, everybody sort of closes ranks and shuns them. There's only three or four of them versus a whole village of Residents, who are not going to be threatened by a few Transients. If it comes down to a street fight, the Transients are going to lose."

A few years later, one of Naomi's colleagues, Canadian whale scientist Graeme Ellis, would witness something close to what she imagined. Graeme, working at the Pacific Biological Station in Nanaimo, was alerted to a Transient pod nearby. He jumped in a boat to locate them and came across ten members of a Southern Resident group called J pod, tearing southward at high speeds toward Descanso Bay on Gabriola Island. Graeme saw convulsions in the water ahead, just outside the bay. It was the rest of J pod, acutely agitated, milling about like big, black torpedoes. Once the rest of the pod caught up with them, they charged into Descanso Bay (ironically *descanso* means "rest" in Spanish) heading straight for the beach.

Graeme spotted three other orcas, swimming away quickly about a hundred yards ahead of the pack. He recognized the trio: members of the T20 Transient group—a mother, her daughter, and her presumed son. The Transients made a run for it, trying to skirt out of the bay along its rocky shoreline, but the twenty or so members of J pod would have none of it. They pursued the aliens with ferocious determination, pushing them up along the rocks. Graeme observed white-water churning, fins and flukes flailing, heads pressed hard against the flanks of other animals, the nipping of skin. The screeches coming from all the whales reverberated through the boat's hull.

It only lasted a few minutes. After a ferryboat cruised by, the three Transients made another break for it, and this time they got away. Graeme followed them, only to discover fresh bloody teeth marks, or rakes, on their skin. "Whatever the reason," he said after the incident, "the T20s definitely got their butts kicked, in my view." Still, after nearly forty years of continuous research on the local orca population, this was the first act of aggression among wild orcas he had ever witnessed.

Resident orcas did not like hanging around Transients, but Naomi could tell they enjoyed the company of each other. All orcas are social animals, but the close-knit bonds between Resident whales are the most precisely documented. Routine socializing among Northern and Southern Residents might involve the entire pod, or just a few individuals, Naomi observed. It usually included play behaviors, especially among juveniles, such as spy-hopping, breaching, fin and fluke slapping, head standing, swimming upside down, rolling over other orcas, or playing with toys such as kelp or even jellyfish. Physical and sexual contact often occurred in these behaviors, especially among males. "When the males get together," Naomi said to incredulous friends back home, "the penises come out and start flying all over the place." A mature sex organ can reach eight feet in length.

Socialization consumed about 12 to 15 percent of a Resident orca's "activity budget," more than that for Transient whales, and it varied between the Northern and Southern communities. Southern Residents were more likely to perform exuberant aerial displays while socializing. They commenced their annual start-of-the-summer reunion with other pods by staging elaborate "greeting ceremonies." The whales approached each other in two facing rows until they were 30 to 150 feet apart, then halted and waited at the surface for about half a minute. Then, perhaps following a signal from the ranking matriarch(s), they rushed toward each other and merged underwater in a felicitous riot of body rubbing and excited, high-pitched vocalizing.

Resident orcas maintained their tight social bonds through additional gestures of trust and cooperation. Naomi occasionally saw them sharing food. She read in one study of Northern Residents that prey items were shared in 76 percent of 235 feeding events—though adult males were much less likely to share food than females and juveniles.[1]

Orcas were also known to assist pregnant mothers in giving birth. In one case, a female was seen swimming rapidly in circles in a thirty-second period before giving birth beneath the surface. Immediately after that, three other pod members were seen lifting the baby up and out of the water for a number of seconds. Over the next two hours, other members were seen in unusual swimming patterns, making celebratory tail slappings and continuing to lift the calf above the surface.[2]

Resident whales also provide succor to their wounded. In 1973, a calf was severely gashed by a ferryboat propeller in the Strait of Georgia, off British Columbia. An older male and female swam close to the injured youngster, cradling the calf between them to ensure it did not turn upside down and suffocate. The male repeatedly repositioned himself during the complicated maneuvering to maintain his support.[3]

Naomi was fascinated by how caring and compassionate these huge animals were. We sure have a lot to learn from them, she thought.

8

OrcaLab

Early summer on Johnstone Strait always ushered in a season of socializing. Boaters crisscrossing the waters often stopped to chat with each other, or called at Telegraph Cove, Alert Bay, or West Cracroft Island, where those intrepid American students were camped out. Life grew more festive as the weather warmed. The area is rich with First Nations' traditions dating back centuries: gift-swapping potlatches, canoe races, or salmon celebrations. Besides that, most Canadians are friendly people, and British Columbians are friendly Canadians.

Naomi did not have much free time, but she did try to socialize with people such as Jim Borrowman and his wife, Anne, and Bill and Donna MacKay. They were kind and hospitable, almost more like family than friends. Bill and Jim would check in with the students almost every day by radio. The *Gikumi* and later the *Lukwa* regularly idled off Cliff or Boat Camp, usually accompanied by camera-wielding tourists. Hot soup was always on board, and the students welcomed the meal. On cloudy days, it could get damp, cool, and raw out there. Warm soup was like student bait. "It's just a little bribe to keep you kids coming back each year," Bill joked. "We don't want to lose you. You're friendly . . . and you're foreigners. *And* you're teaching everybody a whole lot about our killer whales."

Naomi and other students often climbed aboard the tour boat to deliver impromptu talks about orca behavior and society. For the tourists, it was a definite bonus. The young people truly enjoyed sharing their knowledge, not to mention the hot soup.

One day Bill pulled up with a small group of Americans, including Naomi's father and stepmother. They had traveled from Los Angeles for a brief visit to the area, and now Raymond Rose was heading out to West Cracroft to see his daughter at her "office." Naomi got in the *Sparrowhawk* with two of her assistants and putted out to meet the *Gikumi*. She stood in the rear of the inflatable, steering the outboard engine as they approached, eager to see her dad. Naomi had a red cap with a visor on, with her hair braided into pigtails. As she got closer, one of the tourists, a middle-aged woman, reached for a camera and elbowed Mr. Rose in the ribs. "Look at that!" she cried. "A little Indian girl's coming out to greet us!"

The social event of the season on northern Vancouver Island in the late 1980s, from an orca researcher's perspective, was the annual summer gathering on Hanson Island, a craggy outpost crowded with evergreens at the far northwestern reach of Johnstone Strait, tucked between West Cracroft Island and Telegraph Cove.

Hanson had no permanent residents, though it *was* home to one of the most innovative killer whale research facilities in the world: OrcaLab, run by Dr. Paul Spong and his second wife, Helena Symonds. The main building, a high-tech, off-the-grid whale observatory, is a genuine 1970s Pacific-coast contemporary: weather-bleached timbers set on a rustic lattice of cedar piles astride a black-rock outcropping. Its solar panels, slanted, dark skylights, and sloping, shingled roof make OrcaLab look like an Earthship built by Marin County hippies during the Nixon era.

The location could not be more ideal: a tranquil cove often visited by Northern Residents that opens onto Blackney Pass. The pass is a narrow channel, an orca superhighway of sorts, linking Johnstone Strait and points south with Blackfish Sound and northern destinations such as the Queen Charlotte Islands, now known by their First Nation name, Haida Gwaii.

Paul Spong is a five-foot-six New Zealander with a receding line of wispy hair. During the sixties and seventies he looked more like a hippie than a gifted and visionary scientist. Naomi would occasionally drop in on Paul at OrcaLab. By this point, he was a bit older, a little more clean-cut, and very much a legend in his time.

Naomi enjoyed spending time with Paul and Helena, and the whale researcher was happy to regale grad students with stories of captive killer whales back in the early days—the 1960s—when research on the social behavior of *Orcinus orca* was in its infancy. Paul spoke in a soft New Zealand accent, almost flattened by decades in North America.

Paul had quit law school in New Zealand to pursue a BA and a master's in experimental psychology at the University of California, Los Angeles, where he worked on groundbreaking investigations that mapped the effect of different stimuli on various parts of the brain. While in California, Paul also met and married his first wife, Linda.

In 1967 Paul interviewed for the principal investigator position at the University of British Columbia's neurological lab. But there was an unexpected catch: The Vancouver Public Aquarium had just acquired an adult female killer whale. She was the first orca to be made available for scientific research. The aquarium needed a scientist, preferably with a neuropsych background, to study her. If Paul wanted the lab job, he would have to spend half his time at the aquarium doing whale science. Paul knew virtually nothing about whales, but he accepted the offer.

He began work that fall, spending mornings at the neurological lab implanting electrodes into the brains of live cats. The lab housed a collection of mammal brains in glass jars, including a twelve-pound brain from a killer whale. He'd never seen anything like it. Its sheer size and heft, its multitude of convolutions—far more than a human brain—told him that this was an exceedingly complex and sophisticated organ. Clearly this animal had a lot of intellectual power. Paul stared at the jar and wondered, What on earth does a killer whale *do* with a brain like that?

Paul quickly found that his aquarium work with the female orca was much more enjoyable than implanting electrodes in felines at the university. The whale was named Skana (from the Haida language for "supernatural one"). Skana had been caught in Yukon Harbor, Washington, a peaceful cove on the Olympic Peninsula about seven miles west of Seattle. She was netted with other members of her Southern Resident group, K pod, in a large purse seine. Purse seines are nets used to catch large schools of fish that aggregate near the surface, such as salmon, herring, or mackerel. Along the bottom of each net is a series of rings connected by a rope. When a school of fish swims into place, the rope is pulled and the net closes into a "purse," trapping all living things inside.

Purse seines were also ideal for catching orcas. Killer whales travel and forage near the surface and they fear nothing—including nets, at least on their first encounter. Some orcas swim right into purse seines, especially if they are full of salmon or herring.

That day in Yukon Harbor was the first time an entire pod had been captured at once. It was a cold afternoon, February 16, 1967. Ted Griffin, the young, rambunctious director of the Seattle Marine Aquarium, was looking for killer whales. Seven months earlier, Griffin's cherished pet, Namu—the second orca ever to go on public display—had developed a bacterial infection that racked his nervous system. Namu became disoriented, got stuck in the net that formed his sea pen, and drowned.[1]

Griffin was devastated. But he also learned that killer whales were profitable—both in terms of selling them to marine parks and selling tickets to see them. He had spent weeks trying to catch some whales. Now he had gotten a tip from the coast guard: Killer whales were spotted swimming off Port Angeles, along Juan de Fuca Strait, the broad and beautiful waterway that separates the Olympic Peninsula on the US side from Vancouver Island in Canada. The whales were headed east, for Puget Sound.

The following morning, Griffin amassed a platoon of men and machines, including a purse seiner, a helicopter, and a prop plane, to go find the pod. It took all day to locate the whales and corral them into a purse seine at sunset near Yukon Harbor.

In the morning, Griffin realized he had caught fifteen specimens. The next few days were spent sorting the animals into different pens separated by netting and deciding which whales would go to Seattle and which would be released. Three whales died while awaiting their fate. Seven were let go. That left five—all infants or juveniles. Griffin towed them across the sound in a floating sea pen to the Seattle Marine Aquarium.[2]

Two of the males, Ramu and Kilroy, were sold to the new SeaWorld theme park in San Diego. The smallest female, Katy, who was about a year old at capture, lived for just two months in Seattle. Her cause of death was not revealed. The largest female from Yukon Harbor, a young female named Kandu, stayed in Seattle as a popular tourist attraction for three years before Griffin sold her to SeaWorld.

"That left Walter, the fifth calf from Yukon Harbor," Paul explained. "Griffin sold him to a regional boat show, which brought him to Vancouver for display as a publicity stunt." But the fourteen-foot-long, three-thousand-pound animal refused to eat. His new owners knew nothing about cetaceans and tried to force-feed Walter seventy-five pounds of dead fish and hamburger meat a day. He lost weight and developed a terrible skin rash. When Vancouver Aquarium director Murray Newman saw the ailing orca, he offered the boat show $25,000 and took Walter off their hands.

Only then did officials realize that Walter was a female. They renamed her Skana. She was given live fish and finally began eating again: seventy-five pounds of cod, salmon, herring, and flounder every day. Skana regained her health and her skin cleared up.

In the spring of 1968, about six months after Paul started work, another large orca roundup took place, this time in Garden Bay, one of the many picturesque inlets of Pender Harbour, a deeply crenellated anchorage fifty miles northwest of Vancouver. One evening, some local fishermen were attending a bowling league banquet in Secret Cove, about ten miles from Garden Bay, when they heard that orcas had been spotted. They were acutely aware of the swift profits that Griffin and others were amassing by snatching whales from local waters. They raced back to the marina at Garden Bay, loaded their boats with nets, and went out to trap the killer whales.

The men spotted the pod at nine that night, catching seven orcas in their nets. They immediately called SeaWorld, Marineland, and Marine World Africa in California, in addition to the Seattle Marine Aquarium and the Vancouver Aquarium.

Paul's boss, Murray Newman, flew up to inspect the catch. Paul made a trip as well, armed with recording devices to capture the animals' vocalizations. The roundup had begun when a large male, later named Hyak (Chinook for "swift" or "hurry"), was caught and held in a netted pen in the harbor. Before long, members of his extended family, believed to be the Northern Resident A5 pod, entered the harbor and swam up to the net. The captive whale and his free-swimming relatives vocalized wildly and relentlessly. The pod seemed unafraid of the men, machines, and nets that surrounded them. They wanted to be near Hyak, regardless of the consequences. It didn't take long before *they* were captured, too.

Over the next several months, Paul shuttled back and forth from Vancouver to record the whales' sounds. In May, two males named Corky I and Orky II were flown to Marineland of the Pacific in Palos Verdes. Paul watched as they were separated from the rest of the pod. "The vocalizations were the most frequent and excited when these two whales were being prepared for transportation to Marineland," he said. "Their pen was moved across Pender Harbour so that they were separated from the others by more than a mile. I recorded the continual exchange between the groups of whales. During the final stages of removing them from the water, the vocalizations became increasingly frantic. They seemed to be associated with emotion and trauma."

That trauma hardly deterred Marineland officials. "Do we want to buy one? You bet we do!" Marineland curator John Prescott gushed to the *Vancouver Sun,* which noted that the quote was offered "gleefully."

Marineland already had one killer whale—a female named Orky I, a Northern Resident from I pod—when the males Corky I and Orky II arrived from British Columbia. Orky I only lasted about two years longer at Marineland—she died of pneumonia in July 1969 at the approximate young age of seven. The following year, Corky I died from a severe abscess in his body cavity, also at age seven. Corky I was replaced by Corky II, a female A-pod Northern Resident captured in Pender Harbour, in 1969 (Paul had returned to witness that chaotic scene as well). Corky II was the orca that Naomi saw perform with Orky II at Marineland when she was a teenager in Los Angeles. Corky II and Orky II would produce seven calves, none of whom would survive infancy.

The male calf from Pender Harbour was purchased by the Vancouver Aquarium for $5,000. He would soon become the aquarium's second orca, after Skana. Paul's boss, Murray Newman, was already thinking of the dual roles—touristic display and scientific subject—the new whales could play. "We think this is a unique opportunity to have a look at these huge creatures at close proximity," he told the *Vancouver Star*. "We will study the sounds the animals make at the same time."

The young male was hoisted onto a boat and ferried down to his new home at the Vancouver Public Aquarium, which named him Hyak II. It was a highly painful leave-taking from his mother. "When they were separated, she would occasionally turn in his direction and emit a series of about twelve distinct loud clicks," Paul said. "The calf responded with one or two separate vocalizations, and the mother would turn away." During his separation, transfer to Vancouver, and the first two weeks at the aquarium, Paul said, the baby whale "emitted nothing but a single plaintive cry again and again."

Paul had the new arrival placed in a small tank about fifty feet long, thirty feet wide, and eight feet deep—isolated from the other orca, Skana. The two whales could not see one another, but they could clearly sense each other. Before long, Skana and the youngster Hyak were breathing together in perfect synchronicity.

Paul focused much of his early work on Skana, when she wasn't performing in the main pool. He wanted to test her visual acuity. He designed a system of two levers and two cards that could be lowered into the water: One card had a single vertical line and the other had two lines. Each time Skana pressed the lever next to the card with two lines, she received a reward: half a dead herring dropped from a slot on an automated carousel.

Paul and Skana did the experiment seventy-two times a day. When he started, he used a card with a four-inch gap between the two lines. Then he narrowed the gap, to one inch, a half inch, quarter inch, and so on until he reached one thirty-second of an inch. At one-sixteenth of an inch, Skana selected the correct lever 90 percent of the time. At one thirty-second of an inch, she could no longer tell the difference. So Paul went back to one-sixteenth inch, expecting Skana to achieve 90 percent accuracy once again.

"On the first try, Skana pushed the incorrect lever and got no herring," Paul explained. "I was surprised but not alarmed, especially since Skana got it right on her second try and received her positive reinforcement." But then came the third try, and the fourth, fifth, sixth, and seventh. Each time, Skana pressed the wrong lever, even though it meant no reward. "Even more strangely, she began vocalizing forcefully before and after each wrong selection, like she knew it was wrong but chose to do it anyway." Skana kept selecting the single line, no matter what.

He tried repeating the experiment with the other cards, but Skana consistently picked the lever with just a single line, vocalizing loudly each time. Overnight, Skana had gone from scoring 90 percent to zero. Paul was despondent; his project appeared to be an utter failure. Or was it? "I began to suspect that Skana was pushing the wrong lever on *purpose,*" he said. "She was, in effect, telling me, 'Hey, I've got ideas and opinions too, buddy. Don't think you can make me do whatever you want to do just because it means I'll get a dead fish."

Paul came to realize that Skana was probably bored. "I started to think about the issue of motivation. I wondered, why are we giving this whale half a dead herring?" he recalled. "That may not be great motivation. And I said, 'Well, this is kind of a boring environment for the whale, she doesn't have much to do.' So I started thinking, okay, instead of fish, I'll let you do some other things. I said to her, 'You can swim around the pool. And if you get it right, I'll toss a bowl into the pool and let you retrieve it.' Stuff like that."

Under those circumstances her performance came up to about 50 percent, but it never returned to her previous level, Paul said. "So I still was thinking about the question of motivation. And I thought, 'Okay, orcas are acoustic animals. Their primary sense is sound; it's their main way to get information about the world, and their society.' "

Paul started producing a variety of sounds for Skana to hear: bells, flutes, whistles, even spoons on wineglasses. She seemed intrigued by them, perhaps even pleased. She often showed this by swimming in the center of the pool and rolling over onto her back, then approaching the source of the sound. When Paul lowered an underwater sound box into the tank, Skana swam to it and rested her lower jaw (essentially, her outer ear) directly on top of it.

As much as Skana liked new sounds, she also grew bored with them quickly, and she always let Paul know when she'd had enough. After tiring of a certain sound she would swim over to the gateway to the holding pool and stick her head in it, refusing to come out until Paul stopped the noise. "It was another imaginative way for her to speak to me, to tell me no," he explained. "Just like she did with the cards."

It was a whale's way of putting her hands over her ears and going, *Nyah, nyah, nyah*.

During this time, Paul had spent little time with the male calf held alone in the other pool. He did this deliberately and ordered aquarium staff not to spend time with the infant. (Years later, he discovered that some staff would sneak down after he was gone for the day and keep the orca company. A baby whale was just too cute to resist.)

Paul was trying to be scientific, not cruel, in keeping Hyak II isolated. He wanted to work with a "naïve" whale, one that had minimal interaction with humans. Skana was fully trained and performed in the aquarium's show, but not Hyak. "I was working with the thought that, okay, maybe I won't use food for a reward with him. Maybe I'll use sound instead," Paul said. Hyak had been left alone for several months. He had grown sullen and seemingly depressed. He was sedentary and barely moved around, preferring to simply lie still and stare into one corner of his small, rectangular tank. It was exceedingly unnatural behavior.

"I thought, okay, what I'll do is see whether I can train the whale to swim around the pool for an acoustic reward," Paul said. He fashioned a crude underwater loudspeaker out of an empty paint can with a small stereo speaker in it and hooked it up to an amplifier. He then obtained a sound-wave oscillator that could synthesize a huge variety of tones. He also got a record player and a stack of albums—everything from Beethoven to the Beatles.

"What I did was say to the whale, 'Okay, you're sitting in that corner of the pool. If you move out of the corner, I'll provide some acoustic stimulation for you,'" Paul explained. "'And if you keep going, I'll continue to provide this acoustic stimulation. But if you stop and go back to the corner of the pool, I'll stop.' It was a pretty simple situation."

But how did he get Hyak to start moving in the first place? "I just waited," Paul said. "He had to move at *some* point. And when he finally did, I played the music. You know, normal behavioral stuff: You wait until you get a component of the behavior that you want, then you reward it."

Paul was not with Hyak by the pool, but rather hidden inside a building behind glass, where he could observe the whale anonymously. "He would move just a little bit out of the corner of the pool, and I'd turn on the sound," Paul recalled. "And he would stop or go back to his corner. So I'd turn it off. I did this a few times, and within just two sessions, he is swimming around and around and around the pool. That was such a contrast to what had happened with Skana, where she took over six hundred trials to learn what I thought was a fairly easy task. He had solved the problem in less than two." Paul had been amazed.

"I said to him, 'Okay, I'll remove the requirement that you have to swim around the pool, and I'll reward you for *anything* you do that isn't just sitting in that corner. And under those circumstances his behavior underwent this incredible transformation. This little whale, he'd just been sitting there so quietly, all alone, and then I played my records and he became this unbelievably energetic little guy."

Paul always began a musical experiment by playing an underwater tone for Hyak, letting him know they were about to begin. From the corner Hyak would turn his head around and wait. When he heard music, he began to back out of his corner, tail first. Paul played everything for the little whale: Mozart, Miles Davis, the Moody Blues. Hyak seemed to like it all. But what he liked more than anything was music that was *new*.

One day Paul put an album on by the famous Indian sitar player Ravi Shankar. "He was so interested in that, he responded to it so enthusiastically," Paul said. "So the next day, I went down and played it to him again. And we didn't get more than just a few seconds into it when he stopped and went *straight* back to his corner. He sat there and waited for me to put on something else. He'd remembered enough of what I'd played him the previous day, and he didn't want to hear what he'd already heard, again." Hyak wanted something new.

This little whale had an extraordinary acoustic memory, Paul realized. And as with Skana, he believed that Hyak was trying to use operant conditioning to elicit a desired response from the *human,* not the other way around.

Hyak was now constantly demanding new music. Fortunately, in 1968, there was no shortage of it. The Rolling Stone's album *Beggars Banquet* had just been released, in addition to a number one single, "Jumpin' Jack Flash." Hyak seemed to love rock and roll. The aquarium's conservative management frowned upon the racket of English longhairs blasting their electric guitars, but Paul figured, what the hell? He'd give the Stones a go. He got the LP out of its sleeve and put it on the record player.

"All of a sudden, Hyak swam at me and charged around the pool making great waves that washed over the edges," Paul recounted. "He went down one side, leapt, and then turned the corner and r-r-r-aced down the other side. Then he shot his body out of the water once again, did a barrel roll, and dove back in. After that he slapped his pectoral fin on the water. He'd sit there and spray great plumes out of the side of his mouth. It was such an amazing transformation of behavior. And I said to myself, wow, he *really* digs the Stones!"

Hyak had other tastes in music. One day Paul played Beethoven's Violin Concerto in D Major, sending Hyak into a beautiful and bizarre ballet. "His whole body sort of vibrated to the tone of the music, and he slapped his pectoral fins in time with the melody of the concerto," Paul recounted. "His flukes as well, they were totally in time with the music. And there were fountains of water spouting from his mouth, all of it timed together. I was so amazed by this."

Paul ran to find other aquarium officials to observe the surreal waltz for themselves. By the time he returned with witnesses, however, the music had stopped and Hyak had gone back to sulking in his lonely little corner.

Clearly, sound was stimulating to Hyak, but Paul thought there might be another positive reinforcement at play: human companionship. "I believed that *I* was becoming the positive reinforcement," he recalled. "The company that I was giving him was his reward. And in fact, I did an experiment one time to test that idea out."

Paul would go down to the aquarium at night, where he had designed a test to see if he could get Hyak to vocalize in exchange for Paul's company. "I had a light on in the little office I was in, next door to the pool," he explained. "And I said okay, if you vocalize, I'll stick around. And if I heard him on the hydrophone vocalizing, and if he *kept* vocalizing, I would stay there and leave the light on. But if he stopped for more than the limit I had established—three minutes in a row—I'd turn the light off and go home. And under those circumstances he would just continually vocalize, hour after hour." Hyak quickly learned that he could take occasional breaks and stop vocalizing, but never more than the three-minute limit, lest the light go out.

"The light was a symbol that he had some company," Paul said. "It was a sign that the poor, lonely guy had somebody paying attention to him, that he had company during the night."

Paul had been in Vancouver for about a year and was starting to have a surprising and uncomfortable emotion: The dispassionate and highly trained neuroscientist was beginning to have doubts about his own research. Not the methodology, not the scientific rigor, but rather the ethics of the entire enterprise. Was captivity in the best interest of these large and intelligent creatures, no matter how far it was advancing knowledge?

Paul needed a good deal of soul-searching to answer this tough question. But once he did, the answer was simple and resounding.

"I began doing this work with orcas in captivity as a very traditionally trained experimental psychologist. I was interested in behavior and behavior alone. And I was deliberately not paying attention to that brain that I had seen in the glass jar," he said. "I had the thought that captivity might be stressful, but I pushed it way to the back. I was aware of all the speculation at the time about what dolphins *do* with their brains, but I deliberately pushed all that aside. I wanted to focus on finding out some really basic things about these whales."

Years later, as Naomi looked back on Paul's extraordinary experiences, she realized she had not given nearly enough thought to orcas in captivity during those early days of research. The orcas she saw at SeaWorld and other parks looked healthy and relatively content to her. And they seemed to enjoy performing, which they did with unwavering precision and drive.

And as Paul had illustrated, much of what we know about orcas came from studying them in a tank, not the open ocean. His work with Skana and Hyak would have been impossible in Johnstone Strait. The language acquisition studies that Naomi had worked on with Phoenix and Ake, back in Honolulu, could likewise only have been done in captivity.

To properly study animal behavior, scientists simply *had* to observe a species both in the field and in a controlled lab environment, she believed. That necessarily involved captivity, even if temporarily. Naomi was training to become an academic researcher, not a Save the Whales activist wielding leaflets on street corners. Fighting captivity was all about policy and politics; Naomi was interested in science and data. Besides, she wasn't going to work at SeaWorld. Captivity was not her issue.

But within the last year or so, things had begun to change. It had started to dawn on her that captivity might be hard on the marine mammals under study. A slight rumble of disquiet, a tiny pang of guilt over her prior enjoyment of seeing orcas on display had started tugging at her conscience. Naomi was beginning to think of captivity from the whales' point of view.

After all, she used to watch Corky and Orky at Marineland when she was young and she still visited the Vancouver Aquarium when she came here. She had recently seen Hyak with Bjossa and Finna. They were terrific. But Naomi had to admit she hadn't thought about where the whales had come from. She had no idea that Corky was taken from Pender Harbour, or that Hyak had been stolen from his pod at such a young age in the local waters near Vancouver.

Many years later after hearing Paul's entire history, Naomi wondered how he reconciled his emotional feelings with his scientific work.

He didn't. "It's not possible. I had no option but to reach the conclusion that what we were doing to the whales in those tanks was unfair. I had a little bit of experience seeing what they were like in their natural setting, the ocean, because I'd been present at two captures, in 1968 and 1969, both in Pender Harbour. That was enough to understand just a tiny bit about what they were possibly like in the wild. And it wasn't a very large step to say that it was wrong what we were doing, that taking them from their natural setting was unfair to them."

This was the reason why whale watching in the area was becoming so popular, he reckoned. "By experiencing these animals in the wild, people really feel as if they're in tune with something that is meaningful. It's a lovely way of life these whales have organized. And we come along and take a whale out of that environment and put her or him into a concrete tank. It's such an unnatural thing to be doing that you have to conclude that it's just not fair play."

Paul had started to think about the idea of releasing the captive orcas back into nature and reuniting them with their families. The idea did not go over so well with Murray Newman and the conservative management at the aquarium.

That's when Paul gave a lecture at UBC that changed his life forever.

The lecture hall was packed. Paul gave a talk about the biology of orcas, about echolocation, and about the experiments he had conducted with Skana at the aquarium. "This was the first time in the history of behavioral science that a lab animal so obviously refused to do something it knew how to do," he said.

"My respect for this animal has sometimes verged on awe. *Orcinus orca* is an incredibly powerful and capable creature, exquisitely self-controlled and aware of the world around it, a being possessed of zest for life and a healthy sense of humor." Paul could see a number of professors rolling their eyes at this point. Assigning human attributes like a sense of humor to animals was beyond frowned upon, it was taboo. Paul continued anyway. "Moreover, they have a remarkable fondness for and interest in humans." Then he dropped his bomb:

> *It has been my feeling, since observing the semicaptive whales at Pender Harbour, that* Orcinus orca *in the wild, in the company of family, is a decidedly different creature than the* Orcinus orca *that we observe in the aquarium. . . . I am not convinced we get a full or accurate picture of* Orcinus orca *in an aquarium setting. Perhaps an alternative would*

be to work in semicaptive environments like the one at Pender Harbour, where it might be possible to train the whales for release and recall. The feasibility of releasing trained cetaceans into the ocean and having them return upon command has been well established in studies with dolphins by the US Navy.[3]

Most of the audience burst into applause, though Paul realized that some faculty members and aquarium staff were sitting on their hands, looking glum. What Paul didn't realize, not yet anyway, was that a reporter from the local newspaper had been in the room all along, taking notes.

"And the next day there was a headline in the paper saying that this guy wanted to set the aquarium's whales free," Paul said. "I hadn't intended to make that a campaign or public event or anything like that. But I got caught in the fallout and ended up in a fight with the aquarium." His two-year contract was about to expire and management politely let him know it would not be renewed. He still had his job at the university, but things were rather tense there as well. Eventually, Paul accepted a $4,000 research grant as severance and left academia for good.

He was devastated to leave his friends Skana and Hyak behind in Vancouver, and he made an effort to wish a special farewell to each. (Skana would live eleven more years at the aquarium before succumbing in 1980, at about age nineteen, to a systemic fungal infection. Hyak would pass away from pneumonia, at approximately age twenty-four, on February 16, 1991—just four days before Keltie Byrne died in the orca pool at SeaLand.)

Paul used his grant money to start the Killer Whale (Orcinus Orca) Foundation, or KWOOF, whose whole mission was to halt all whale captures in the waters of British Columbia. But he also wanted to keep studying orcas, preferably in the wild. The best place to do that, a friend of his said, was Johnstone Strait and Alert Bay, the "Home of the Killer Whale."

The Spongs arrived in Alert Bay just before Christmas in 1969. Paul set about speaking with locals, fishermen, and First Nations people in other locations. Everyone told him how large numbers of orcas came through the area every summer, staying through the end of the salmon runs and then disappearing again. Eventually, Paul and Linda came across a peaceful cove on Hanson Island, tossed a hydrophone in the water, and established a camp there.

The marriage with Linda did not last, but Paul's fascination with killer whales did. He was hooked. Paul remained on Hanson Island, continuing to build OrcaLab from the ground up. The lab is dedicated to the belief that one can closely study the wild without interfering with it.

"We don't go out in boats to observe the whales," Paul explained to Naomi. "We do land-based work as a matter of philosophy. It's research without interference." In addition to recording the whales via hydrophones and analyzing the vocalizations for individual and family patterns, staff and volunteers also take visual sightings of orcas as they pass OrcaLab, and reports from land observation sites staffed by OrcaLab volunteers.

Paul met his second wife, Helena, at an end-of-summer party and salmon roast on a nearby island in 1979. She was a schoolteacher from Vancouver, beautiful with blue eyes and long strands of blond hair running down her back. In 1978 she took a teaching job at the Little School in Alert Bay, whose student body was about 80 percent First Nations. After marrying Paul, she quit the school to work with him full-time at OrcaLab. "It was impossible not to get involved," she told friends. Helena developed a deep interest in orca acoustics, society, and behavior, as well as the impact of human activity on orca habitat.

The idea for a network of hydrophones was born from one of the summer gatherings that Paul and Heleña hosted each year at OrcaLab. The informal powwows attracted fifty to seventy-five people from the local area and as far south as Puget Sound. It was a who's who of *orcaphiles* (sometimes affectionately teased as "orcateers"). They slept on the living-room floor, camped in tents in the garden, or crashed in one of the outer buildings in sleeping bags. Helena would ply the hungry mob with gourmet meals that she somehow coaxed from a wood-burning stove, much to the astonishment of just about everybody.

The summer gatherings brought everyone together to brainstorm and discuss ideas on advancing orca science. Naomi always looked forward to the rendezvous, and to taking in the musings of the cast of characters who showed up every year. Jim Borrowman and Bill MacKay would ferry people out to Hanson Island from Telegraph Cove and join the orca talks themselves. They knew the Northern Residents as well as anybody. Paul and Helena were the celebrated hosts, but one guest was the center of attention: veteran orca researcher Michael Bigg.

To Naomi and many others, Mike's surname was appropriate. He was a big, sandy-blond guy with, everyone agreed, an outsize heart and generous spirit. He is still revered as the father of contemporary killer whale field research, credited with developing the photo-ID system that revolutionized it. Naomi adored Mike and considered him to be a friend and mentor.

Mike was born in London in 1939, but his family moved to the Pacific coast of Canada when he was young. He soon grew to love nature, especially marine mammals. He received a PhD in biology from the University of British Columbia with a dissertation on the reproductive ecology of harbor seals. In 1970, Mike accepted the directorship of marine mammal research at the Canadian Department of Fisheries and Oceans station in Nanaimo.

One of Mike's first assignments was to conduct an official census of the killer whale population in British Columbia. At the time, it was widely believed that thousands of orcas populated local waters because so many had been spotted in the region. There was no way to identify individuals. It didn't occur to anyone they were seeing the same whales over and over.

Mike ingeniously enlisted the public's assistance. He printed up fifteen thousand surveys and had them distributed to fishermen, ferrymen, tour-boat pilots, lighthouse operators, and private boaters, asking them to record every killer whale they saw—all on the same day: July 27, 1971. This would help to eliminate the problem of counting the same whales more than once.

The results shocked just about everybody. According to the survey, there were not thousands of orcas in British Columbia. The estimated population was, at most, 350 whales.[4]

Mike loved to share his knowledge, and he worked closely with colleagues such as marine mammalogists John Ford and Graeme Ellis (who witnessed the Resident whale rampage against the Transients in Descanso Bay). The team knew they needed to devise a foolproof system to identify each member of the orca population. They also needed to delineate the varied family patterns that make up killer whale society. No one even realized the difference between Resident, Transient, and Offshore whales, let alone the subdivision between Northern and Southern Residents.

Early in the survey, Mike noticed that one whale had a distinctively mutilated dorsal fin, probably after meeting a propeller blade. When he got back to the lab, he wondered to his team if he could find and identify her again. They did. Mike named the female "whale A1," though he also called her "fifty-year granny."

By taking photos of all the whales and comparing them, the researchers noticed that scratches and scars on the body and the unique gray saddle patches on the back could also be used for positive identification, in addition to fin nicks and notches. Saddle patches are typically divided between "open" saddles—with streams and fingers of gray in their patterns—and "closed" saddles, which are more solid in appearance.

The small team of photographers and orca spotters soon grew into a small army of whale-watching volunteers. Bigg, Ellis, Ford, and their team sorted through the thousands of black-and-white images that had been sent in. (Each whale was photographed from the left side, an arbitrary choice that standardized the system.) Mike and his team compiled an exhaustive catalog of photos of each whale. It numbered and lettered each animal according to pod and listed known or estimated gender and birth year. The constantly updated catalog remains the most comprehensive orca family history in the world, documenting the matrilineal relationships of every killer whale that frequents the shores of British Columbia and Washington.

The official census was a scientific and policy triumph. Now for the first time, the killer whale population could be better managed. Scientists could actually *count* the animals rather than guessing their numbers and whereabouts. The study of live killer whales in their natural world was founded.

Armed with individual orca IDs, researchers set out to collect and catalog data on longevity, reproduction, social relationships, travel patterns, births, deaths, and even diet—crucial knowledge for the full understanding of any species. The census found that the 120 or so members of the Northern Resident community were divided into sixteen pods and three distinct "clans" (clans are closely related pods with similar vocal dialects). The more tightly knit and much smaller Southern Resident community, with 71 individuals, was composed of just three pods within a single clan.

Pods were further divided into subpods, closely related and composed of two to eleven matrilines, followed by the most basic orca social unit of all: the matrilineal group. Matrilines consisted of a mother and her offspring, who almost always traveled together and usually rested side by side. When mature females had offspring of their own, their children became part of the matriline. These groups typically had three and sometimes four generations of whales, usually numbering about three to nine animals in size.

Mike and his colleagues also discovered that the orcas of Washington and British Columbia had a big problem: captures at Pender Harbour, Yukon Harbor, Penn Cove, and elsewhere—by industry giants—had left a significant reproductive dent in certain pods, especially among the Southern community.

In 1976, Mike submitted his official report to the Canadian government. It included data on the rate of captures, and how the unbridled "culling" of such a small population was unsustainable and endangering to the whales. He urged Ottawa to enact binding restrictions on the taking of orcas from any waters under the jurisdiction of Canada.

That same year, funding for his work dried up." He was ordered to work on other projects, mostly pinnipeds, the seals, sea lions, and walruses. The move did nothing to deter his killer whale work. He continued the research in his free time, usually at his own expense, often with contributions from private citizens infected by his excitement and love for *Orcinus orca*. In the late 1980s, Mike summarized his work in two papers on the photo-identification of orcas, to be published by the International Whaling Commission. But in 1990, he was diagnosed with leukemia. Naomi was crushed.

"It sounds as though Mike Bigg will die soon, a matter of weeks rather than months," she wrote in her journal on October 13, 1990. "How sad. And of course how unfair . . . the heart will go out of the project when Mike dies. He is the center, the lodestone—when he is gone, we will all mill aimlessly, no one to send data to, no one to send pictures to, no one to stop in to see along the road to Telegraph Cove."

A few days later, the final proofs for Mike's two papers were couriered to him from the publisher in England.[5] Mike Bigg, a true scientist until the end, reviewed the opus on the afternoon of October 18, 1990, then passed away a few hours later. He was fifty-one.

The news devastated the northern island community, including Naomi, who was working on her dissertation at Dan Kirby's fishing lodge when she heard the news. "Mike Bigg died last night," she wrote in her journal. "And the whales went by, perhaps even as he was dying—I'd like to think their slow passage past Hidden Cove was connected with Mike's passing from this world." She added, "You'll never get to see my thesis now! How will I know it's right?"

When her dissertation was completed, Naomi would dedicate it to Dr. Bigg. A few years after that, the whale sanctuary and rubbing beaches across from West Cracroft Island would be renamed the Robson Bight (Michael Bigg) Ecological Reserve, after intensive efforts by colleagues and friends up and down the coast.

Mike's death was not the only bad news to rock the tight-knit community. In 1986, Robin Morton, a Canadian naturalist and filmmaker, along with his American-born wife, Alexandra, were in the midst of a long-term field study on the islands of the area, including the daily acoustic recording of cetacean activity. Robin was also working on an orca documentary.

Alex had studied dolphin communication at Marineland in LA, then worked with the park's killer whales, Corky II and Orky II. One day she dropped a hydrophone into their tank—and pioneered the study of killer whale vocalizations. Alexandra recorded the sounds of mating, childbirth, and grieving after a calf was stillborn. She wanted to study how infant whales learned language from their elders, but none of the calves at Marineland survived for more than a month and a half. Alexandra would later go on to fight Norwegian salmon farms in the region, which she said were killing wild salmon populations and threatening the survival of the Northern Residents.

One morning Robin, Alexandra, and their four-year-old son boated into Robson Bight so Robin could do some underwater filming at the rubbing beaches. His gear included a high-tech "rebreathing" apparatus that captures exhalations (which contain oxygen), scrubs them of carbon dioxide, and recycles them back into the air supply. It prevented the release of bubbles that might distract marine life. But on this morning, the mechanism failed. Carbon dioxide filled Robin's lungs and he gently passed out. Alexandra dove into the freezing sea to find her husband lying on the stony bottom, faceup and arms outstretched, as if reaching toward her through the dark green water. She dragged him back to the boat. It was too late.[6]

In the late 1980s, SeaWorld included the tragedy in its educational material on orcas, "Killer Whale Fact Sheet," saying that "killer whales have attacked boats and bitten scuba divers. A diver in Canada was killed while photographing killer whales in 1986." The implication was that orcas were no more dangerous in captivity than in nature.

But the whales had nothing to do with Robin's drowning.

9

Happy Talk

Soon after settling into his new job at SeaWorld, Jeff Ventre realized just how much of his social life centered on the park. Many careers offer opportunities for friendship and romance, but something comfortable and insular about SeaWorld seemed to encourage it. Company executives tried hard to infuse the workplace with an infectious sense of camaraderie, fealty, and mutually supportive teamwork. Smiling was an on-the-job requirement.

Socializing among coworkers was the norm, much like students at a small college. People on staff often shared housing and living expenses. There were beach parties, pool parties, golf games, and nights out on the town. Weddings between SeaWorld employees were not unheard of.

Jeff made friends, even though many on staff were not his type. A large number seemed to be politically reserved, if not conservative, and more fervent in their Christian faith than he was used to. An oddly high ratio sold Amway products or participated in network marketing schemes on the side, to make ends meet.

In Jeff's early years at SeaWorld (he worked there from late 1987 through 1995), his core group of friends included Mark Simmons and Carol Ray—who were hired at about the same time as he was—and Samantha Berg, the smart, wisecracking firebrand from New York.

Carol is five feet six with honey-colored hair and sea-blue eyes. She combines the classic loveliness of Grace Kelly with the next-door warmth and humor of Meg Ryan. Born in Pennsylvania and raised in Connecticut, Carol went south for her higher education to study at Rollins College, Florida's oldest college and a respected liberal arts school in Winter Park, north of Orlando. Like Jeff and Sam, she was an excellent swimmer and loved the water.

Carol loved psychology as well and excelled in her courses. In her senior year, she took a specialized track on behavioral studies and did research on human behavior, and especially the concept of classical and operant conditioning and the work of B. F. Skinner. Carol knew all about the pigeon films. But she never considered working in the field of animal behavior or training. She had never seen a killer whale in her life.

After graduating in 1987, Carol decided to find a temporary job while weighing what to do for graduate studies. She thought she might like working with kids, perhaps those with developmental disabilities, though she wasn't sure. Then one day a friend mentioned a job interview she had at SeaWorld, for a tour-guide position in their education department. "Wouldn't it be fun if we could do it together?" her friend asked.

Carol wasn't so sure. She had always been conflicted about animals in captivity. She grew up with an intense desire to see wild animals up close because they were so beautiful and remarkable. But even as a young girl, she would leave zoos and aquariums feeling depressed, even a bit guilty for having enjoyed herself so much. But it sounded like a nice summer job, and it didn't involve contact with the animals. It wasn't as if Carol would be exploiting any of the creatures herself. She got an interview and drove down to SeaWorld.

During the meeting, the HR people kept glancing down at Carol's résumé as she answered questions. Something was up. "You know," a woman said, "you have a psych degree, you specialized in behavioral studies, you're scuba-certified, *and* you've been a competitive swimmer. You really have the background of an animal trainer. Would you be interested in any positions in *that* department?"

Carol was shocked. She assumed people required specialized instruction and hands-on experience working with animals to even think about applying for a trainer job at SeaWorld. She had none of those things. She declined the offer. Besides, Carol wanted to make some money, travel, and then go to grad school. This sounded like too much of a commitment.

Carol was hired as an educational tour guide and found that she liked the work. She enjoyed taking groups around the park to have a peek behind the scenes at areas not normally viewed by the public.

Carol was liked by guests and staff alike. She was careful to follow the cues for proper on-the-job terminology, and how to deflect difficult questions from tourists. A training manual given to guides contained a list of offending terms to avoid. "Certain words and phrases have negative connotations," the manual said. "At SeaWorld, we call these 'buzzwords.' Avoid buzzwords and use more positive words—you'll give guests a better overall impression."

The booklet listed alternatives: the animals had not been "captured," but "acquired." They were not kept in "cages" or "tanks," but "enclosures" and "aquariums." Meanwhile, "captivity" was to be called "a controlled environment" (today the term seems to be "in human care"). And no matter *what* guests saw, animals never had "sex" at SeaWorld. They engaged in "courtship behavior."[1]

Other loaded terms were to be avoided as well. Some of them required lying—or promulgating scientific ignorance—to get around:

Dead, die—If people ask you about a particular animal that you know has passed away, please say "I don't know."
Kill—This word sounds very negative. Say "eat" or "prey upon."
Play, talk, enjoy—Anthropomorphic, they give human traits to animals.
Evolve—Because evolution is a controversial theory, use the word "adapt."

In November, Carol was contacted by the HR department again. A few trainer positions were opening up, due to some Orlando staff heading to the new SeaWorld in Texas. Did Carol want to try out? She got the transfer, to Whale and Dolphin Stadium, remaining there for five months with the belugas, dolphins, and false killer whales. In spring of 1988, Carol was kicked "upstairs" to Shamu Stadium, where she joined Jeff and other rookie bucket scrubbers.

"I was really enjoying my time at Whale and Dolphin," she told Jeff. "But the bigwigs showed up this morning and called us into the office one by one for a talk. I had no idea what was going on." Several people were transferred into and out of Whale and Dolphin that day. Some were sent to Sea Lion and Otter, only one or two went to Shamu. "I have mixed feelings—it's kind of exciting but I'm bummed it looks like it will be a while before I can do much up here," she told Jeff. "What do you think?" Jeff said it was the transfer that most SeaWorld trainers dreamed of. "You're supposed to be excited!"

Once at Shamu, Carol was handed another set of manuals and booklets to read, this time with information geared toward trainers rather than guides. One was a primer on training marine mammals. Another was a glossary of preferred and prohibited terms to be deployed while fielding questions during a show, titled "Difficult and Unusual Questions and Answers."[2]

One of the first phrases every trainer learned to repeat was the animals ate only "restaurant-quality fish." Jeff, Sam, and Carol could utter the mantra in their sleep. It never occurred to them that there is no actual definition of what constitutes "restaurant-quality fish." One would presume it meant fresh fish that had been inspected and deemed fit for human consumption.

One would be wrong. On page 4 of the "Difficult and Unusual Questions and Answers" handout was the question "If the fish that we feed our animals is 'restaurant quality,' why are the boxes stamped 'not for human consumption'?" The answer: "Fish going for consumption by the public must be federally inspected." All the fish at SeaWorld came from the same source as seafood served in restaurants, the booklet said. "It is of the same quality, but inspection is not required. All uninspected fish sold must be labeled 'not for human consumption.' "

Processing, moving, and feeding all that restaurant-quality fish to the killer whales was a major part of everyday life at Shamu. Tons of the stuff were everywhere. Adult killer whales at SeaWorld ate between 140 and 240 pounds of food each day, mostly thawed herring and capelin (a small smelt), but also some salmon, mackerel, and other species.

Fish was batch-tested for calories, fat, protein, moisture, and other values. The whales were weighed and carefully monitored for changes in eating habits and other signs of nutritional problems. On command, the orcas would slide up onto a special scale and stay calm while their weight was read. The animals were given random batches of food at random times throughout the day, whether during shows, training sessions, relationship sessions, or just periods of relaxation. "To provide a high level of mental and physical stimulation," SeaWorld's materials explained, "it is important to vary feeding times and amounts for mental stimulation."

SeaWorld would not reveal how much it spent on animal food, nor would it say how much cash it took in by selling tourists little pieces of thawed fish to feed to dolphins, stingrays, and other animals. It was a shrewd revenue stream: Have your visitors open their wallets to feed the animals they have already paid top dollar to see.

The corporation must have bought its fish at high-volume, wholesale prices (who else would consume tons of smelt in central Florida?), but implied it paid retail like everybody else. "The price of fish fluctuates too widely to give you a consistent figure," the "Ask Shamu" educational bulletin explained. "We feed our adult bottlenose dolphins between 25 to 35 pounds of fish every day (fish such as mackerel, smelt, capelin, and herring). You can go to your local supermarket and investigate how much 25 to 35 pounds of these types of fish would cost in your area to get an idea of the daily cost to feed just one dolphin."[3]

Other "difficult questions" that trainers had to prepare for included:

Q) Are any of our animals treated for ulcers or other stress-related illnesses?

Ulcers are found in both wild, as well as captive, cetaceans. In cetaceans that are ill, it is common to use medications similar to those used in the human field to prevent formation of ulcers. It appears that ulcers may be caused by many factors and are not related solely to stress.

Q) What is the difference between sedatives and tranquilizers?

A tranquilizer is often given to an animal who is acting normal but is about to undergo some type of procedure. In this case the veterinarian may be using a drug which allows the animal to be handled safely. A sedative might be considered a drug used to allay *further* excitement in an animal which is reacting to a change in circumstances.

Q) How long do killer whales live? Not 100 years like in the wild?

The most current scientific documentation (Heyning and Dahlheim, 1988) indicates Killer Whales have a *maximum* life span on the order of 35–40 years. The *average* is less and may be about 25 years.

Q) Are they happy here?

While it is difficult to qualify any human emotion in animals, we believe our animals are happy. We base this on behavior and social interactions with other animals, reproduction, health, and longevity.

Carol got some tips on the proper use of SeaWorld jargon from various trainers. "You always want to turn a difficult question—no matter what the topic is, or what the crux of the question is—into some type of approved sound bite," one of them explained. "So, if someone came up to me and said, 'The whales seem unhappy,' I would say, 'Well, what we've found is that whales in captivity live as long as they do in the wild, they're fed restaurant-quality fish every day, and they're given the best care possible.' "

Carol understood. It was like politicians who deflected treacherous inquiries by giving another answer. It was 1989 and the Iran-contra scandal was winding down. No shortage of verbal jujitsu had been emanating live from hearings on Capitol Hill, land of the misleading sound bite.

10

Mama's Boys

Naomi Rose knew she wanted to study the male killer whales of Johnstone Strait. Now she needed to figure out *how*. Over her first two summers on West Cracroft Island, she began honing her thesis topic to focus on the evolutionary biology of Northern Resident males and the possible selective advantages of their uncommon, if not singular, stable family bonds.

The pioneering work of Naomi's supervisor, Robert Trivers, was her constant guide and scientific muse. Dog-eared and heavily annotated copies of his book *Social Evolution* and his original papers went with her each summer to Canada. (Trivers was ending his time at UCSC—he eventually ended up at Rutgers University.) Naomi moved to the lab of Burney Le Boeuf, the pinniped expert and leading authority on the social behavior and migratory patterns of elephant seals. From 1986 onward, Naomi would spend her winters studying these seals.

Social Evolution had nothing about orcas, or even other cetaceans, but the general theory of evolutionary biology was universal and eminently applicable to *Orcinus orca*. As far as Naomi knew, no one had yet applied it to study social behavior in male killer whales.

She realized from reading Trivers's book that the different thing about killer whales versus the rest of the animal kingdom was this social structure where the males don't disperse from their mothers. "Male dispersal is the norm in most mammals, though I do know some human males who never leave home," she would often joke to others.

Naomi had been playing with the idea of trying to answer why. Why did this dispersal pattern evolve? What were the forces acting on the evolution of this social structure? Others thought it was a superb line of inquiry. If Naomi could reach some conclusions, her work would be groundbreaking. But her plan had a blemish. Naomi was at a loss when it came to the type of data she could collect to answer the question why. "It's too big a question," she lamented to colleagues later.

The problem was insurmountable. Naomi altered her question from why Resident males don't disperse to what are the consequences of their staying with their mothers? After all, having a posse of outsize, testosterone-charged males stay close by their moms and siblings can create stress and social disruption. How were those consequences mitigated? And how might a female-dominated society of apex predators, focused on raising calves, benefit from the permanent presence of rambunctious and libidinous bulls?

Tension often exists between males in many species—they are hormonally wired to compete for access to females. The disruption this causes is a key reason why most mammalian males disperse and live in bachelor groups or alone. "But with killer whales, you've got all these males living with their moms, so how do you keep that tension from boiling over?" Naomi asked her adviser, Burney Le Boeuf. "How do they manage it? Why don't we see male orcas going at it the way that male elephant seals or bighorn rams or wolves go at it, inflicting real damage to each other? How do they *avoid* that?"

Naomi was excited when she finally settled on her dissertation topic. "I'm going to investigate how this community works out ways of living together that minimize the disruption that males and their hormones normally cause," she told her buddy Janice. "And I want to study how they adjust for the added pressure of competing for food." Competition for all kinds of resources occurs in a group environment, she said. "That's why men often play violent sports. It's a way to ritualize natural aggression that group living can cause. It's a testosterone reliever."

Orcas don't play sports, but the males did display all sorts of male-only behaviors that seemed to mimic some of the ritualism, drive, and energy of football or basketball. Was that how males minimized the inevitable pressure and tension of remaining in such large and tight-knit family groups?

In the seasons that followed, Naomi set out to observe as many male Northern Residents as she feasibly could. She wanted to discern patterns of male behavior and write about what those patterns might say about the species. With whom did these males hang out? What did they *do* all day? How much time did they spend traveling? How much time did they spend foraging, socializing, or resting? When their mothers were resting, were they also resting nearby? And who was initiating all these behaviors?

Naomi began refining her research ideas, running them by her adviser, Burney Le Boeuf, colleagues Graeme Ellis and John Ford, and friends such as Jim Borrowman and Bill MacKay, Paul Spong and Helena Symonds, Janice Waite and Dave Bain, among others.

"What is the payoff for the male? Why does he stick around?" she wondered. "That's what I'm trying to get at. If you're a grown man and you're living with mother, there are definite disadvantages to that, like being subject to the house rules, right?"

Naomi thought she might have an answer. "For resident males, there's an advantage to staying with mom. And that's the fact that females are very gregarious. When multiple pods get together, the females gravitate toward each other and have their own sewing circles, or whatever." That intensive socializing gave their sons instant entrée to all those unrelated girls.

The son might hook up with the daughters of his mother's friends, or even with his mothers' friends themselves, she continued. Naomi had seen several adolescent males who were not socially mature yet—they were sexually mature, but not socially mature—hanging out with post-reproductive grandmas unrelated to them. "I think it's literally a Mrs. Robinson situation," she said, referring to the 1967 hit film *The Graduate*.

"That female may be past menopause, but it doesn't mean she doesn't like to have sex. And this young guy, who's got lots of get-up-and-go but no reproductive female that's going to give him the time of day, she'll hang out with him."

If the sons get dates, then what do the mothers get out of this unusual arrangement? The trade-off couldn't be so one-sided, Naomi knew, or it wouldn't be stable. Naomi suspected the benefit mothers enjoyed from letting their grown sons stick around was free babysitting.

Alloparenting occurs in both sexes in several cetacean species, but with orcas, males tend to look after their own siblings. Being able to count on older sons to babysit allows a Resident mother to be more reproductively successful. "It lets her concentrate on her newest born and not worry about the five-year-old calf that's still full of beans and is potentially going to run off and do something stupid, because the older brother's looking out for them," Naomi said. "It even allows her to get some 'me' time, which no doubt recharges her batteries and improves her health, making her a better mom."

Over the ensuing seasons, Naomi observed that adult males often traveled in the middle of the strait, where the water ran deep and fewer salmon swam, while females and younger calves hugged the shoreline, where shoals and kelp beds were more likely to shelter prey. "The males are sort of getting the dregs, but I think they're forced into it by their mothers," Naomi explained to the other students at the West Cracroft Island camp. "I think she's basically saying, 'Get out of my face. Go off in the middle there. I'm going to take the best pickin's here along the coast." Not only did the mother ensure more fish for herself and her calf, she was also protecting the youngster from any threats, keeping her back against the wall, basically.

When the pod was not foraging, but traveling or resting instead, adult males stayed much closer to their mothers, usually within a body length or two.

Young calves were virtually inseparable from their mother, though many showed definite adventurous streaks. Naomi saw some of them testing their mother's tolerance for letting them go explore. Killer whales are born precocious. They are exceptionally curious about their world before they grow up and learn to exercise more caution. Naomi noticed that most adults preferred to avoid an unknown object rather than swim up to it impulsively. "But the young ones will indulge as much of their curiosity as their mothers and siblings allow."

One day Naomi was following a large pod of Residents traveling south along the Vancouver Island side of the strait near Robson Bight when a young mother and calf, perhaps two years old, appeared on her starboard. Naomi followed behind them and to the side, conducting an observational session for her database. She maintained a distance of about a hundred yards, noting the pair's speed, diving intervals, distance from each other, and other data on a special sheet she had modified from her colleague Janice's earlier work on alloparenting.

Soon the *Sparrowhawk* moved a bit ahead of the pair. But the calf vanished. The mother was swimming just behind where he had been. Then he resurfaced, splashing around just feet from the churning propeller. What struck Naomi was not the calf's dangerous curiosity, but his mother's measured response to it. She wrote about the experience later in a cetacean anthology titled *The Smile of a Dolphin: Remarkable Accounts of Animal Emotions:*

> *Now that he insisted on examining my boat, she came nearer, overcoming her caution. . . . The whole time the calf frolicked in my engine wash, she was a steady presence. She didn't herd him away from the potential danger; she tolerated his curiosity. But she stayed within striking distance. [Somehow] this mother indulged her offspring's playful spirit and overcame her own natural caution in order to allow him some behavioral latitude. [But] she was keeping pace, keeping vigilant.*

The mother was permissive, yet also protective. The special relation between mother and child was palpable. Years later, Naomi would realize that killer whale calves taken from their mothers at a young age were denied this permanent bond. She knew that intelligent mammals, such as Tilikum, were subject to depression and neurosis, just like humans. Could an animal like that fully recover from suffering such a life-changing separation?

By August of 1990, Naomi was satisfied with the amount of data she had collected on Northern Resident males. She needed to start crunching the numbers and outlining the basic chapters of her manuscript, so she accepted the caretaker job at Dan Kirby's Hidden Cove Lodge. It would be a long, cold six months, a period that would mark the deaths of Mike Bigg, Naomi's grandmother, and the trainer at SeaLand of the Pacific, Keltie Byrne.

That winter there was an orca incident, too. It happened in early December. Naomi first heard about it one evening on the two-way radio she kept on in the fishing lodge—her only constant contact with the outside world. A female orca had washed up dead on a nearby beach. The sad news deeply moved Naomi and many of the people in the area. She waited until early morning before boating over to Telegraph Cove to get more details from Jim Borrowman and Bill MacKay.

When she arrived dockside, the necropsy was already underway. The whale's belly had been slit and her dorsal fin removed from the body. It would later be cast into a fiberglass mold.

She was A9, otherwise known as Scar because of ghastly gashes that a boat propeller had ripped into her back, just behind her dorsal fin and just before her tail. That accident, years earlier, nearly severed her flukes from her body. It was a wonder she not only survived but lived so long. The orca was estimated to be about sixty when she died. Scar was the mother of a well-recognized male, A5, better known as Top Notch because of a visible crescent-shaped nick near the top of his dorsal fin.

Back when Mike Bigg first began categorizing the whales, he believed the oldest males were the dominant members of each pod, traveling with their presumed "harem" of females. Mike had identified Top Notch as A5 and named his group the A5 pod. Only later did he realize that the females were dominant, and in a group with an adult male, he was related to them—most likely the son of the oldest female.

Scar, Top Notch, and the rest of the A5 pod had been captured in Pender Harbour, British Columbia, in 1969. Eventually the adults were released, including Top Notch and Scar. They remained outside the nets for days, waiting for the younger whales to be released, but that never happened. Aquatic theme parks snatched up the calves and juveniles, including Corky II, who was flown off to Marineland in Palos Verdes, California. Corky was almost certainly a member of the A5 pod.

Years later, Scar gave birth to another son, dubbed A26. She, Top Notch, and A26 were beloved and well-recognized fixtures up and down Johnstone Strait. Naomi would often see the female and her boys resting close together in the quiet waters of a protected inlet. They were inseparable.

Naomi was saddened by Scar's demise: She had been following the whale and her sons for five years. But now there was work to do. On that cold, wet morning, she assisted where she could, helping sort through the stomach contents of the poor animal. Even this late in the year, Scar had been finding salmon. Remains of nineteen fish were identified in her belly, seventeen of them chinook. Scar was approximately sixty-four years old and no cause of death could be determined. She had not been injured anytime recently. Hours after the necropsy, her sons Top Notch and A26 were spotted swimming aimlessly around the strait, vocalizing in deep, forlorn cries, seemingly lost in a sad and fruitless search for their mother. Naomi described the day in her journal:

> *How strange to be up so close to those scars that so distinguished her. How oddly unpoetic for such a beautiful animal to be reduced to literally nothing more than a bag of bones. . . . She seemed so much smaller than I would have thought—and she was, really, only 20.5 feet. . . . I sure hope the boys are all right. They did seem to be looking for her that day.*

The skeleton was sent to the maritime museum in Sidney, north of Victoria, where it remains on display. Shortly after Scar died, both of her sons left the strait. No one knew if they had also died or swum off somewhere to mourn.

When adult male Residents lose their mother, their lifelong companion, some have been known to die within months. Naomi believed it had something to do with the stress of being alone. "Some people think their heart's broken, and maybe there's something to that," she said. Once cleaved from his mother, a male might easily grow despondent. "But I think it's more than that," she said. "They don't have as much status once their mother is dead, and they were already low status to begin with. If there isn't any female relative who's ready and willing to take them in—it's going to be tough for them."

Two and a half months later, Naomi came across Top Notch and his brother. It was a great relief, though somewhat bittersweet. "I went on the water because it is flat glass calm this morning and there was a report of whales," she wrote in her journal on February 15, 1991. "Found A5 and A26 slow milling just in front of Bauza Cove, heading in a generally east direction, around 10:30 or something . . . it looks like these two guys are wandering around all by themselves—they haven't attached themselves to any female. It just seems so sad, like they're looking fruitlessly and listlessly for mom."

Naomi ended her field studies in the summer of 1990 and, after her stint at Hidden Cove, labored on her dissertation back in Santa Cruz, though she did make it up to Vancouver Island for a two-week visit in the summer of 1991. By now, Johnstone Strait felt like home, and Jim, Bill, Paul, Helena, and others felt like old neighbors. The following summer she led two four-week field courses for college students on West Cracroft Island for the School for Field Studies, a Massachusetts-based study-abroad organization centered on conservation, sustainability, and ecological health. The teaching fee helped Naomi pay bills as she finished her dissertation—a full year and a half later than the fastidious student had planned.

11

Breakfast at SeaWorld

Mornings at SeaWorld were hectic. All trainers starting out at Shamu as "bucket scrubbers" learned the routine well. They often showed up at 6:00 a.m. to find a pool of hungry whales anxious for fish and attention. A bit earlier, the animal-care staff had driven up in a flatbed truck to drop off hundreds of pounds of fish, semifrozen in giant cubes packed in cardboard boxes in the "fish room," leaving them there to thaw. The facility, big and brightly lit, was lined with gleaming tiles and outfitted with refrigerators, sinks, and stainless steel countertops, like an industrial-scale sushi bar for Goliaths.

A whiteboard in the busy room listed each orca's name and the breakdown of fish it would eat that day. Katina, for example, might be given eighty-five pounds of smelt, sixty-five pounds of herring, fifteen pounds of salmon, and a few pounds of mackerel and squid. Orlando's only male at the time, Kanduke, ate the biggest breakfast of all. Jeff was vaguely aware that Duke was a different type of orca, but he wasn't sure what the term Transient meant. Carol and Sam (who never worked together at Shamu) had no idea that Duke was a subspecies apart from the girls, and nobody knew Duke was meant to prey on marine mammals. What they *were* aware of, however, was his bellicose and unsettled disposition.

Duke was kept by himself much of the time in one of the back pools, where he often floated motionlessly at the surface for hours, his flopped-over dorsal fin warming in the subtropical sun. His main function was to breed with the females in Orlando and provide the big splash finale at the end of the show, after everyone else had cleared out of A Pool. But Duke was unpredictable and ornery.

No one was permitted to get in the water with Kanduke. He was also prone to self-injurious behavior, including banging his head against the pool wall or grinding his teeth on the bars of the gates that separated the pools. On several occasions, Jeff saw blood running from Duke's battered jaws. At night, clouds of mosquitoes would settle on his back.

The main task before breakfast was to stuff fish gills with vitamins, antacids, and sometimes other pills that the apprentice trainers figured were antibiotics because they came in capsules and were only given when an animal was sick, or acting "slow" in SeaWorld parlance. On the counter each morning they found individualized freezer bags with the whales' names on them, filled with packets of pills each animal was to receive. Trainers were not told what was in the pills, other than vitamins. Nobody asked. But they could see the little packets were marked with letters such as TAG, which they assumed stood for the antacid Tagamet.

At other times, when a whale was ailing and dehydrated, the bucket scrubbers were given syringes and instructed to inject herring and smelt with freshwater from the tap. Whales, like all mammals, cannot drink salt water and rely on prey as their source of freshwater. The hydro injections were tedious, and by the time the fish were carried out to the pool for breakfast, much of the water had leaked out.

Each whale had its own set of metal pails for the day, marked with the first three letters of their name: KAT, GUD, and KAN, for example. Each bucket weighed about thirty-five pounds when filled with fish. A male such as Kanduke could easily go through seven or more pails a day.

Throughout the morning, trainers assessed each animal's health, mood, and response to signals and tried to evaluate their social interactions for signs of precursors that might send them "off behavior." Just before each show, the trainers huddled to discuss which animal would perform which behaviors, including wet and dry work. They also assigned responsibility for one whale to each trainer while that whale was in the show pool, though a trainer might "hand off" the animal to another trainer for a specific segment or behavior.

The main primary reinforcement at SeaWorld, food, was meted out during the day and into the evening, most often during training sessions and shows: It added motivation for the whales, who quickly came to associate training and performing with getting fed. Buckets were also spread out physically around the stadium at strategic points where trainers could reward a whale between behaviors. Sometimes a whale ate another whale's fish or two, but everyone assumed it would all even out in the end, and it usually did. No orca ever went hungry at Shamu Stadium.

But no one got a free lunch—or breakfast. "They have to comply with our program in order to get that food in the morning," Jeff explained to Carol. Even when feeding an animal breakfast, you had to get and keep it under behavioral control. That meant separating them, so each animal could be fed from its proper bucket. Usually, the whales were anticipating breakfast and peered over the ledge looking for buckets. Most mornings, they complied, Jeff said, but not always. "Sometimes something whacky is going on in the social environment and you can't get them under control right away. When this happens, we walk away and come back and try again a few minutes later."

The bucket scrubbers put medicated fish on top of the breakfast pails so they could be fed to the whales first. It was the only way to ensure that the right vitamins and meds went to the right whales. Jeff gave some medicated food first, followed by a "chaser" of fish not stuffed to the gills with pills. The animals knew when they were getting pill-stuffed herring and would sometimes spit the spiked food out. The chaser usually prevented that from happening. It was sort of like with a cat, when you try to disguise a pill in the rest of its food.

Any animal that consumed up to 250 pounds of food each day was also going to leave a lot of stuff behind. Orca feces enter the water in a murky greenish-brown liquid, rather than a solid mass, and quickly dissipate into trailing clouds behind the animals. Salt water, bromide, and state-of-the-art filtration take care of the rest. Moreover, the bulk of water work was done near a whale's head or back. Direct contact with jetsam oozing from the anal slit, luckily, was minimal. The whale poop habitually drew bays of laughter from the tourists who saw it.

When cetaceans urinate, it's usually not detectable in the water. While Jeff was working at Whale and Dolphin Stadium, one of his colleagues noticed that when false killer whales hauled themselves up onto the slide-out area, they would often pee. It was easy to figure out why: They had lost buoyancy and their urinary bladders grew pressurized. It was only natural for them to let it spill out. The trainer quickly learned to reward that behavior with a whistle bridge, telling the animals they had done a "good job." After just a few times of doing this, the false killers urinated when asked. Trainers waited near their genital slits with jars to collect samples for testing. They would then reward the animal with fish and show it the jar of urine.

When the bigwigs at Shamu Stadium saw this, they adopted the procedure for killer whales as well. Before long, the trainers simply had to call the orcas into the slide-out and show them a collection cup, and they would pee on command. They were also trained to present their flukes to extract blood through a plastic tube.

If not on breakfast duty and the morning shift, junior trainers were sometimes assigned to "night watch," which was essential when newborns or sick whales required round-the-clock attention. Night-watch staff also observed the whales resting or milling about in their respective pools. Cetaceans do not fully sleep the way other mammals do. They do not breathe automatically and must remain at least partially conscious to avoid drowning. Instead, one-half of their brain "sleeps" at a time, and never for more than ten minutes.[1]

The main task during the overnight shift was counting respirations. Every thirty minutes the staff would time the breathing patterns of a whale, noting how many times the animal breathed in five minutes. Any noticeable change, faster or slower, indicated a potential problem, and they were told to then contact the on-site vet. Night watch was boring, the trainers thought, but at least they could catch up on their reading or grab coffee in the lounge.

The junior staff members at Shamu Stadium were not truly "trainers," but really more like performers and behavior "maintainers." They read the lines and went through the motions and even began to give hand signals to whales during performances. But whale training had already been completed by senior staff. Infant whales needed less training: They learned show behaviors by watching and mimicking their mothers and remembering which hand signals or underwater tones were associated with each movement. But younger animals did need training to learn certain husbandry behaviors, such as presenting their flukes for blood tests, and the apprentice trainers were sometimes assigned the task. They conditioned the young whales by pinching their flukes with two fingernails, to simulate a needle prick, then offered a reward for not flinching.

Rookie animal trainers at the time received surprisingly little instruction in how to work with animals. They mostly learned by observing, much like the baby orcas. They were also given reading materials, including the booklet "Animal Training at SeaWorld and Busch Gardens." It covered the basics, such as operant conditioning, positive reinforcement, and stimulus discrimination (SD), in which animals learned to distinguish between different stimuli (usually hand signals, but sometimes underwater tones) and then to display the correct behavior.

The philosophy behind animal training at SeaWorld was straightforward: "1) Trainers create an environment that is fun, interesting and stimulating for the animals; 2) Trainers reinforce desirable behavior with a variety of rewards and do not draw attention to undesirable behavior; 3) Trainers build strong and rewarding relationships with the animals based on a history of positive and stimulating interaction," the manual stated.

Positive reinforcement for desired behavior, in addition to food, included "back scratches, rub downs, grooming, toys, favorite activities, squirts with a water hose, ice cubes, puzzle games and one-on-one time," the manual said. All behaviors were *variably reinforced,* meaning rewards came at different times, so the animals never knew exactly when the next reinforcement was coming. "Experience has shown that a random schedule of reinforcement is more effective than a fixed one."

Training of animals at SeaWorld took place in six types of interactive "sessions." *Husbandry* sessions were for weighing animals, taking urine and blood, and other procedures; *exercise* sessions were for stimulation and recreation; *learning* sessions were how trainers conditioned the animals to perform specific behaviors; *play* sessions were downtimes for trainers and animals to interact with games such as ring-around-the-rosy and such toys as balls, cones, and floats (even though staff were told to avoid use of the word *play* when talking about the animals); *relationship* sessions were intimate moments between animal and trainer to develop and build mutual trust; and *show* sessions were the live performances, in which show components and positive reinforcements were constantly switched around to make them more stimulating for the animals.

SeaWorld called it the HELPRS system, for husbandry, exercise, learning, play, relationships, and shows.

Of all the terms to know, and methods to master, the most fundamental concept was *target recognition*. Each behavior was divided into a series of discrete steps, so animals needed to be presented with a "target" that they could recognize and physically touch. Trainers "use their hands as a focal point," the book explained. "Animals are trained to come to the trainer's hand, hold on it, and await the next signal. This behavior is called 'targeting.' "

If a behavior required the whale to be out of reach of the trainer, another "target" was used instead of the hand. "Just as a flagstick is a target that directs a golfer toward a golf hole, a target directs an animal toward a position or direction," the handbook said. Targets were fashioned from long fiberglass poles with a foam float at the end. Others included slapping the water to call a whale to stage, tapping the poolside glass to call a whale to the edge, or lobbing a chunk of ice into the water. Whatever tool was used, "trainers teach an animal to 'target' by touching the target gently to the animal," the manual explained. "The bridge signal is sounded, and the animal is reinforced. This is repeated several times."

Once a whale learned to target the object, it was moved a few inches away. The trainer waited. It didn't take long for the orca to figure out she was supposed to move over and touch the target, and she was reinforced for doing so. This was repeated several times until the whale successfully targeted the focal point on each round. Next, the target was positioned even farther away, perhaps a foot or two. When the animal touched it, the trainer again reinforced her. Before long, the orca knew she was supposed to follow the target wherever it moved.

"The target may then be used to lead the animal through a series of steps to gradually perform complex behaviors," the book said. "Eventually, the target is replaced by a hand signal. As with other stimuli in animal training, the hand signal stimulus is learned by introducing it along with the target."

Hand signals were used to send whales on any number of behaviors in their large repertoires, including bows, triple bows, "fast swims" around the perimeter, fluke splashes, back dives, spy-hops, "hula" spins, head shakes to "say" yes and no, rollovers, side breaches, pectoral-fin waves, "double-pec waves," water squirts, slide-out poses, tongue kisses, and the wildly popular "raspberries"—when whales make farting noises from their blowhole.

The book explained other terms that were key to animal training, including the *least reinforcing scenario (LRS)*. The LRS, often referred to as a "neutral response," was divided into two parts. "The first part is a consequence for incorrect behavior," the handbook said. "The second part is a stimulus providing an opportunity for reward—for two or three seconds the trainer is relaxed and attempts no change in environment. This brief time period is a stimulus to the animal to remain calm and attentive." After the LRS, the animal is rewarded for remaining calm and attentive. The trainer might then offer the animal a chance to execute another behavior, one that if completed will produce a reward.

By reinforcing an animal after the LRS, it will "learn from its mistakes," the book said. But a neutral response must be given properly: "An animal is never forced to respond to a situation, nor is it ever punished." Over time, the consistent application of LRS "decreases undesired behavior and increases calm and attentive behavior." It can help lower animal frustration that might come about from a lack of reinforcement (even a "neutral" response is still a response). The LRS "teaches the animal to respond without aggression."

Though she enjoyed learning these techniques, Carol Ray only lasted a little more than two years at Shamu. She grew disillusioned with SeaWorld after being transferred to the stadium. No single event made her say, "I want out of here." But a series of smaller incidents niggled at the back of her mind. She didn't discuss it much with coworkers, but the doubts and questions mounted.

First there were the dangers. Carol was still unsettled by what had happened to John Sillick in San Diego, in November 1987, when Orky II crashed on top of him as he rode on Nootka. It happened a few months after she started working as a tour guide. Sillick had sustained a broken pelvis, femur, and ribs among other injuries.[2] He was permanently disabled. Carol was told it had been an accident, chalked up to "trainer error." In other words, as long as trainers didn't make mistakes, they had no fear of being injured in the killer whale pool. That didn't sit well.

When Carol was first transferred over to Shamu, she began nosing around for more details on the Sillick affair. She figured people there could offer more of an inside scoop on what had happened. She was dismayed to get the same response as she had in the Education Department. "Trainer error. Nothing more" was the mantra repeated at Shamu Stadium.

Carol had experienced her own frightening killer whale moment, with Kalina, though she blamed herself for the scare. One day in a side tank, B Pool, Carol was playing with Baby Shamu, keeping her engaged with a variety of small behaviors until it was their turn in the show in A Pool. Carol had leaned down close to the whale, alternating between asking for an "open mouth," a "tongue out," a "jaw pop" (snapping her teeth together), and so forth. She reached into Kalina's mouth to give her tongue a pat.

"I was moving her quickly between behaviors," she told Jeff after work that day. "After asking for an open mouth with tongue out, I reached in her mouth to give her a tongue pat. I think she anticipated I was going to ask for another jaw pop, and as I reached in to pet her tongue, her jaw popped right down on my hand."

Carol sustained a gash between the first and second knuckle on top of her right hand, and on the same location of her palm. Her whole hand was swollen and throbbing. Fortunately no more skin was broken. Carol pretended nothing had happened. Despite the pain, she smiled at the audience and proceeded to work Kalina from the stage. She periodically dipped her hand in the water to wash off the blood, before giving Baby Shamu another signal.

Carol realized this was more than a love nip. She could not grip anything with her right hand, not even a small fish, nor could she open her palm completely. The other trainers spotted blood on her hand. They knew something had happened, but they put on their game faces and ignored it. The show had to go on.

Immediately after the performance, Carol's colleagues called for the medics, who bandaged her wounds. "It was my fault. Really," Carol told them. "Don't blame Kalina. It was a stupid novice-trainer's mistake. I should've been more clear with her, especially with my hand in her mouth! I shouldn't have been moving her so quickly between those specific behaviors." It took Carol's hand two weeks to fully heal.

In other incidents, Kanduke took a swipe at someone, either trying to grab the person's hand or else rising from the water, jaws agape, in a threatening posture. Carol noticed that these incidents, though not injurious, were seldom reported in the Animal Profiles or incident reports. She had seen Duke's chart. It didn't mention any of the close calls she had witnessed or heard about.

Two babies were born at Shamu Stadium during Carol's tenure—Katina's calf Katerina and Gudrun's daughter Taima, half-Transients sired by Kanduke. Both of them—but especially Taima—had also aggressively tried to jump on top of people in the water. At other times they mouthed the hair, arms, or shoulders of trainers. Most staff shrugged off the acts as roughhousing by rowdy juveniles. No one thought they were cause for alarm. They told each other that such incidents were something to be aware of when in the pool, but little more.

Trainer safety was not the only thing tugging at Carol in those moments of doubt. She was troubled about the whales themselves. Why did they require so many vitamins, antacids, and fish shot up with water? Something was fundamentally wrong with that. Something was lacking in the daily life of these whales.

The abrupt removal of the young Kalina was also disquieting. When Baby Shamu was four, the trainers were told she would be moved out of Orlando to SeaWorld Ohio. Carol was shocked. She was no expert on killer whales, but she knew enough about their social structure and strong bonds to believe that ripping Kalina from her mother and siblings—the only family she knew—was wrong.

"I thought we had a responsibility to provide these whales with as much of a normal life as possible, despite being held in captivity," she complained to friends on the outside. "Now I feel like we are letting them down by needlessly disrupting their social group."

Carol raised her concerns with a supervisor in private. She was distraught by the reply. "Forget about it," her boss told her. "Kalina is disrupting the shows. Won't it be nice to have her gone and not distracting the other whales?" Any further efforts by Carol to complain to superiors about the pending removal of the calf were met with derision.

Carol was assigned to work the night of Baby Shamu's departure, February 12, 1990. She labored to help separate the calf from her mother and the other females. They had been training for this moment for some time. But Kalina was still not used to being separated from her mom. It was difficult to compel her into the medical pool, and nets were often required to separate her. (While originally assigned to stay near and work with Gudrun during the removal of Kalina, Carol was ultimately pulled off that duty after raising questions about the appropriateness of the impending move. Instead she was told to take observations from the catwalk above the stadium.)

Carol was also told to stay on for night watch that evening, after Kalina was hoisted on a sling, lowered into a box of ice water, driven to the airport, and flown off to Ohio. Carol made observations regarding Katina throughout the night. What she heard and saw shattered her.

Katina was never a vocal whale, but that night Carol listened in sorrow as the distraught mother remained immobile in a corner of the pool, emitting wretched cries into the night. The harrowing din withered Carol. The gate to Katina's pool was open to the other females, including her infant, Katerina, but no one entered her tank. Carol thought the girls were leaving Katina alone with her grief. The bitter moaning reminded her of when Gudrun had first been shipped to SeaWorld from an aquarium in the Netherlands, in November of 1987. Carol was still a tour guide back then, but Gudrun's haunting distress calls could be heard emanating from Shamu Stadium for weeks.

Even the usually adoring press questioned the wisdom of removing such a young calf from her mother. Park officials conceded that some research on wild killer whales demonstrated "strong bonds" between mothers and their children. But they insisted that "many questions about the animals' social structure remain unresolved," the *Orlando Sentinel* reported when SeaWorld made provisions to send "the little celebrity" off to Ohio.

Ed Asper, SeaWorld's zoological director, did not believe that all killer whale offspring stayed with their mothers throughout their lifetime. And, he told the *Sentinel,* the bonds between "Shamu and her mother" had weakened since Katina gave birth to her next calf, Katerina.

Not long after Kalina left, SeaWorld updated its trainer handbook, "Difficult and Unusual Questions and Answers," with the following: "Q: Why did you take Baby Shamu away from her mother? A: Baby Shamu was no longer a baby. She was five years old and quite independent from her mother, who was socially pushing her away from her and was already raising a new baby."

Of all the whales at SeaWorld during Carol's tenure, the most morose and ornery was Kanduke. Duke spent much of his time alone in C Pool, floating motionlessly at the surface. He seemed to stay there by choice: The females would often gang up on him, picking fights with him and raking their teeth over his skin. The trainers did not spend a lot of time interacting with Kanduke because he was not a water-work animal—and he was unpredictable. From what Carol observed, he got a half hour, tops, of exercise during the day from doing the splash segment of the show, or from a training session now and then, before heading back to skulk in C Pool.

Other times, however, Kanduke swam free with the girls, out front in the main pool. They would circle the tank harmoniously and, for some reason, counterclockwise.

In September 1990, trainers reported that Duke had been acting "slow." SeaWorld vets agreed that he seemed a bit sluggish and noted a mild drop-off in his eating habits. No one thought the symptoms were serious, and they certainly did not expect the ten-thousand-pound male's health to worsen. But it did. On September 20, 1990, the great Transient died.

SeaWorld plunged into mourning. Despite his savage reputation, everyone loved the Duke. Upon his death, at 5:40 p.m., the final evening performance was canceled. Guests were not told why. The next morning, it was back to business as usual.

A few weeks before, Kanduke had scraped a pectoral fin and bled during a performance, but SeaWorld officials said that accident had been minor and was unrelated. "His death was sudden," spokesman Nick Gollattscheck told the *Orlando Sentinel,* adding that Kanduke was in his "middle to late twenties." But that was not possible. Kanduke had been captured as a young calf in Pedder Bay, British Columbia, on August 16, 1975. He was believed to be about eighteen-to-twenty years of age.

Gollattscheck also "cited three scientific studies showing killer whales generally live to be 25 to 35," the *Sentinel* reported, even though "other scientists have estimated that male killer whales can live to be 60 while females can live to 80 or 90."

The paper reported, "Kanduke's death leaves SeaWorld of Orlando with four killer whales, all female. It leaves the company, which also has parks in California, Texas and Ohio, with 13 whales—12 of them female." The *Sentinel* said the cause of death was undetermined pending a necropsy, adding, "The procedure will give officials an idea of why the animal died, but a full report that includes tissue analysis will take four to six weeks."

Kanduke's skeleton was to be featured as an educational tool, though a display site had not been chosen. Some of his remains would be buried on the Orlando property, the *Sentinel* said. The rest would be sent to a rendering plant, to be turned into fertilizer and pet food.

Meanwhile, the "lopsided sex ratio" left in Kanduke's wake "may cause problems for SeaWorld's killer whale captive breeding program," the *Sentinel* reported. "To prevent inbreeding and ensure healthy calves, animal care specialists prefer having several mature males and females. Capturing wild killer whales is controversial, and many animal welfare groups protest such actions." The company had no more permits to catch killer whales in the wild, according to the federal government, but that was not a problem. "It prefers to import whales from foreign marine parks, particularly parks that are 'substandard,'" the paper said.

Animal activists were infuriated. A group called the Dolphin Project, led by former marine mammal trainer Ric O'Barry, who once worked on the *Flipper* set, said Kanduke's demise proved that killer whales cannot adapt well to life in a concrete tank. O'Barry announced that his group would stage a consumer boycott of all Anheuser-Busch products until it promised to stop capturing wild marine mammals and to close dolphin petting pools. "We're not doing anything unrealistic, like saying, 'Set all the dolphins free,'" O'Barry told the *Sentinel.*

Two months later, SeaWorld issued a necropsy report stating that Kanduke had most likely died of a viral infection, but tests would continue. Though Duke had been found with a collapsed buoy in his stomach's first compartment, the report asserted that the float did not harm the whale. "While the presence of the object . . . was visually striking, it was not related to the cause of this animal's death," it said.[3]

In an extraordinary move, company officials met with staff members from the National Marine Fisheries Service (NMFS) and the federal Marine Mammal Commission to present the report in person. Necropsies were typically just mailed to Washington, DC, but SeaWorld took the measure because "we knew the government was getting a lot of calls for information," park spokesman Gollattscheck said.

Earlier, in the summer, Carol had requested a transfer out of Shamu Stadium. When Kanduke died, she felt the move could not come fast enough. It was approved and she was sent back to Whale and Dolphin. But by then she knew she would be leaving SeaWorld altogether. Carol was ready to move on. And it was time for some soul-searching.

In December of 1990, Carol met with Chuck Tompkins to ask for a six-month leave of absence. As she expected, it was denied. She turned in her notice. "I'm really going to miss the animals! It's so easy to become attached to them. But I won't miss seeing the problems that are created by having them in captivity, like the stress of manipulating their 'families' into unnatural groups." Carol could not imagine that the ensuing "struggle for dominance" would *not* lead to aggression and violence in some of the animals.

Jeff and Samantha were sad to see Carol go. But her departure fertilized seeds of doubt that had already germinated in their own minds, seeds that would one day flourish into a full-frontal assault against their employer.

12

Whale for Sale

A decisive public outcry followed Keltie Byrne's death when the three killer whales dragged her under the water at SeaLand of the Pacific in February 1991. Protesters appeared daily outside the theme park's gates, rattling the quiet, upscale community of Oak Bay. SeaLand was seriously considering ending the orca shows, after twenty-three popular years.

One thing was clear: If the whales remained at SeaLand, continuing the shows would be costly and cumbersome. In June, the Coroner's Court of British Columbia had issued a verdict in the case: Keltie died as a result of "drowning due to or as a consequence of forced submersion by orca (killer) whales, due to or as a consequence of falling into the whale pool." Even though the whales had "thwarted" her escape, the death was ruled an accident.

Among the expert witnesses at the coroner's inquest was Paul Spong, who came down from OrcaLab to testify about his experiences with killer whales at the Vancouver Aquarium. Paul stated that the SeaLand whales had been subjected to "severe sensory deprivation" for fourteen hours at a time in the dark metal module, and this "may well be the key to understanding the behavior of the whales" in the moment they killed Keltie.[1]

The five-member jury was headed by Dr. David Duffus, an associate professor at the University of Victoria and founding director of the school's Whale Research Lab. (Duffus would become the US government's expert witness in the legal wrangling with SeaWorld over Dawn Brancheau's death, two decades later.) The panel recommended that the provincial Workers' Compensation Board implement a list of seventeen different safety reforms at SeaLand to better protect not only staff, but visitors as well.[2]

SeaLand would need to appoint at least one person whose sole job would be to watch for any breach in safety precautions "by the general public or SeaLand staff," the panel determined. The park would also need to implement a system for quickly isolating one or more whales in the pool; install hand- and footholds around the edges for easy exit from the water; conduct safety and emergency training for all staff; and require personal floatation devices for anyone who went onstage, amid several other improvements.

The jury seemed to accept Paul Spong's assertion that the stress of captivity and cramped spaces contributed to the incident. "Confinement of whales in the 'module' should be limited to veterinary or animal husbandry functions only," they asserted. "Separation of the whales for behavioral reasons should be accomplished using partition(s) in the main pool."

Making matters worse, SeaLand had been warned about aggression among its whales. The alarm had been raised by a staff member in April 1990, just ten months before Keltie was killed. Animal trainer and marine biology student Eric Walters had written to the Canadian Federation of Humane Societies outlining his complaints and concerns about safety lapses at the facility. Walters was hired in May of 1987, but quit two years later after SeaLand management ignored his calls for improvements to animal welfare, and to protect the life and limb of staff and guests alike. Among his concerns he noted:[3]

- The three killer whales were "housed from 1730 hrs until 0800 hrs the following day in what is called the 'module.'" Lights were kept off all night and no form of stimulation was provided. The animals cut and scraped themselves on the metal sides of the small, dark tank. "I have seen the male, Tilikum, with the ends of his flukes abraded and bleeding." The tight space "leads to conflict between the whales, which have no options for avoiding confrontations. Often the whales' skin shows teeth marks from aggressive action between the three, which are not just superficial tooth rakes."

- Nootka was kept on a constant dose of antibiotics the entire time Walters worked there. "She receives exceedingly high doses of tetracycline every day: 56,000 mg. (about 2 ounces). She has been receiving this drug for over 3.5 years on a daily basis." Walters claimed that one of the veterinarians repeatedly warned against the practice, but his pleas were ignored by management. One supervisor reportedly asked, "Why stop if the animal appears healthy?" Nootka was also diagnosed with anemia, and the antibiotic seemed to boost her red blood cell count, he said.

- Noise from the harbor at Oak Bay was entering the tanks through the nylon-mesh nets, disturbing the whales. "When a hydrophone is lowered into the pool there is a constant roar from the props and pumps in the marina." In addition, sewer runoff and low salinity levels in the bay were likely causing health problems. Oil slicks from pumped-out boat bilges would routinely enter the pool.

- The orcas were becoming increasingly aggressive with each other. One time Nootka chased a tankmate into the module and ended up smashing her head on the metal side. Blood ran from her blowhole, but "no veterinarians were called until the next day, demonstrating negligence on the part of SeaLand."

- Park staff members were underpaid, underqualified, and undertrained. "SeaLand's policy seems, ironically, to specifically hire those with little or no experience with marine mammals. SeaLand, for the most part, ignores all conferences that are offered and provides very little in the way of public education. Therefore, most of the information that is expressed to the public is either inaccurate or out of date."

- SeaLand was a hazardous place to work. "I have rescued two trainers from the killer whales," Walters charged. On two occasions Nootka had pulled people into the water. But even after those incidents, no precautions were taken other than banning staff from physically touching any of the whales "for a few weeks."

"I feel that sooner or later someone is going to get seriously hurt," Walters concluded with chilling prescience.

Walters did not write a second, I-told-you-so letter following Keltie's death, though it must have been tempting. On April 1, 1991, a few weeks after the incident, he sent a nine-page memo to the British Columbia coroner's office, which had asked him for an assessment of the facts as he saw them. This time, he provided more details about conditions at SeaLand and the acts of aggression, especially by Nootka, he had witnessed. They were anything but "accidents."[4]

In the first instance, Nootka grabbed a camera from the hands of a startled tourist. Lead trainer Steve Huxter ran over and tried to pry the camera from the orca's jaws. "The whale pulled back when Mr. Huxter was holding onto the camera strap and I watched as he was pulled into the pool," Walters wrote. "He then exclaimed to me, 'For Christ's sakes, Eric, pull me out!' I then reached out my hand and he grabbed it, exclaiming, 'She's got hold of my leg!'"

Huxter was yanked back toward the stage, and Nootka ultimately relinquished his leg. Walters lifted his boss from the water and onto the floating deck. Huxter stood up and "nonchalantly continued his conversation with some of the visitors," Walters wrote. Nootka's act had been deliberate, but gentle. There were "no visible marks on his leg in the area where the whale had been holding him."

The next assault was more serious. Walters was working the killer whale show along with two other trainers, including a young woman named Henriatte Huber, who was stationed on the far end of the stage. Huber was speaking to the audience through a wireless mike while working with one of the whales. Then Walters saw her fall into the pool.

"I immediately ran from my position to where she was holding onto the stage from within the pool. I pulled her out and noticed that there was blood coming from her hand/wrists." According to Walters, Huber had been feeding the whale (presumably Nootka) and speaking to the audience when she took her eye off the animal. "It grabbed/bit her wrist which caused her to lose her balance and fall into the pool. There was a lot of blood coming out because the whale's tooth had punctured her artery."

No report was filed in the case. "Therefore NOTHING was done as a result of the particular incident," Walters wrote. Meanwhile, the no-touch policy was heeded for just a few weeks before Huxter himself returned to physical contact with Nootka.

Other incidents were also unreported, Walters wrote. "I can remember giving Nootka the cue for a kiss, and she came up with her mouth open and attempted to grab my hand. On another occasion, a blind member of the public was brought down to pat Nootka's tongue and she closed her mouth on the woman's hand." Still another time, Nootka clamped her teeth down on Walter's own hand. It took some time to convince her to release him, and the bite left his hand bruised, though not seriously damaged. He reported the incidents to his superiors, who once again forbid anyone from touching Nootka or performing the "kiss behavior" with her. Once again, the ban only lasted a few weeks. "In the case of the blind lady being held, [management] told the trainers to not let members of the public down on the stage for the time being. This policy stood except when Mr. Huxter had someone down that he knew and he would allow his 'friends' down to the stage."

SeaLand was not alone in dealing with aggressive orcas. "At Marineland in Niagara Falls, I was told by Mr. Huxter that one of the whales held a woman at the bottom of the pool long enough that she suffered minor brain damage," Walters continued. "At SeaWorld in San Diego, trainers have been grabbed by killer whales and others have been held at the bottom of the pool. Still others have been hurt as a result of inexperienced trainers issuing cues to whales at the inappropriate time." One trainer at the Vancouver Aquarium suffered a broken leg "thanks to the head of a killer whale."

Given all those incidents, "SeaLand should have taken responsibility to inform their staff of the threat and dangers in working with such intelligent and large animals," Walters charged. "Caution and tact should have been used when working with these animals at all times. I feel that I was not properly trained to appreciate the dangers, nor were safety measures in place to deal with any unexpected accidents."

Walters alleged that the whales at SeaLand were understimulated and just plain bored. Ironically, an outside consultant hired by SeaLand, Bruce Stephens, had written an animal behavior handbook warning against cetacean tedium. The handbook had been distributed to park trainers. "On page 1 it states that 'constant change in the environment of trained animals is critical to maintain their attention and positive attitude,' " Walters noted.

"Consider the dolphin. Pressed into service as an entertainer, an object of fascination for those fortunate enough to enjoy his antics, he is sometimes thought of as a financial asset rather than a thinking, feeling organism," the manual went on. "It may be assumed by some thoughtless humans that given enough to eat and an occasional new behavior to learn, all will be well."

Just like people, trained animals become bored with daily routines, Stephens wrote in his manual. "If you fail to provide the animals with the excitement they need, you may be certain they will create their own excitement."

Finally, Walters told the compensation board that SeaLand's fatal mistake had been a failure to "desensitize" the killer whales to having people in the water. "Animals will become comfortable with a new situation quickly and with little stress if they are exposed to it in small steps," he wrote, quoting the Stephens manual. A foot in the water will quickly draw the attention of a naïve whale. But by putting a foot in the water over and over, it will "desense" the animal to accept the foot as normal. Slowly, if more and more of the person is put into the water, ultimately having a whole human in the tank is no longer a novelty. Walters predicted that after desensitization, whales would "no longer react the way the three SeaLand orcas had reacted."

That didn't mean that water work was safe, Walters took pains to clarify. "I am not saying that performances should include trainers in the water. What I am saying is that, in the event a trainer should end up in the pool, the whales would be experienced with respect to the appropriate response they should elicit under such a circumstance."

Based on what he knew about killer whale training at SeaLand, Walters concluded, "The death of Ms. Byrne could have been prevented."

Soon after the verdict, the city council of Oak Bay voted to make removal of the orcas by the end of 1993 a condition of renewing SeaLand's lease with the city.

SeaLand had had enough: It was time to dump its whales. But what does one *do* with three killer whales who have been implicated in the death of a young trainer? Animal welfare advocates said the whales should be set free in local waters. But SeaLand vice president Dean Strongitharm rejected that idea, saying the Icelandic orcas would be "completely defenseless" trying to survive on their own in the ocean off British Columbia. Complicating matters, both females, Nootka IV and Haida II, were pregnant by young Tilikum.

Two months after the jury verdict, Strongitharm announced, "Part of the future of SeaLand does not include the display of killer whales." The show would go on for now, but "we will be phasing out the killer whales." He acknowledged that the official inquest into Keltie's death hastened the decision. He declined to speculate on who might buy the killer whales. A year later, SeaLand would close its doors and disband altogether.

Paul Spong of OrcaLab had his own idea for how to handle Nootka, Haida, and Tilikum. They should be returned to Iceland, he said, and set free. Selling them to another marine mammal facility "would, because of their history, entail enormous risks," Paul told the Associated Press. While generally opposed to the trade in orcas, he was particularly against the export of these whales. "That is, they killed a trainer," he said. The three whales could be retrained to catch fish and learn other skills to survive in the ocean. "They're extremely adaptable animals," he told the AP. "If they can adapt to one way [to live in captivity], they can adapt to another."

Paul noted that at least one orca, a large male named Charlie Chin, had successfully been released after extended captivity. Charlie Chin was held for more than a year in a sea pen at Pedder Bay, west of Victoria. "And he's still out there today," Paul said. He added that Nootka, Haida, and Tilikum would first need to spend time in a type of cetacean "halfway house" if they were ultimately going to be released into the ocean—where perhaps they could be reunited with their pods. Such a facility, a sea pen in a protected bay, would be used to train the whales for "release and recall." Under that regime, researchers could periodically call the whales to shore to monitor their health and see how they were adapting to the wild. "We could teach the whales how to make a living again in the ocean," Paul said. "I'm convinced it could be done."

SeaLand ignored Paul's advice and continued to look for a buyer. After all, no profit was made from an ocean release. Quite the opposite: Such a project would be monstrously expensive, and SeaLand wanted its money back and was not interested in investing more dollars into a halfway house for wayward whales that had killed an employee.

But who would want the animals even if two unborn calves were thrown into the bargain? The Vancouver Aquarium had no interest in Victoria's notorious Icelandics.

That fall of 1991, SeaWorld indicated it might be willing to take in the rogue beasts.

"SeaWorld has to study the whales before determining how they would be used in shows," Brad Andrews, vice president for zoological operations, told the media. He downplayed Keltie Byrne's death: "I think we have to talk about what really happened. The person drowned. It was an accident."[5]

In the end, SeaWorld said Tilikum, Nootka, and her calf should be sent to Orlando, while Haida and her calf should go to San Antonio. SeaWorld San Diego, meanwhile, wanted to import another orca—a captive-born two-year-old male from Marineland in Ontario, Canada.

SeaWorld needed whales. It had recently lost its seventh and eighth orcas in just five years. They were tragic deaths, and a public relations catastrophe for the company. The two deaths of 1991 did much to raise the disgust of its friends and foes alike.

On May 14, 1991, a female named Kahana had died in San Antonio at about the age of fourteen. She was found floating in a back pool with a male and a female orca. Some outside observers wondered if she had been killed. Kahana suffered multiple fractures of the mandible and skull, lacerations exposing the brain, and severe loss of blood.[6] "There was no indication the animal was in distress or had any health problems," Bob McCullough, PR director for the Texas facility, told reporters. SeaWorld had been trying to breed Kahana, but she miscarried her calf just six months before her death. She might have been pregnant again, but McCullough did not know. Little information was released about the incident. A necropsy report later said the whale had a cancerous tumor in her brain, which is presumably what caused her to smash her skull on the side of the pool, an act of panic driven by the pain. NMFS listed the cause of death in the marine mammal inventory as "severe trauma."

Kahana was given a cold and businesslike sendoff: "Shamu has not died. Shamu will never die," McCullough said. "It just happens to be one of the whales who portrays one of three roles at any given time."[7]

Kahana's violent demise left SeaWorld with just twelve killer whales in its entire collection, only one of them a male. SeaWorld Florida was seeking yet another import permit for a female orca named Winnie, from Windsor Safari Park in the UK. But after Kahana died, activists vowed to block that sale. They called for Winnie to be released in Icelandic waters, where she had first been captured. "To shift an animal from Windsor Safari Park to SeaWorld is like shifting it from Leavenworth to Sing Sing," Ben White, founder of the anti-captivity group Dolphin Rescue Brigade, told the *Orlando Sentinel*. "There is one option always available that is always ignored and that is to bloody well set them free."

Winnie's import application also brought unwanted scrutiny from one of SeaWorld's fiercest and most well-organized critics: the Humane Society of the United States (HSUS), the influential animal-protection group based in Washington, DC. The well-funded outfit contacted NMFS in April 1991 to request a public hearing on the import permit for Winnie and to express concern over

Keltie Byrne and two whales that had died the previous year: Nootka I in San Diego, and Kanduke in Orlando.

HSUS officials also questioned the "peculiar" circumstances surrounding Kahana's death. They had obtained and analyzed her necropsy report and complained to NMFS that it was "deficient and contradictory." Many details on Kahana, such as the discovery of her carcass, were omitted, HSUS said. The microbiology report was missing, and a diagnosis listed as "positive" in one supporting document was described as merely "possible" in another.

Three months after Kahana's, another death occurred at SeaWorld. In August 1991, the pregnant female Kenau died of pneumonia during a stillbirth at SeaWorld Orlando. Technically, two whales died that day, the mother and her stillborn calf.[8]

HSUS went on the attack again, complaining about an incomplete bacteriology report, inconclusive determination of the primary pathogen, "peculiar circumstances" surrounding a tooth infection, and the "sudden and serious nature" of the onset of pneumonia. "This tremendous loss of life is unacceptable, especially in light of the fact that SeaWorld is widely considered 'the best' in the public display industry," the HSUS wrote to NMFS. "This series of events indicate that systemic difficulties exist in the maintenance of orcas in captivity." HSUS urged the feds to investigate the recent spate of orca deaths and their necropsy reports, and to look into SeaWorld's orca husbandry in general.

Anti-captivity groups reminded the media that since 1965, when SeaWorld bought its first killer whale for display, the company had owned a total of thirty orcas. Now, nineteen of them were dead. "I think it's time the public wakes up and realizes these animals are not meant to be kept in captivity," warned Jerye Mooney, the marine mammal coordinator at the Fund for Animals.[9]

SeaWorld's McCullough rejected the criticism; his company was proud of its record. "I think we do have outstanding animal-care facilities and outstanding animal-care staffs," he told the *Orlando Sentinel.* Most of the orcas had died of "old age." He told the paper that SeaWorld believed killer whales lived twenty-five to thirty-five years. But as the *Sentinel* noted, "Biologists who have studied wild whales believe female killer whales can live 80 to 90 years, and males, 60 years."

On November 7, 1991, SeaWorld filed a formal application with NMFS to import Tilikum, Haida, and Nootka from SeaLand in Canada, despite the growing public concern on both sides of the border about keeping the deadly animals in such close proximity to people. The deal, which included two unborn calves, was estimated to be worth some $5 million.[10]

SeaWorld intended to apply for an emergency permit to fly Tilikum to Florida right away. Rapid action was needed "to eliminate his potential interference in the birth, bonding or nursing process" of his two sired offspring, said a letter from the company to NMFS. The feds pushed back. It would take several months, and a period of public comment, before any decision was made on the applications, including the emergency permit for Tilikum. Part of the reason for the delay: Federal officials wanted to carefully consider the death of Keltie Byrne *first*—especially since SeaWorld trainers entered the water with their killer whales.

If it was a tough time for SeaWorld and the display industry, things were about to get more troublesome. On November 22, NMFS and its parent, the Department of Commerce, held the federal government's first ever public meeting on captive killer whales. Officials sat for eight hours listening to "conflicting views about . . . *Orcinus orca*," the *Orlando Sentinel* reported. Captive killer whales were portrayed "as everything from 3-ton teachers to exploited prisoners, from scientific gold mines to depressed, aggressive orphans."

The Humane Society sent its vice president for wildlife and habitat protection, Dr. John Grandy, to testify. "The public is changing its attitude and challenging the ethics of additional orcas in captivity and the well-being of current captives," he said. Grandy asked the fisheries service to ban the capture and importation of any more killer whales, conduct a thorough investigation of currently held orcas, and "analyze the suitability of orcas for captivity and evaluate educational and research programs utilizing captive orcas."

Eight killer whales had died at SeaWorld in the past five years, Grandy continued. Wild whales lived fifty to eighty years; captive orcas rarely made it past thirty. "SeaWorld officials dispute the 50-to-80-year life span and say it is closer to 30 or 35 years," the *Sentinel* noted. Grandy said the three SeaLand whales should not be allowed into the United States.

And then he brought up the dreaded R-word—*release:* "We request NMFS, together with the public display industry and humane community, to explore the option of rehabilitation and release of captive killer whales into the wild."

Dr. Paul Spong was also in Washington to testify. He warned that the separation of a young calf from its mother and tight-knit social group was "a perversion of the natural order of things. We should not be surprised by the bizarre behavior of orcas in captivity." He talked about the death of Keltie Byrne and said it would be dangerous to send Tilikum, Nootka, and Haida to perform at SeaWorld.

But the company's director for research programming, Dan Odell, "downplayed safety concerns during a break in the proceedings," the *Sentinel* reported. "How many people get killed riding horses every year?" Odell said. "Whales are big animals. They deserve a lot of respect."

Brad Andrews, vice president for SeaWorld's zoological operations, said the whales in the company's collection received world-class care and helped to "present marine life in a manner that is meaningful to the public." He added that SeaWorld's $130 million breeding program had resulted in a number of scientific advances, such as the discovery that the gestation period for female orcas was not twelve months as believed, but actually seventeen months (it has since been revised to eighteen months). He said that would never have been learned from studying wild orcas. Andrews predicted that SeaWorld's application to import the Canadian whales would be approved "because there is no reason why it shouldn't be."

The public airing of problems with captive killer whales did nothing to deter SeaWorld's pursuit of more animals, including the three orcas in Victoria. But the federal government persisted in its due diligence, especially given the animals' notorious past.

On December 17, the chief of NMFS's permit division, Ann Terbush, wrote to Brad Andrews requesting "a discussion of SeaWorld's consideration of the tragic incident at SeaLand of the Pacific," and a list of "actions SeaWorld has taken, or will take, to prevent recurrence," before and after importation.[11]

SeaWorld was handed a lot of homework by the feds. "This discussion should include: the factors which SeaLand, the involved Canadian governmental agencies, and SeaWorld believe were involved in the incident (e.g., please submit copies of reports, findings, etc.)," Terbush wrote in a stern bureaucratic tone, " and the recommendations made by each of these entities."

Andrews refused the government's request. He put the onus of investigating the Byrne affair squarely back on the shoulders of NMFS. His attitude seemed to be: If you want that information, go get it yourselves. "Although we are generally familiar with the circumstances surrounding the accidental death of a trainer in the killer whale pool at SeaLand," he wrote back to Terbush on December 30, 1991, "we do not have any of SeaLand's records or reports on this matter. SeaWorld urges NMFS to request whatever documents NMFS believes necessary for NMFS's information from the appropriate authorities of SeaLand and the Government of Canada."

SeaWorld did not believe any safety issues were involved with importing the killer whales. The company, Andrews implied, provided a better, safer work environment than did its Canadian counterpart. "SeaWorld understands the historical interest regarding the incident at SeaLand," he wrote, "but SeaWorld believes its present employee training and safety program should be judged on its merits," especially since conditions were so much better at SeaWorld. "We believe the accident was unique to SeaLand and due to the combination of a poor pool design which prohibited exit from the water, inadequate emergency life saving procedures, and interference by whales unaccustomed to the presence of people in the water."

In contrast, all animals at SeaWorld were "highly trained" and used to interacting with trainers and veterinarian staff, Andrews claimed. The whales from SeaLand were "essentially untrained," however, and would be managed as untrained animals. Moreover, SeaWorld had "implemented an enhanced employee training and safety program," following the 1987 accident with John Sillick, in which Orky crashed down on him while Sillick rode the back of another whale. "Since implementing this program, there have been no accidents at SeaWorld," Andrews assured NMFS, adding that "safety for both employees and killer whales is paramount in all our work."

There were no other issues with importing the whales, Andrews insisted. "These animals have no apparent medical condition requiring quarantine upon arrival, and SeaWorld personnel who have examined the animals believe the animals are behaviorally normal."

Just four days later, on January 3, 1992, SeaWorld applied to NMFS for an emergency permit to immediately import Tilikum, citing "medical reasons." Haida's calf, Kyuquot *(ky-YOO-kut)*, had been born on Christmas Eve, and Haida and the pregnant Nootka had chased Tilikum into a small medical pool and threatened him each time he tried to leave. SeaLand decided to lock him up full-time in the dark, enclosed overnight module, away from the newborn and the females, so as not to "interfere with the nursing and bonding process," said James F. McBain, SeaWorld's corporate director of veterinary medicine, in a letter to NMFS sent with the permit application.

The government granted the permit five days later, on January 8. But the Commerce Department was not pleased with the strong-arm tactics and self-imposed "emergency."

"Both SeaLand and SeaWorld had at least two months advance knowledge of the imminent birth of at least one, and possibly two, killer whale calves," NMFS official Nancy Foster chastised Andrews in a letter.[12] "SeaLand is responsible for these animals and should have taken steps to ensure that arrangements were made to hold the adult male killer whale, 'Tilikum,' at or nearby the SeaLand facility or at another facility in Canada following such births."

Presumably, Foster added, SeaWorld shared a "significant interest" in the well-being of the animals. As an adviser to the Canadian park on their proper care, SeaWorld "should have taken such steps even if SeaLand had elected not to do so. Such temporary holding arrangements, whether they involved construction of temporary pens or other enclosures, should have been possible, at least for the few months necessary to consider and decide upon the permit application pending from SeaWorld for the import of these killer whales for public display." Had SeaWorld done that (and not held Tilikum's well-being for ransom), "the present need for an emergency permit authorization could have been prevented. However, such reasonable and prudent precautionary steps necessary for the health and welfare of Tilikum were not taken by SeaLand or SeaWorld."

SeaWorld rejected that criticism. Brad Andrews protested that it was not SeaWorld's fault if Haida and Tilikum could not share the same tank with their own offspring. "It's easy to point the finger and say, 'You should have made plans,'" Andrews told the media. "We knew what we needed to do, but sometimes the animals don't follow all of the plans."[13]

He added, "'When you stop and think about it, he's in a miserable place and he needs to go to the largest facility in the world to be part of a nice breeding program."

Tilikum required a remedy. NMFS sent agents to Victoria to verify Sea-World's claims that his situation was, indeed, an emergency. "NMFS has concluded that the best interests of all the killer whales concerned, particularly that of Tilikum, the newborn calf, and the calf expected to be born in the next few months . . . would be best served if an emergency authorization is granted for Tilikum's importation," the agency said.[14]

The agreement would permit SeaWorld to temporarily import Tilikum "for the purpose of providing medical treatment and care that is otherwise unavailable in Canada at this time." Tilikum could remain at SeaWorld while the company went through the normal procedure for a public display permit. But NMFS attached several conditions.

First and foremost, if the permanent display permit was denied, SeaWorld would assume all costs "for the return of Tilikum to Canada and placement at a suitable facility." If no facility was found, then SeaWorld would need to arrange for "the return and release of Tilikum at the original location of capture." In other words: Iceland. SeaWorld would also be banned from displaying Tilikum before a display permit was issued, and NMFS reserved the right to revoke the emergency import at its own discretion.

SeaWorld agreed to the conditions. But even as the application was under review, Brad Andrews wrote to Iceland's minister of fisheries, Thorsteinn Palsson. He wanted to know if Iceland would agree to let a killer like Tilikum be released into its national waters.[15]

Andrews did not shy away from stating his corporation's view on the issue. Tilikum's possible return to Iceland required "an analysis of the likelihood that Tilikum will survive, the possible impact of Tilikum's release on the fishing industry . . . and the potential impact of Tilikum's release on the marine environment," he warned. Such an environmental impact study would entail an "in-depth analysis of the incidence and distribution of disease and disease causing organisms in the fish and marine mammal populations in Icelandic and Canadian waters, as well as an analysis of any possible latent pathogens being carried by Tilikum."

Andrews reminded the Icelandic minister that Tilikum had been "maintained in an ocean pen surrounded by a 500-boat marina and occasionally consumed fish native to the region." Even with the most thorough medical examination possible, he cautioned, "it is possible that the presence of some subclinical organisms not native to Icelandic waters might not be detected."

Andrews did not give the full picture. He failed to mention he had informed the US government that all three whales at SeaLand had no "medical condition requiring quarantine upon arrival," and that the animals were "behaviorally normal."

SeaWorld deemed it unfeasible to return Tilikum to Iceland, Andrews concluded in his "query" to the minister. "However, SeaWorld wishes to consult with the Government of Iceland and obtain its views on these matters, since any such release would be subject to the laws and jurisdiction of Iceland."

Palsson agreed with Andrews. It was too risky to take Tilikum back. "The history of this killer whale and the best available scientific information concerning the possible consequences of a return of the animal to its place of collection have been carefully considered," he wrote back to SeaWorld.[16] "This Ministry, with the concurrence of other interested Ministries, has concluded that the return of the killer whale, Tilikum, is not feasible. It is evident the animal may carry diseases that are both undetectable and alien to killer whales and other animal populations in Icelandic waters. Furthermore, it is the view of the Icelandic scientists that the survival of the animal in the wild would be highly questionable."

Palsson did not stop there. He took the opportunity to issue a preemptive blanket ban on any captive killer whale from Iceland returning to its native waters.

"The factors that render the return of the killer whale, Tilikum, infeasible would apply to other killer whales that have been maintained in captivity outside of Iceland. Consequently, this Ministry and the other interested Ministries have concluded that neither the return of this killer whale, nor the return of others now in captivity, will be authorized. We trust that our response fully and finally disposes of any questions regarding the return of killer whales to their point of collection in Icelandic waters."

It was a blow to the anti-captivity movement and a victory for the display industry. The vast majority of orcas in captivity were of Icelandic blood. Now they could never go home. Activists cried foul. They charged SeaWorld with sending a precipitant missive to prime the political pump in Iceland, bias the government, and stave off the return of any captive orcas.

Andrews scoffed at the suggestion. A few years later he would argue that SeaWorld had been required by NMFS to seek the opinion of Reykjavík. "We don't play games, and we don't play baseball," he said. "The National [Marine] Fisheries Service asked the question. We didn't. The Icelandic government responded. We didn't. . . . So whoever thinks that we lobbed anything into the picture is absolutely off base."[17]

It was official: Tilikum would be sent to Orlando on an emergency basis. Permits for the other whales at SeaLand would be considered more deliberately, NMFS said, as would a permanent display permit for Tilikum. This time, public comment would be allowed. The activists had lost a battle, but were thrilled that Tilikum's emergency import permit even mentioned the possibility of his return to Iceland. "The fact that it's beginning to be considered on a serious level is extremely important and exciting to us," said Paula Jewell, the marine mammal coordinator for HSUS. The group, which was quickly becoming the world's chief opponent to keeping killer whales in captivity, would soon make an offer to marine biologist Naomi Rose, one that she would find impossible to turn down.

13

Dissertation

In the fall of 1992, Naomi the doctoral candidate returned to Santa Cruz after leading the School for Field Studies courses to finally complete her overdue dissertation, "The Social Dynamics of Male Killer Whales, *Orcinus orca,* in Johnstone Strait, British Columbia." It was the culmination of seven years' work. Naomi was proud and relieved. The long journey was almost over; she was about to become a bona fide scientist, a *doctor,* ready to go out in the world and make her difference in it.

Still, Naomi felt that she had *already* made a difference, no matter how modest, in the scientific knowledge of orcas. "This study was the first attempt to examine systematically the behavior of male killer whales, in an effort to describe the behavioral consequences of their unusual social structure," she wrote in the dissertation.

Her study period had extended over three summers from 1988 to 1990, comprising a total of 131 attempted observation days, of which she observed whales on 118 days, or 90 percent of the time. Observation days began at 7:30 a.m. and did not end until 6:30 that evening.

At the time, the Northern Resident community had seventy-eight males, though only forty-five of them visited Johnstone Strait regularly enough to study. Of those, Naomi gathered enough data on thirty-two to include in her sample. Those whales represented nine pods and sixteen matrilineal groups.

Male orcas are unusual among mammals in that they are the *philopatric* sex of the species (from the Greek *philo,* for "love," and *patra,* for "fatherland"). That is, they never emigrate away from their home territory—in this case, their mother and her matriline. In many mammal species, having grown males hang around the females and offspring would be disruptive—the mother does not tolerate it and pressures young males to leave. In other species, a male simply feels the urge to move beyond his birth territory when he matures. This was clearly not the case with Resident orcas.

Naomi wrote about the deep and lifelong bonds that form between Resident mothers and their sons and noted that adult males spend at least 40 percent of their time within one body length of their mother—approximately nine and a half hours every day, 365 days a year.

The adult male is so dependent on his mother, Naomi observed, that if he loses her, he may try to transfer that bond, in a slightly weaker form, to another close relative, usually a sister, grandmother, aunt, or even a younger niece. Older sons who survived their mother's death often traveled, foraged, and even rested up to a half mile away from their sisters, implying that adult males without mothers "are most peripheral to and the least integrated into the matrilineal group," Naomi wrote. (This could explain why Tilikum tried—unsuccessfully—to bond with Haida and Nootka when he arrived at SeaLand, Naomi would later realize. And though Icelandic whales do not share identical social traits with Pacific Residents, some scientists now believe the two fish-eating populations exhibit at least some cultural overlapping.)

For the adult male of any species to become nondisruptive and thus socially fit for philopatry (nondispersal), his evolutionary path must develop in ways that mitigate, or even eliminate, aggressive behavior. We now know that male humans, for example, experience a precipitous drop in testosterone production after becoming fathers. For Resident male orcas, domestic docility meant greater acceptance within a female-dominated society: Keeping their cool around the pod was better for the mothers, easier on the calves, and ultimately more beneficial for the males themselves. "This almost total repression of aggression," Naomi wrote, "may be the most significant consequence of the mother-son bond."

Much of her paper dealt with the question of how male Northern Resident social behaviors fit harmoniously into the larger matriarch-dominated society. Naomi wanted to know, what did these males *do* all day, and how did that help them socially integrate so well?

The answers were engrossing and closely tied to the evolutionary suppression of male orca bellicosity. Two social behaviors, in particular, helped explain how and perhaps even why Resident males evolved to become the subdominant, philopatric sex—and how such big bulls remained socially fit to stay integrated in the matriline. The consequences of these two behaviors conferred striking benefits to mothers and sons alike.

In the first case, small groups of Resident males had been observed during the social summer season engaged in temporary "bouts" of intensive physical contact. Before Naomi came along, no one had collected enough data on these all-male encounters so it was not understood what was going on. The scrimmages, which she coined MOSIs, or "male-only social interactions," usually got going in the afternoon. Anywhere from two to four males would wind themselves into a ball of roiling black-and-white energy, slapping the surface with fins and flippers, churning up white water as they rubbed and pushed against each other or leapt in the air like forwards at an NBA game.

Naomi spent a total of twenty-five hours observing MOSIs, which lasted anywhere from several minutes to a few hours. During the group behavior, males were far less attentive to what was going on around them, meaning Naomi could get as close as ten meters away; they were utterly unperturbed by boats during a MOSI. Encounters always involved at least one adolescent male (age six to fourteen), and adolescents were four times more likely to participate than adults. MOSIs usually involved non-kin, though members of the same matriline did sometimes participate.

The most common type of behavior observed during a MOSI was sociosexual, including "frequent body contact, percussive and aerial behaviors and penile displays," Naomi wrote. Cetacean penises are prehensile—the animal can bend and wrap it at will, like an elephant with its trunk or a monkey with its tail. During MOSIs, Naomi observed male killer whales extruding their phalluses for several seconds and sometimes, though not always, inserting them into the genital slit of another male. Other behaviors she described as being in "beak-genital orientation."

The contact sport had odd variations. In summer 1988, two different pairs of adolescent males—A32 and A26, the little brother of Top Notch; and A33 and A38—were seen several times engaged in their own unique MOSI. Naomi described one encounter: "The two males swam slowly side by side, dove simultaneously, surfaced to float facing each other about 20–25 meters apart, slowly approached each other, picked up speed (essentially surface-swimming), and finally butted heads together, sometimes at high speeds (a 'ram'), sometimes more gently (a 'bop'). After impact, they rubbed past each other, dove simultaneously, and surfaced seconds later again side by side, heading in the same direction as before."

Liaisons between individuals changed from moment to moment during a MOSI, as some males joined the fray and others moved on, either to return to their mothers or spend a brief time milling around alone. The most common groupings were in pairs, though encounters of three and four whales were seen about one-third of the time. On average, males spent 10 percent of their activity budget in MOSIs. These hyperphysical, energy-devouring bouts played "an important role in the male behavioral repertoire," Naomi wrote.

But why would males exert so many hard-earned calories on these summertime tangles? What was the evolutionary advantage of MOSIs? Were they "agonistic" in nature, Naomi wondered, meaning they were designed for males to establish dominance and reinforce social hierarchies? Or were they "affiliative," evolution's way of bringing unrelated males into a bonding experience that reduced aggression and built mutual trust?

Naomi concluded the latter. In the dissertation, she stated that her own data analysis "supports the hypothesis that MOSIs are play interactions," and that the behavior "creates high levels of tolerance and affiliation." One solid piece of evidence that all-male sociosexual encounters were not meant to establish dominance and hierarchy: 88 percent of all bouts were reciprocal in terms of who was the actor, and who was the recipient.

According to Naomi, juvenile and adolescent males might have distinct sociobiological motives for taking part in MOSIs. For the younger whales, the activity probably brought the benefit of motor-skill training. For adolescents, it likely reinforced cognitive and social skills, while offering practice for courtship with females and the honing of reconciliation skills with potential male rivals. MOSIs also helped "relieve active libidos with fellow males, since access to reproductive females may be restricted, either by adult males, or by the females themselves."

A key evolutionary benefit of male-only social interactions, then, was likely a profound diffusion of tensions—sexual or otherwise—among males in the community. This ritualized abatement of aggression helped ensure their continued integration with the matrilineal group. MOSIs helped keep the peace, not only among related males, but more importantly among non-kin bulls.

Affinity among unrelated males is highly unusual among predators that hunt in packs. Male wild dogs and chimpanzees "are very aggressive when they encounter unrelated neighbors," Naomi wrote. Not so with killer whales, whose "overt physical aggression between matrilineal groups and pods is not only attenuated, but apparently nonexistent."

As Resident male orcas grow up and participate in fewer MOSIs, they seem to retain the lessons learned from the sociosexual romps. "Adolescence is a critical period in which to acquire certain skills," Naomi wrote in her dissertation. "MOSIs are a critical social arena in which to acquire them." Older males did sometimes join in (perhaps as a refresher course in pod relations and anger management) "to strengthen ties with 'old friends' or reconcile with erstwhile rivals."

The second behavior that allowed male Residents to integrate more closely into their matrilines was a seeming willingness to alloparent siblings under five years of age. Most calves with older relatives were babysat once or more each observation day as their mother foraged, socialized, rested, or looked after her youngest, nursing infant.

The benefits of babysitting are obvious to humans, but hard to quantify in terms of animal behavior. But at least one scientifically demonstrable benefit of alloparenting could be measured in orca mothers. When matrilineal groups travel, they typically swim in "echelon formation," where the youngest calf sticks next to the mother, and her other children spread out, each one to the side and slightly behind the other. The youngest calf gets the coveted spot not only for protection, but also propulsion: slipstreaming alongside the mother helps the infant save energy. But it's also a drag, literally, on the larger animal.

Janice Waite, Naomi's graduate school colleague and friend, discovered that respiration rates were much higher when females were "carrying" an infant than when they were swimming alone. Carrying a youngster that way at all times might wear out even the strongest swimmer. But if the mother had a grown son or two to share the burden, she could conserve precious energy, helping her stay healthy and increasing her chance of raising the calf successfully.

Having a son to alloparent younger offspring conferred other advantages on the mother, and on her larger group. Leaving older offspring with babysitters allowed mothers of newborns to focus on nursing their infants and provided "temporary relief from constant vigilance or from rough-and-tumble play" with juveniles in her brood. Naomi speculated that having older sons around contributed to the overall physical and mental health of childbearing females. She had observed ten females with older males in their matrilineal group whose reproductive rate was "very successful." On the other hand, one female without any potential allofathers had lost two offspring out of four.

Staying at home has its costs, however. Just ask parents with unemployed adult children who "remain in the natal territory." Philopatry by its nature increases competition for food and other resources, so the "repayment model" comes into play. The philopatric sex must be able to offer something valuable to offset the drain on resources their permanent presence entails. Babysitting was evolution's way of charging adult orca males room and board to remain at home with mom.

"The philopatric sex pays back some of the cost of having it there by caring for its parent's subsequent offspring," Naomi wrote in her dissertation. "Over time it may be 'cheaper' to produce a philopatric sex, leading to a bias toward that sex in the ratio."

This intriguing hypothesis was fodder for further investigation, Naomi mused. Was it possible that females had more male calves earlier on, in order to lock in future alloparenting services? It was hard to say. "In a social system where having sons around to alloparent subsequent offspring increases a female's reproductive success, then having more sons, and/or having sons earlier in life, could be favored," Naomi would later explain to a friend back home. Some unidentified prenatal mechanism might be at play, she speculated, some kind of chemical or biological favoritism toward male-producing sperm.

Another potential, but far less likely, explanation was that a mother was more attentive to sons early in her reproductive life, and less so to daughters. "But that's a wasteful way of doing it, and less likely to be favored by natural selection," Naomi said. "To put in all the physical resources and time to gestate and bear a daughter, only to give her sub-optimal care to the point of her dying? That doesn't seem like a mechanism evolution would favor, but who knows?"

Naomi handed in her dissertation in late December 1992. Other than for a brief visit in 1994, she would not return to Telegraph Cove and Johnstone Strait for more than sixteen years. But in her thoughts she would never leave the hauntingly beautiful place. At the end of her acknowledgments section, she thanked the location itself, and the remarkable creatures that summered there each year. "I am fully cognizant of the great privilege I have been granted, to gain entry at last, after long hoping, into the magnificent world of wild killer whales," she wrote. "This was a privilege indeed, and whatever comes after in my life, I will always remember with awe that in these waters of the Inside Passage, I experienced five truly wondrous years."

Now came the daunting task of making a living—Naomi was not keen on returning to her own "natal territory." During the coming months, as she began pondering her new career search, and over the years that followed, Naomi would think back on the unusual social behaviors of Resident male killer whales, and the critical role the behaviors played in turning these powerful beasts into cooperative, nonaggressive, contributing members of their society.

But if MOSIs, alloparenting, and other natural behaviors benefited Resident orca society and individual animals alike, what would be the consequences of *denying* such behaviors to killer whales, ones who don't live in nature?

By her third year in the field, Naomi had begun to ask herself many related questions: What happens to a female in captivity without relatives to assist with birth, or older offspring to alloparent a calf? What becomes of young males who don't learn through MOSIs or other ritualized exercises how to burn off aggression, energy, and libido? What stress might it cause when adult males are prevented from foraging or resting alone, or just being apart from the dominant females for a short time? How does an adult male cope when he suffers from low social status and has no female relatives around to replicate the bond with his mother?

Naomi was finally starting to question the ethics of keeping killer whales in captivity. By the time she handed in her dissertation, she had progressed to full-blown discomfort with seeing them in tanks.

It was good timing. In late April of 1993, she applied for the just-created position of marine mammal scientist at the Humane Society of the United States, which she was offered in early May. Almost overnight, she would be transformed from a scientist to a new and rare breed, that of scientist-advocate.

Years later, looking back on her epiphany that moved her from dispassionate researcher to actively engaged campaigner against cetacean captivity, Naomi wrote about her memories of the big male Hyak, and how he helped convince her that orcas and aquariums do not mix. In the 2003 anthology *Between Species: Celebrating the Dolphin-Human Bond*, Naomi wrote a chapter titled "Sea Change," noting:

> *Every time I saw Hyak after that first summer, he seemed bigger and his tank smaller. I started noticing how much time he and the other two spent floating motionless at the water's surface, something the Johnstone Strait orcas rarely did. I started wondering how the three Vancouver whales filled their time when the aquarium was closed and all the people had gone home. They did not need to forage and they could not travel anywhere; their tank never presented them with any challenges or changes. I wondered how it must be for Hyak to live with two orcas from an entirely different ocean, and I wondered if all three of them remembered their families. At long last, I was thinking critically.*
>
> *By the time I finished my doctoral work, I was not so sure about captivity anymore. For me, several summers of experiencing the real thing—cetaceans in the wild—stripped captivity down to its basic element: cetaceans trapped in empty concrete boxes. [Then] I learned details about what goes on behind marine park shows that once and forever convinced me that keeping these socially complex, long-lived, intelligent, far-ranging creatures in tanks is wrong. Orky and Corky were captured in 1968 and 1969, respectively. In my years in Johnstone Strait, I came to know their families.*

PART TWO

DARK SIDE

14

Arrival

SeaWorld received its federal permit for the emergency importation of Tilikum on January 8, 1992. The next day, he landed at Orlando International Airport.

The hasty exodus from SeaLand of the Pacific was carried out with little fanfare. SeaWorld's publicity department, which normally secured glowing headlines each time any new whale came to one of its parks, apparently kept Tilikum's arrival as low-key and reassuring as possible. The park's hometown paper, the *Orlando Sentinel,* buried the news in its "Around Central Florida—Briefs" section. Tilikum was not mentioned by name, nor was any reference to the killing of Keltie Byrne included in the short account.

"The whale arrived at the airport about 6:40 p.m. and 'is doing quite well,' said Brad Andrews, corporate zoological director," the *Sentinel* reported. Tilikum, who was now twenty feet long and about eleven years of age, had been flown inside a specially built crate aboard a chartered cargo plane that left Victoria, Canada, at 9:30 a.m. Eastern time. No explanation was given for why the journey took more than nine hours. All told, Tilikum was likely in that box for more than eleven hours.

One Associated Press photo ran on the wire. It showed a docile killer whale floating on his stomach, pectoral fins calmly extended from each side, as a male trainer in a wet suit stood by in waist-deep water, gently stroking the animal's back. It was a touching portrait of man and beast in a bonding moment of mutual trust.

Opponents of the transfer were quick to point out that Tilikum's permit was only temporary—his fate had not been fully sealed. To them, Orlando was a mere way station to more appropriate destinations, preferably one in the waters off Iceland. Members of the Canadian House of Commons, for example, began a letter-writing campaign to the US and Canadian governments, urging a ban on putting whales up for sale and advocating for some captive orcas, including Tilikum, to be released back into the wild.

Erich Hoyt, the author of the seminal book *Orca: The Whale Called Killer,* who was now working with the UK-based Whale and Dolphin Conservation Society, also warned the US fisheries agency that allowing any of the Victoria whales into the United States "poses a thorny problem to NMFS as well as SeaWorld."[1]

Although Keltie's was the first death at an aquatic theme park, there had been "dozens of well documented injuries, many of them serious and many of them at SeaWorld," he wrote. "It is amazing that many more orca trainers have not been killed. Certainly, now that there has been a death, it calls into serious question the suitability of exhibiting orcas in captivity."

Hoyt said that SeaWorld should provide a "full review" of the death at SeaLand, charging that SeaWorld's brief explanation of the account had been "simply not acceptable—it's difficult to believe that SeaWorld has not investigated this matter, but if it hasn't, it should be required to."

Hoyt had spoken, off the record, with killer whale trainers in Orlando. "Some are very concerned about the SeaLand orcas coming to SeaWorld," he wrote. "They say that there simply is not enough room for them, and that their previous behavior makes them dangerous." SeaWorld itself had told NMFS that the three whales were "basically untrained," he said. If true, it was incumbent on the agency to "find out what they've been doing for the past nearly 10 years in captivity, and the implications to all concerned for making the transition to SeaWorld."

Sadly for SeaWorld, the protests arrived in more than written form. On January 19, about twenty demonstrators, including several children, appeared at the main gate of the Orlando park, picketing the entrance with homemade signs that read: ANOTHER DAY, DOLLAR, DOLPHIN; CAPTIVITY KILLS; SEAWORLD IS A WHALE JAIL; and FREE ME! across the image of Shamu.

The *Sentinel* ran a photo of the protesters with a caption saying only that they were demonstrating "against SeaWorld's recent acquisition of Tilikum, a killer whale from Canada." Again, no mention that this killer whale had actually killed somebody. The activists were pictured outside the facility on a busy Orlando roadway, "after troopers, hired by SeaWorld, shooed protestors off park property," the caption said.

Inside the park, trainers at Shamu Stadium were only vaguely aware of the controversy; even less so of the danger warnings from Erich Hoyt, NMFS, and others. New Yorker Samantha Berg was working at Shamu Stadium the day Tilikum arrived. (Jeff Ventre was on a tour of duty at Whale and Dolphin, but would return to Shamu in 1994; Carol Ray had left SeaWorld one year before.)

Tilly settled into one of the back pools, where he would be kept alone for the time being, until senior trainers felt he was ready to meet the girls: the queen, Katina; the lower-caste Icelandic females, Winnie and Gudrun; and her headstrong, half-Transient daughter, Taima.

Sam did not think there was any reason to fear Tilikum. The way she heard it, Tilly had refused to give Keltie Byrne's dead body back for hours because he was in possession of a "toy" and it was rare for the females to part with any toy and let him play with it. She was never told how the whales blocked Byrne's exit from the water while she was still struggling for her life. Perhaps some of the senior trainers knew, but those who were lower on the totem pole did not. "It's a very different scenario," Sam said years later, "thinking of her passing out quickly from hypothermia and drowning and Tilikum carrying her around like a souvenir; versus the whales aggressively trying to keep her in the water, whether they knew they were killing her or not."

Sam was told that Tilikum had not hurt anyone, but that message did not dovetail with what she was seeing around her. On Tilly's first day at SeaWorld, one of the female trainers was standing on a gate between two pools, making cooing noises at the big whale and trying to get him to respond to her. Her hair was down, her wet suit unzipped and dangling around her waist. When Chuck Tompkins and chief animal trainer Thad Lacinak saw her in that position, "their heads almost exploded on the spot," Sam told Jeff that afternoon. The two men screamed at the trainer to get away from the gate.

Shortly after that, Chuck and Thad announced a new set of rules that would pertain only to Tilikum. Sam listed them for Jeff: "Absolutely no unzipped wet suits around him; nothing hanging for him to grab, no hair or whistles hanging down; and no one is to lie down next to him—*ever*." All sessions with Tilikum were to be conducted from a standing position only, and exclusively by senior trainers, and only then when another senior trainer was on-site to act as spotter. A perimeter line was established around Tilly's pool, and only senior trainers authorized to work with him or spot those who did were allowed to cross that line.

Sam was a bit confused about the fuss. After all, what happened in Victoria was a tragic accident. Chuck and Thad, she assumed, were just being overly cautious in announcing the new rules for Tilikum. It didn't occur to her they might know more about Tilly's past than they were telling at least some trainers. Then again, many things were like that: You were told things on a need-to-know basis. Sam was never approved to work with him, so she didn't need to know.

Sam also overheard other staff talking to management about whether they should desensitize Tilikum to being with humans in the water. Jeff's friend Mark Simmons argued in favor of "desense" for Tilly, as did several other trainers. Sam felt conflicted about the debate. To begin with, who would volunteer to go in the water with him? She supposed they could have pulled it off if there was a gigantic pool, with Tilikum on the far side, and someone got in the water on the other side, so the person could slip out of the pool if anything went wrong. Even then, the trainers would have to make ever-closer "approximations" with the whale, until they were in an indisputably vulnerable position. In the end, Sam could see why management declined to take that risk.

Sam said nothing. She loved her job, and she had learned that the best way to keep it, and the quickest way to move up through the ranks, was to simply do as she was instructed and not ask questions. It could take six months to a year before a new trainer at Shamu was allowed in the water, depending on his or her previous experience. Sam had a vested interest in going along with what she was told, even if she suspected that something was wrong.

A few weeks after Tilikum was flown to Orlando, a new member of the training staff began his apprenticeship at Shamu Stadium—John Jett, a five-foot-eight-inch Midwestern boy with light blue eyes, buzzed head, and sturdy wrestler's build. John had started at SeaWorld six months previously and had now been transferred to Shamu. Before long, John and Jeff Ventre became good friends.

John grew up on the Kansas side of Kansas City and graduated from the University of Kansas in 1989 with a degree in environmental science. He got hired at an environmental testing laboratory after college, where he did a lot of fieldwork taking samples from woods and streams, but also in buildings and other sites.

One summer John was sent do to water testing at a local amusement park called Worlds of Fun, which included a pitiful little dolphin show. The animals were kept in what looked like a big bathtub and made to perform tricks for the crowds. On his first day there, John arrived early, before the park opened, to begin collecting samples undisturbed. There was no sound, except for the squeaking of the dolphins. Nobody was around to stop him, so John walked up to the small tank and interacted with them, petting their flanks and laughing at their antics.

"It changed my life," he told Jeff one night over beer and hot wings at a south-Orlando BBQ joint. "And I said to myself, 'Man, I wouldn't mind working with cetaceans.' But I realized this was a really crappy situation for the dolphins. I knew there had to be a better way to do things." John began researching job openings working with whales or dolphins. In 1991, he sent a résumé to SeaWorld. "I thought, if you want to work with cetaceans, then this was the place to do it."

Jeff sat and listened to his friend tell a similar tale to his own. John considered himself an aspiring scientist—he wanted to get a master's degree. SeaWorld seemed like the ideal place to begin conducting original research on cetaceans. "And when I started work, I went to the PR training, and I remember going home that evening thinking, this is not right."

The "training" session surprised John. Nothing in the talk was about the natural history of the animals at SeaWorld, nothing about dolphin biology or killer whale social behavior. Instead of learning cetacean science, John felt that he was being spoon-fed corporate sound bites. The new staff was taught to repeat certain approved phrases: "Our animals live longer in captivity than they do in the wild"; "The dorsal fins of male killer whales also bend in the wild, it's perfectly natural"; "These whales receive the finest veterinary and dental care in the world"; and so on.

Most disturbing of all to John, the new hires were told to avoid contact with outside scientists who came to the park to observe the animals.

John left that first meeting with an unsettled feeling in his stomach and walked out into the thick, hot air. Was this really the right move for him? He decided to stick it out. After all, John was lucky to have landed a job at SeaWorld. His new bosses never let him forget that: Plenty more people were in line, just waiting to take his job if he didn't want it.

In the coming weeks and months, John would hear tour guides and animal trainers tell visitors about "all the important scientific research" that went on behind the scenes at SeaWorld, but John was certainly not seeing any of it.

"That's because there isn't much going on here, scientifically speaking," Jeff told John. The only "research" the trainers were aware of was the rudimentary data collection about diet, weight, length, dorsal fin bending, and respirations. Those last were taken day and night. "What are you going to do with that breathing data?" John said. "It will never be used by anybody. Putting all these respirations into a log is just a way to make sure that the whales are reasonably healthy. There's no *science* here."

After several months at Shamu, John Jett was assigned to Tilikum's team. He helped prepare Tilly's food and monitored his respirations; he assisted with training and exercise sessions, usually by holding the extended target—a ball on a pole—over the water or providing Tilikum with fish as a positive reinforcer.

Because John arrived at Shamu after Tilikum did, he was told even less about the whale's past than Samantha had been told. John was not in the least bit fearful of the male. He knew this wasn't a water-work animal: Tilly was being trained to come out at the big splash finale of the show, once his public display permit came through from NMFS. But that was all. John was only vaguely aware that being close to Tilly might pose any sort of danger. The whale always struck him as a gentle and sensitive giant—an overgrown, misunderstood guy.

Tilikum wasn't getting much affection in the pools of Orlando. Much as

Nootka and Haida had clobbered him at SeaLand, some of the Florida females—Katina in particular—often raked the newcomer with their teeth. One morning John came in early to find streams of blood running behind Tilikum as he moved through the water. John saw it happening fairly regularly, and not just to Tilly, but also the subdominant female Winnie. John frequently witnessed her being brutalized by the other females before she was moved to Ohio.

John called the deep red trails of blood streaming through the blue water "skywriting." It quickly dissipated and people in the audience almost never noticed. But on two occasions while John was there, the killer whale show had to be canceled because the bleeding was discernible from the stands.

By the fall of 1992, John came to accept that he was working in show business, not science.

That October, NMFS granted SeaWorld a permit to import the remaining SeaLand whales into the United States and put them on public display. Haida and her calf, Kyuquot, would go to San Diego; Nootka would fly to Florida to join her old tankmate and torment victim, Tilikum. Nootka's calf, a son sired by Tilikum, was born in February 1992, but for some reason would not nurse from his mother. He died a month later.[2]

In addition to granting a permit to import the three whales from Canada, NMFS also gave SeaWorld permission to keep Tilikum for permanent public display in Orlando. He had been flown to Florida as an invaluable breeder; now he would be part of the show as well. Conditions were attached to this permit as well. Most notably, SeaWorld was required to provide the complete "animal/human history" of each whale to all staff who would be handling, training, or in any way involved in the "direct interaction with or management of the animals."[3]

On a freezing morning in January 1993 a screeching Haida and Nootka were hauled out of their sea pen of ten years, along with Haida's calf, Ky, and placed into separate water-filled crates lashed onto flatbed trucks. The two females would never see each other again.[4]

Nootka's arrival in Orlando was treated with even more nonchalance than the advent of Tilikum. There were no news accounts of the arrival, no protests; inside the park walls, most staff were told even less about this animal's involvement in Keltie Byrne's death than they had been about Tilikum's. John had no idea that she was linked to the incident. He didn't know she was involved in any aggressive acts at all. Still, as with Tilikum, SeaWorld management deemed Nootka unfit for water work.

John began working with Nootka almost every day in a support role (she was not one of his primary animals), in addition to his Tilikum duties and performance work. He helped to keep the new female busy with exercise and husbandry sessions in the back pools. Despite the requirement of her import permit that all staff interacting with her were to be informed of her "animal/human history," that was not the case. Only years later would John come to believe that SeaWorld trainers—other than those at the top—were deliberately kept unenlightened about a wide variety of safety issues in killer whale work.

"A lack of detailed information was the norm whenever accidents happened at other parks," he said. "I remember one incident when all of us were pulled from water work for a short time. To this day I don't know what happened."

John said it would have been "nice" to receive detailed incident reports from other parks "so we could have possibly prevented it from happening in our pools." Senior management, he alleged, "purposely kept us in the dark for at least two reasons. First, they didn't want trainers questioning the overall safety of their program; after all, the show had to go on. Second, they probably assumed that if we didn't really know how dangerous the job was, then we wouldn't be inclined to demand more than near-minimum-wage earnings."

15
Humane Society

When Naomi first set out to get her doctorate from UCSC, she expected to remain within the secure confines of academia after completing grad school. She envisioned herself teaching marine mammalogy at one of the leading schools in the field—the University of Washington perhaps, or maybe Texas A&M or UC San Diego.

But a few years into the program, she began to change her mind. Naomi found the petty competition for grant money and tenure to be distasteful. She grew disillusioned with the political backstabbing so common in academic settings and was disturbed by the offbeat and unprofessional behavior she had witnessed from some of the male faculty at school. She was not patient with students and began to see she didn't have the temperament to teach full-time. Naomi began to think that life on the tenure track at a big university would probably be more stressful than fulfilling.

By the time she got her PhD, she was thinking she would like to work in the environmental protection field, perhaps at the Natural Resources Defense Council, the World Wildlife Fund, or a government agency such as the National Marine Fisheries Service (NMFS)—anything with an aquatic connection.

In early 1993, Naomi was back in LA living with her dad and getting desperate about finding a job. She took on some temp work—a lot of typing, mostly—but so far her PhD was getting her nowhere. Most of the career openings were in fields for which she had no training. None of the job descriptions said "marine biologist wanted" much less "specialist in killer whale social behavior sought." Naomi went to just three job interviews in three months. Her mother was getting worried.

Then in April of 1993, she saw the ad. Someone was looking for a "marine mammal biologist."

"Well, what do you know?" Naomi said, flabbergasted, as she circled the announcement. "What are the odds of *that*?" The position was with the Humane Society of the United States (HSUS), at their new headquarters in Gaithersburg, Maryland, a leafy suburb outside Washington, DC. The influential organization was looking for a trained biologist to head up its fledgling marine mammal program. The new hire would be expected to respond to government policy and other actions, educate the public on species protection and habitat conservation, write articles for the national newsletter, and conduct advocacy campaigns on behalf of all marine mammals—cetaceans, pinnipeds, sea otters, and even polar bears—including those in captivity.

Naomi didn't even know there *was* a Humane Society of the United States. She only knew about the local shelters where people went to adopt a puppy. HSUS is not affiliated with those shelters, a point that critics—and there are many, mostly on the political right—often make, though HSUS does provide assistance to animal shelters and related programs.

Established in1954 as the National Humane Society by the journalist Fred Myers and others, the group's mission is to fight abuse and cruelty and promote the welfare and protection of all animals, wild or domestic, around the world. While far more mainstream than People for the Ethical Treatment of Animals (PETA) and other "animal liberation" groups, HSUS has still managed to attract its share of controversy. Many leaders in the fields of agribusiness, animal testing, and the zoo and aquarium industries loathe the organization for its combative opposition to their less humane practices. Though unaffiliated with local humane society animal shelters, HSUS is the world's largest animal advocacy group.

Given Naomi's extensive experience with killer whales and seals, and her gradual conversion from aquatic theme park fan to opponent, this did seem like the perfect job. And she was the perfect candidate. She mailed in her résumé and waited.

The phone call came from John Grandy (the HSUS vice president who testified at the NMFS meeting on captive killer whales). "I'm going to be in LA for a meeting," he told her. "Can you meet me at the airport?" Naomi was almost unnerved by her own excitement. On the appointed day, she drove out to LAX to have lunch with John in the dining room of the airport Hilton on Century Boulevard.

John told her about the recent HSUS efforts against SeaWorld and how the group had tried to halt the import permits for Tilikum, Haida, Nootka, and other killer whales. He said the group's president, Paul Irwin, wanted to intensify HSUS's campaign against captive marine mammals. "Paul wants to do more. Much more," John said.

He also talked about a movie that Warner Bros. was releasing in the summer of 1993 called *Free Willy,* about a troubled teenaged boy who bonds with a captive killer whale and helps return him to the sea. Irwin wanted to piggyback onto the *Free Willy* publicity to launch a public campaign against whales and dolphins in captivity, John explained. The president wanted somebody who could be a marine expert, someone with a PhD who could do media "and actually talk with authority on the subject." Naomi, a recent graduate, a doctor in biology, and an expert in killer whales, could not have come along at a more opportune time. John was impressed with her knowledge—and demeanor.

It was a good lunch.

Naomi drove back to her dad's house almost certain she had got the job. A few days later, the formal offer arrived by mail. The pay was abysmal, but a great deal more than she had ever made before. Besides, it was a good job in her chosen field, and it was definitely time to move out on her own. Naomi accepted. Only later would she realize that Grandy mistook Naomi's unfailing politeness at lunch to be a sign of agreeability. "It's not what he got," Naomi told friends, "because I talk back to people all the time, as you know."

Naomi had a couple of weeks to prepare for the big move. She arranged to have her belongings shipped East, while she, her Newfoundland/border-collie, Huey, and her cat, Jeb, drove to Maryland together in her car. Once there, she would stay with an aunt and uncle while looking for an apartment. Naomi planned a leisurely ten-day trip across country, stopping at the major attractions, but also poking along lesser-traveled byways and exploring the roadside oddities that dot the American landscape. She wanted to see the world's biggest ball of yarn.

A couple of days before taking off, Naomi's father announced, "I'm coming with you."

"No, Dad. You're not."

"Yes, I am. I'm too uncomfortable with the idea of you driving all that way by yourself."

Naomi lost the argument. Making matters worse, her father had a meeting back in California within the week. They were going to have to drive fast—the trip would take three days. An hour outside LA, Naomi's dad complained about the U2 music on the stereo. It was unmitigated hell: three eighteen-hour days of straight driving. No music, no byways, no big ball of yarn. Maryland was more of a relief than a final destination.

In late May, after finding a small apartment in Gaithersburg within walking distance of the office, Naomi reported to work at HSUS headquarters—a squat, 1980s, brick-and-glass two-story building, hidden behind shade trees in an anonymous corporate office park. She had no idea about the amount of work that was about to be thrown at her.

Free Willy was scheduled for nationwide release that July. Buzz was growing about the picture, which was looking to become the feel-good family hit of the summer. Paul Irwin wanted to use the opportunity to slam SeaWorld and the public display industry for keeping whales and dolphins in captivity for the entertainment of paying guests. He wanted the new marine biologist to prepare a report for publication—a white paper of sorts—on the dark side of cetaceans in captivity. Naomi only had a few weeks to get up to speed on the industry: Irwin intended to stage a news conference to release the paper—at SeaWorld's doorstep in Orlando—when the movie opened.

And that wasn't all. The Marine Mammal Protection Act of 1972 was up for congressional reauthorization. SeaWorld and other aquatic theme parks were already working hard behind the scenes to gut the landmark law by softening or removing provisions regulating the export of animals, minimum standards on animal welfare, and requirements for education and conservation programs in order to display marine mammals. Naomi would need to hit the ground running to coordinate the counterattack and keep the MMPA intact. She knew little about the MMPA, having done her research in Canada, and had no clue how to go about protecting it.

One afternoon at work, Paula Jewell, who was handling marine issues up until that point, handed off her files to Naomi—not just about killer whales and captivity, but about all marine mammal species and the issues and challenges that each faced (live captures, hunting, pollution, entanglement in fishing gear, etc.). There were piles of files. The killer whale drawers alone brimmed with thousands upon thousands of pages of memos, letters, reports, scientific journals, news articles, printouts from recently established Internet forums, and copies of the federal Marine Mammal Inventory Report—which listed all the births, deaths, sales, loans, and transfers of every US marine mammal in captivity. There were also necropsy reports on several killer whales that had died at SeaWorld.

Naomi had a lot to learn in a short time. Captivity was more complicated than she had imagined.

As a methodical scientist, Naomi wanted to understand the context of the debate. She pulled the files on the history of the marine mammal display industry and got to reading.

Among the most shocking materials in the files was the history of killer whale captures. It was sickening. The hunting or live capture of marine mammals has never been pretty. Untold thousands of whales, dolphins, and seals perished in nets or at the end of a harpoon line when men came hunting for them—whether for meat, blubber, fur, or entertainment.[1]

In the early 1960s the undisputed leader of marine mammal captures was Marineland of the Pacific. Frank Brocato, the park's chief animal collector and his assistant, Boots Calandrino, had grown famous for capturing a wide range of dolphins, porpoises, belugas, and pinnipeds. But they hadn't given much thought to bagging an orca for Marineland's burgeoning menagerie.

In November 1961, the collection crew were cruising around Newport Harbor in Orange County about an hour down the coast from Palos Verdes, when they came across a sick, disoriented female killer whale swimming alone among the schooners and yachts of the upscale inlet. They corralled the animal and lifted her onto a flatbed truck for the drive back to Marineland. She was the first killer whale ever taken into captivity. It did not go well.

"We'd suspected the animal was in trouble because of its erratic behavior in the harbor," Brocato said in a PBS interview, years later. "But the next day, she went crazy. She started swimming at high speed around the tank, striking her body repeatedly."

Finally, the orca shot straight for a concrete wall, smashed her snout, convulsed several times, and died. She had lasted only two days. The autopsy report found acute gastroenteritis and pneumonia, possibly acquired from pollution in Newport Harbor. Whether her illness was related to her suicidal behavior will never be known.

Now that Marineland officials had captured one orca, however temporarily, they wanted more. No member of the general public had ever before seen a captive killer whale up close. People would line up with open wallets to take in such a horrible beast.

Snagging killer whales out of the ocean was no easy task. Frank Brocato and his crew sailed up to Puget Sound in search of another orca for Marineland. A month later, scouring the waters off the San Juan Islands in Haro Strait, they spotted a pair of killer whales, a male and female, hunting porpoises.

The female chased a porpoise beneath the boat as the men waited for her on the other side with a lasso. They snagged her. But then "everything started to go wrong," Brocato recalled later. The female made a sudden and unexpected turn. The nylon ropes of the lasso wrapped around the propeller shaft, killing the engine and stranding the vessel. The female took off at a frantic pace, only to be stopped short about 250 feet away when the lasso's tether ran its length. The men could see the orca's black silhouette rise above the surface through the distant mist.

Then they heard the noise: an agonizing shriek that shattered the silence of the waterway. The trapped female was calling for the male, whose six-foot dorsal fin quickly appeared next to her. The animals turned toward the boat and charged at it at top speed, rushing the hull repeatedly and thwacking it hard with their powerful flukes. Brocato grabbed a. 375 Magnum he kept on board and fired into the gray water, slamming a single bullet into the male, who either died or fled. It took ten bullets to finish off the female. They towed her body to a dock in Bellingham, where she was weighed and measured for posterity. Brocato extracted her teeth for souvenirs, and the rest was reportedly ground into dog food.

Brocato had tried and twice failed to land a healthy killer whale that could be displayed to the public. That wouldn't happen for three more years. The first orca to be put on live exhibit was at the Vancouver Aquarium in 1964. The aquarium's director, Murray A. Newman, had commissioned a sculptor, thirty-eight-year-old Samuel Burich, to hunt down and kill an orca and stuff it, for use as a life-size model in a special display for the aquarium's new British Columbia Hall.

Burich took a harpoon boat out to Saturna Island, on the Canadian side of the water, and waited. It took two months but it paid off. The crew spotted a pod of thirteen killer whales near the shore. Burich harpooned a young whale. It was wounded, but did not die. Suddenly, two pod members came to the rescue of the young whale, gently pushing it to the surface for air. The harpooned animal tried to free itself, thrashing in the water and issuing a high-pitched screech and shrill whistles that could be heard above the water, three hundred feet away. Burich steered over to the whale and fired several shots into its flank. Still it would not die.

Burich called the aquarium. Director Murray Newman took a float plane from Vancouver out to Saturna Island and, upon arrival, decided that the whale should be spared and towed to the aquarium for display. The crew attached a line to the harpoon in the animal's back. It took sixteen hours through choppy seas to make the journey. The pain and terror inflicted on the whale during that time can only be imagined.

The captive animal was placed in a makeshift pen at the Burrard Drydock in North Vancouver, where she was identified as a female and named Moby Doll. She was an instantaneous sensation. Scientists from around the world flew in to see the specimen up close and record her unearthly sounds. What struck most people, scientists and the general public alike, was how tame and docile the whale seemed to be. She also appeared to be sick and was probably still in shock from the harpoon and bullet wounds.

It took Moby Doll two months to regain weight, but she continued to be weak and developed a severe skin rash due to the low salinity of the harbor water. After eighty-seven days in captivity, Moby Doll died. *The Times* of London noted the media coverage of the capture and display of the orca, calling it "some of the first positive press ever about killer whales." An autopsy on Moby Doll later revealed the whale was, in fact, Moby Dick.

The next killer whale to be captured, in 1965, was a male calf accidentally caught in a fishing net near Namu, a remote coastal outpost in northern British Columbia. Ted Griffin, owner of the Seattle Marine Aquarium, caught word of the capture and immediately offered the fisherman $8,000 for the whale. A special cage was built to tow the calf—now christened Namu—four hundred miles down to Seattle. As the voyage began, the young whale emitted a string of distress calls, attracting a pod of thirty orcas who tried but failed to free him. A mature female and three calves continued to follow the boat and cage for another 150 miles before finally turning away. They were presumed to be Namu's mother and siblings.

Once in Seattle, Namu was placed in an enclosed pen in Rich Cove, where he quickly took on rock star status. Tens of thousands of people lined up to see him eat his daily diet of 375 pounds of fish. Once again a small battery of scientists arrived on the scene to study the whale, measuring Namu's blood pressure, heart rate, and even his brain waves.

Eventually, Griffin came to trust the whale enough to climb into the tank with Namu. That same day, he learned how to ride the animal. Soon after that, he was able to teach Namu a few performance routines for the public. The most popular tricks were Namu's jumping from the water to retrieve a salmon that Griffin dangled from a tower and taking his owner for a ride around the pen. Namu performed five shows a day, alternating with a trained-seal act. "Ahab had his whale," Griffin liked to say, "and I have mine."

Ticket sales to see Namu more than doubled. Marketing the orca became even more lucrative after he starred in the 1966 film *Namu, the Killer Whale*. The rather lame but endearing drama starred Lee Meriwether and Robert Lansing. The film's tag line read, "Make room in your heart for a six-ton pet!!!"[2] People would never look at orcas the same way again: The animals were being transformed from killer to cuddly in the public eye.

Namu and Griffin thrilled crowds for eleven months, but Namu never really adjusted to captivity. He could be heard making loud, strident screams from his pen, at times picking up returned calls from orcas passing by in Puget Sound.

In July of 1966, Namu started to act erratically, ramming his stainless steel nets at high speed, refusing to perform, and "acting mopey," according to the Associated Press. Namu caught a bacterial infection, got caught in his stainless steel net, and drowned. Griffin told reporters that Namu was "lovesick" and may have been trying to escape to mate with females out in the sound. Griffin mourned his loss, but he had been bitten by the show business bug and the allure of ticket sales. A massive effort to trap more orcas began, and an industry was born.

But Griffin wanted more.

The second killer whale to be kept in captivity and put on display was a young female captured in 1965 by Griffin and his business partner Don Goldsberry. They named her Shamu—a combination of *she* and Namu—the *original* Shamu in a long line of Shamus. Griffin had initially intended for Shamu to be the cure for Namu's lovesickness, but she was far too young for mating. Griffin sold her to the new SeaWorld park in San Diego, according to the AP. On December 20, 1965, the twenty-three-hundred-pound female was sent to California aboard a chartered cargo plane. The second orca to be kept in captivity was also the first orca to fly.

Shamu was released into her new tank in San Diego, completely ignoring the bottlenose dolphin, named George, put there to keep her company. "Shamu let out a squeal like a rusty gate when she plunged into the water," the AP reported. SeaWorld refused to disclose what it had paid for the whale, but the AP said the price "was reported to be $75,000" (over half a million dollars at 2012 value). *The New York Times* called her "the ugly killer whale on public display."

Goldsberry and Griffin perfected the purse-seining technique for capturing killer whales in Puget Sound. By the early 1970s, they had taken more than two hundred orcas from these waters. Most were released; the rest went to aquariums, including SeaWorld.

In 1970, the two partners undertook the most notorious orca capture ever attempted: a hugely controversial whale roundup at Penn Cove, a tree-lined bay on crooked Whidbey Island, forty miles northwest of Seattle. The men snared eighty killer whales in a single net, comprising nearly every member of J, K, and L pods, or the entire Southern Resident community. Most were released during the ordeal. The men chose seven whales, all of them young, to ship off to anxiously awaiting aquariums around the world. One of them was Lolita—the only orca still alive from the Penn Cove capture. She lives and still performs at the Miami Seaquarium in the world's smallest killer whale tank.

Many orcas did not survive the long and stressful ordeal at Penn Cove. Some of their carcasses were dragged up by a fishing trawler; their bellies had been slit and their bodies weighted down with steel chains and anchors. Goldsberry and Griffin later admitted to trying to prevent the dead whales from washing ashore.

Two years later Congress passed the historic Marine Mammal Protection Act of 1972, which banned the live capture of any marine mammals in US waters, except for the purpose of public display or scientific research. SeaWorld continued to qualify for the display exemption.

But live captures of killer whales in Washington State waters were about to come to a halt. The last straw was in March 1976, when Goldsberry, working for SeaWorld, overstepped the bounds of what the public would tolerate. Searching for whales off the shores of Olympia, the state capital, Goldsberry spotted a group of Resident whales. He used aircraft and exploded seal bombs to force six killer whales into a waiting net at Budd Inlet. The spring afternoon had drawn many boaters onto the water that day, including Ralph Munro, an aide to Governor Dan Evans, who witnessed the whole thing.

"It was gruesome as they closed the net. You could hear the whales screaming," Munro said later. "Goldsberry kept dropping explosives to drive the whales back into the net."[3]

Washington State filed suit against Goldsberry, alleging that he and SeaWorld had violated the terms of their permits, which mandated that the roundups be humane. It was the worst publicity SeaWorld had ever endured. The company agreed to release the whales being held in Budd Inlet. A Seattle district court subsequently demanded that SeaWorld relinquish its permit to collect killer whales off Washington.[4]

SeaWorld would have to look elsewhere for new animals. It promptly dispatched Don Goldsberry on a worldwide tour in search of promising hunting grounds. He selected Iceland—the whales were plentiful and the government was cooperative. In October of 1976, SeaWorld had its first killer whale flown in from the island nation.

Naomi grew angry at the long list of killer whales that had been yanked from their homes and families since that first ailing female was taken from Newport Harbor. Nearly one hundred and thirty orcas had been shipped to display facilities between 1961 and 1992.

Now, only twenty-four of them were still alive.

16

Backstage Doubts

Nobody likes cognitive dissonance, the itchy, uncomfortable feeling that your previously held beliefs about a person, place, or thing—a job, say—do not conform with what your eyes and ears are telling you. Human nature goes into overdrive to eliminate, or at least tone down, the unbearable internal conflict.

Denial and rationalization are thus highly useful for collecting a paycheck. As the great muckraker Upton Sinclair put it, "It is difficult to get a man to understand something when his salary depends upon his not understanding it."

Looking back at their earlier years at SeaWorld, former trainers such as Jeff Ventre, John Jett, Sam Berg, and Carol Ray marvel at the denials and justifications they used to muzzle the whispering doubts that followed them to sleep at night. They were loyal team players at the world's premier marine-life enterprise. No matter what their eyes and ears were telling them, they still thought SeaWorld was a great place to work—for people and animals alike.

After all, they had been trained to believe many workplace myths: The whales and dolphins in their care were happy, healthy, and pampered, with longer life spans than animals fending for themselves in "the dark, scary ocean," as SeaWorld officials sometimes called the natural environment. The trainers were part of the SeaWorld "family": They were fairly if not richly rewarded for their one-of-a-kind jobs; their bosses, while not always congenial, were leading experts in animal behavior; and above all, they felt *safe,* even while doing water work with killer whales. No one at SeaWorld ever led them to believe otherwise.

Over time, however, cognitive dissonance grew stronger; the soothing balm of denial and rationalization eventually began to wear off.

Jeff and John had become good friends; they spent many Orlando nights over beer at local hangouts, where talk often turned to the more disquieting aspects of their jobs. Each recognized the other as a critical thinker, and both had ethical questions about keeping marine mammals in captivity, and its effect on the animals' mental and physical health. They were also growing skeptical about the integrity and intelligence of their supervisors and the relatively low pay they received.

Such discussions were reserved for quiet corners in dark pubs, far from the earshot of other SeaWorld staffers. Bill Clinton had recently taken office and was concocting a plan to allow gay Americans to serve in the military, as long as they did not reveal their sexual orientation. Jeff and John now felt as if they were living under their own bizarre set of Don't Ask, Don't Tell restrictions. They understood the rules. They didn't ask questions at work, and they saved the "telling" for after hours.

One thing the men noticed was that whales, because they were so smart, easily got bored. The animals needed to invent ways to amuse themselves when humans weren't interacting with them. Frequent targets of the whales' restlessness were birds—usually seagulls, but other Florida fowl as well. Captive killer whales in San Diego had been observed leaving small bits of food on the surface to attract hungry birds, then ambushing and killing them, for fun, not a meal. It is believed that one whale devised the trick, which the others learned by observing. In other cases, whales would float a whole fish on the surface as bird bait. This indicated purposeful intelligence: the willingness to forgo food up front for the potential of a greater reward later on. The use of bait to attract a victim was a form of tool deployment, a hallmark of intelligence in animals.

Birds in the water could also demonstrate how stubborn and unyielding killer whales can be. One day in San Diego a killer whale grabbed a pelican right in the middle of a show, much to the revulsion of the guests. The whale refused to relinquish the new toy and ignored all callback commands to drop the bloody carcass and return to the stage.

The doubts that began to germinate in John's and Jeff's minds were sometimes planted from people on the outside. SeaWorld executives had already anticipated that anti-captivity critics—activists and scientists—who routinely showed up to observe the animals would also try to make contact with staff members. But the company had done a thorough job of infusing its employees with a general disdain for anyone who criticized the business of keeping cetaceans in tanks. All members of groups such as the Animal Welfare Institute, Whale and Dolphin Conservation Society, In Defense of Animals, and, of course, HSUS were almost universally despised at SeaWorld. They were nut jobs to be avoided at all costs.

One outsider did manage to break through the wagons to connect with a few of the trainers in Florida, including Jeff and his friend and fellow hotdogger at SeaWorld Mark Simmons. Her name was Astrid van Ginneken, MD, PhD, a tall, athletic woman from Holland with gray-blond hair pulled back behind the ears and a deep, heavily accented voice. Jeff first noticed her back in the spring of 1988, sitting alone for hours in the stands at Shamu Stadium, during shows and in between, watching one of the whales: Astrid first arrived after the female Gudrun had been brought to SeaWorld from the Netherlands.

Jeff began speaking with the stranger, even though most staff dismissed her as that "crazy Dutch whale lady." Eventually, he became more cautious and met Astrid off-site, to avoid prying eyes and ears.

Astrid, who had designed an innovative computerized patient-record system for the Erasmus University Medical Center in Rotterdam, first fell in love with killer whales after seeing one at the Dolfinarium Harderwijk theme park in the Netherlands. It was Gudrun, whom the Dutch aquarium—together with SeaWorld—had captured off Iceland in 1976, along with the female Kenau. The two whales were brought to Harderwijk, east of Amsterdam. Kenau was quickly sent to SeaWorld in Orlando but Gudrun remained. Astrid found herself spellbound by the giant black-and-white dolphin with the gentle personality.

In 1986, work brought Astrid to a conference in Washington, DC, where she visited the Natural History Museum, which was having a blockbuster exhibit on whales. Again, Astrid was enraptured by the creatures. She purchased a few books on cetaceans at the museum store. Two of them were about whales: Erich Hoyt's *Orca: The Whale Called Killer,* and *Song of the Whale* by Rex Weyler, which tells the story of Paul Spong's metamorphosis from dispassionate scientist to global whale advocate. She was deeply inspired by what she read.

Astrid went back home with a deep desire to see Gudrun again. She arrived at the Dutch aquarium early one Sunday to see if the orca was still there. She was. But it was November now. The park was closed for the season. The staff allowed Astrid some time with the whale.

As they approached, the trainer cautioned, "Gudrun doesn't like strangers much. She can be very nervous in their presence. So please don't look her in the eye, and don't make any wild movements or that sort of thing." But Gudrun acted calmly around the stranger. Astrid felt they had a bond. Using her charm and enthusiasm, her university credentials, and the reference of a curator she knew at the New York Aquarium, Astrid talked her way into a standing invitation to visit Gudrun. She was even allowed to feed the whale.

Astrid visited every week for the next year. The aquarium had a small underwater viewing area, and Astrid would also spend time down there, staring into Gudrun's eye. It was a profound experience. "She dramatically changed my life," Astrid told Jeff one day, "because somehow she broke through my shield.

You see, I was—how do you say it?—I didn't dare show affection in the open, I was very shy about that." Gudrun had changed that.

Astrid became so fascinated by killer whales she began looking into places where she could study them in the wild. She learned about the Center for Whale Research, an orca science compound on San Juan Island, in Washington State, run by Kenneth Balcomb. Ken, a premier expert on wild killer whales, had worked with Mike Bigg on the early surveys of the Northern and Southern Resident communities and coauthored several papers with Mike. Ken also cowrote the seminal field guide *Killer Whales,* with his Canadian colleagues Graeme Ellis and John Ford.

The Center for Whale Research operates out of Ken's home and office on a few acres of land overlooking Haro Strait and across to Vancouver Island. Its ongoing project, Orca Survey, is a long-term photo-identification study of the three pods J, K, and L, which make up the Southern Resident community. Astrid contacted Ken and offered to work as a volunteer in the summer of 1987. He agreed to take her on.

In the fall, it was announced that Gudrun would be transferred to Florida on a "breeding loan." The director of Dolfinarium Harderwijk, F. B. den Herder, had contacted SeaWorld vice president and zoological director Dr. Lanny Cornell, informing him that Gudrun, who was captured in a joint operation between SeaWorld and the Dutch aquarium, had reached "reproductive age." At this point in nature, den Herder added, most females would already be impregnated by males.

"We would like to mimic this natural process which would be beneficial to the well-being of our female," den Herder wrote, "but we have no possibility to provide a male killer whale." He proposed sending Gudrun to one of SeaWorld's "killer whale breeding facilities" on a breeding loan basis for at least four years. "The ownership of the potential young should be decided on in a future stage of negotiation."[1]

But the Dutch were not offering Gudrun for free. She was a "major attraction" at Harderwijk, den Herder reminded his American colleague. "We are afraid that her absence will decrease the number of future visitors." In order to "overcome this potential problem," den Herder requested that SeaWorld send two of its false killer whales to Holland for the same period that Gudrun was to be in Florida. He also wanted two white-sided dolphins, to "give the public a chance to come in contact with other species than the 'Flipper' [bottlenose] dolphins."

The exchange could happen in November, when the aquarium's 1987 season was winding down.

Astrid was allowed to stay with Gudrun in the final hours before the flight, in mid-November 1987, keeping the animal calm and hand-feeding her fish.

Gudrun had a rough time when she got to Florida. Not only did Katina assert her dominance by raking and shoving the newcomer, but SeaWorld began breeding her almost immediately. She was locked in a back pool with Kanduke, who chased her around the tank, trying to penetrate her over and over, and often succeeding. What seemed like serial rape to Jeff produced the birth of Taima, the unpredictable Transient-Icelandic hybrid, in July 1989. Born during a summer storm, her name was a Native American word for "crash of thunder." It would prove to be an appropriate moniker.

In Europe, Gudrun had spent most of her time with bottlenose dolphins. SeaWorld officials were unsure what kind of mother she would be, but Gudrun showed herself to be loving and competent with Taima. Within a year, the two of them were performing daily at Shamu Stadium.

Astrid had been disturbed to see Kanduke's rough treatment of Gudrun. But after the Transient died in 1990 and was replaced by Tilikum in 1992, things began to get better for the Icelandic female. Both Gudrun and Taima took to Tilikum soon after he arrived, Astrid observed, unlike the dominant matriarch Katina, who harassed Tilikum and raked him with her teeth—unless she was in estrus (heat) and wanted to mate.

"Gudrun is different," Astrid told Jeff. "Maybe she comes from Tilikum's clan in Iceland, or maybe their personalities just match better. We'll never know. But they do spend a lot of time together, and very harmoniously."

The two also mated harmoniously. There was no "rape" between Tilly and Gudrun. One day, a trainer informed Astrid that Gudrun was in estrus and invited her to watch the pair mate in a back pool. She watched in amazement and even captured the ritual on her camcorder.

"Tilikum was so gentle!" she marveled to Jeff later that day. "He would swim behind her, and Gudrun would be in the lead, and she would look back at him, as if to say, 'You're still following me, right?' And then he would swim up to her and caress her with his head, or he'd roll over and take her on his chest. It was so romantic. Afterwards, they were completely content, resting side by side. It was totally different from Kanduke."

On New Year's Eve, December 31, 1993, two years after Tilikum arrived in Orlando, his fourth offspring—a three-hundred-pound, seven-foot female named Nyar—was born to Gudrun.[2] It was the second birth in Florida that year. Four months earlier, Katina had given birth to a male calf, named Taku, who was also sired by Tilly.[3] SeaWorld's killer-whale-breeding program was booming. It was now the most productive in the world: Nyar was the ninth successful birth at the chain.

Gudrun's new daughter seemed to fit in well with the artificial pod she was born into. Even the queen, Katina, assisted Gudrun while she was in labor with the calf. The newborn began playing with Taku, and sometimes her older half-sister, Taima, joined in as well.

SeaWorld officials told the media they had been given an "unprecedented" opportunity to study the development of social behaviors simultaneously with two infant killer whales. "This is a very happy occasion for all of us at SeaWorld and the zoological community," Brad Andrews told reporters. Each successful birth, he claimed, was "adding to the scientific community's knowledge of killer whales."

Mother and calf "appear to be doing well—we're all hopeful this is a strong and healthy young whale," Andrews continued. But he cautioned that the days and weeks ahead would be crucial for the calf's survival. Half of all baby orcas died before their first birthday in the wild, Andrews stated. "This successful birth is further evidence that the killer whales are thriving in SeaWorld's environment."[4]

Andrews spoke too soon. Nyar did anything but thrive. The infant girl seemed to suffer from some kind of congenital birth defects. She was physically and mentally unsound. Gudrun rejected her calf and tried to drown her several times before SeaWorld separated the two. Nyar made little progress and had trouble swimming correctly. Blood tests showed she suffered from immunosuppression. She had trouble learning and was unfit for shows.[5] Nyar often spent time in the company of Tilikum, her father, who treated her with great gentleness.

Astrid told Jeff that Gudrun's rejection of her calf would be rare in the wild. The Dutch scientist was on her way to becoming a bona fide killer whale researcher herself, having now spent six summer seasons on San Juan Island at the Center for Whale Research.

Jeff had never seen a killer whale in the wild. He wasn't sure if any of the trainers at SeaWorld had ever seen one.

"You really must come to the San Juan Islands one summer, very soon," Astrid told him. "You will never look at an orca the same way again."

Jeff loved the idea, but he wasn't sure how well his bosses would take it if they found out he went to the Pacific Northwest to observe wild killer whales with the "nutty" lady from Holland. He said he would think about it.

Over the years that she visited SeaWorld, Astrid taught Jeff much about the natural history of killer whales in the wild, something he would never learn at the marine park. The Dutch doctor was not overtly opposed to captivity and was by no means anti SeaWorld. But many of the things Jeff learned from his discussions with her left a lasting impact. However inadvertently, Astrid was helping to change Jeff's feelings about captivity. Jeff's talks with her added more fuel to his nighttime ruminations with John. Both of them were growing more concerned about the whales' health—the drugs in their morning meals, their dorsal fins, the way they kept dying.

And then there was the tooth issue.

Many of the killer whales had developed serious dental problems—mostly chipped and broken teeth, but also teeth that had been removed or fallen out.

Most disturbing of all were teeth that needed to have the pulp drilled out of the center, leaving behind a conical cylinder.

Fighting between the whales could also lead to tooth loss. Astrid told Jeff about one particularly awful conflict she had witnessed. Katina had been acting up with Gudrun, shoving and ramming her repeatedly as they swam around the main pool. Gudrun tried to defend herself by raking Katina with her teeth. But just as Gudrun moved in with her mouth wide-open, Katina thrust her tail at it. The impact was loud and severe. Two of the teeth in Gudrun's lower jaw were driven down into the bone. She was in excruciating pain. Blood and green vomit spilled from her mouth.

An emergency medical crew was assembled and Gudrun was separated into D Pool, the small medical pen. The floor was raised and the water drained, stranding Gudrun on a mat in the pool. She was held by trainers and animal-care staff, who used a four-by-four block of wood to keep her mouth open long enough to allow vets to yank out the two impacted teeth.

Jeff and John were beginning to believe that stress and boredom were adding to the tooth problem. The steel gates that separated the park's pools were made from horizontal bars. These gates were the first line of defense when the whales went "off behavior" and became aggressive and in need of physical separation. Once separated, two whales sometimes bit down on the bars, a display of aggression called jaw-popping. Jeff knew that some animals show more aggression when restrained than when unfettered—the way some dogs will snarl at other dogs when on a leash, but be friendly when off leash. When they know there is no chance for actual physical combat, they exaggerate how tough they are.

Even when they were not challenging each other through the restraints of the gates, some whales passed the time fighting boredom by simply chewing on the bars, or on the corners of the concrete pools. Several times Jeff and John discovered teeth or fragments of teeth on the bottom of the tanks, especially near the gates.

All that breakage left a lot of exposed tooth pulp. If left untreated, decaying pulp can form a large cavity that becomes plugged with food. Impacted food can cause infection and inflammation and possibly harm an animal's immune and cardiovascular systems. Most orcas at SeaWorld were relatively young; the roots of many of their teeth were still too immature to accommodate a root canal, as it was explained to John and Jeff.

John sometimes assisted in drilling Gudrun's teeth. She was trained to submit to the awful procedure just like any other behavior—with a lot of positive reinforcement. First she was asked to put her chin on the deck. Then the trainers would show Gudrun the high-speed drill—a handheld Dremel like those used around the house—then reinforce her by rubbing her fins or giving her some fish. They would touch the drill bit to her tooth, without turning the machine on, reinforcing Gudrun once again. Next they touched the bit to her tooth and turned it on at the lowest speed, so she would feel it and hear it just slightly, followed by more reinforcement. Finally they started to drill.

They then made a hole through the pulp and into the jaw in what is known in dentistry as a pulpotomy. The whales hated it and often refused to submit to the drill by sinking down beneath the surface, shaking their heads violently, or breaking from control and swimming away. The staff knew it was a successful drill when blood started to bubble out from the bore hole. Once drilled out, the hollow teeth needed to be flushed—or "irrigated"—with a solution of iodine and saline water three times a day. Otherwise, abscesses, bacterial infections, and systemic sepsis might set in, and possibly much worse. In humans, poor dental health can lead to heart disease, pneumonia, stroke, or heart attack. SeaWorld called it "superior dental care."

Most whales did not seem to mind the regular tooth irrigation, probably because they were heavily rewarded (fed) for participating. Food and other materials in the cavity might have been irritating them as well. Each time John flushed out an orca's teeth, all matter of detritus came spilling out. Sometimes he extracted strips of killer whale skin from the borehole—curled up like black string—possibly from Tilikum or another subdominant whale subjected to frequent raking.

Jeff and John were slowly coming to accept that life at SeaWorld was just too stressful for killer whales, though some animals seemed to handle confinement better than others. Katina, for example, was always businesslike and ready to follow signals consistently and predictably. But other animals were not so reliable.

Taima was the least predictable water-work whale in Orlando. Strong-willed and independent, she would break from control during sessions far more often than the others, then go off and do her own thing. The impetuous Taima had many "fuck-you" moments, Jeff liked to joke, as in "Fuck you, I'm not going to do the bow you just asked for, I'm going to swim circles on my back instead."

Taima had always been different. As an infant, she had grabbed the ponytails of female trainers, who were subsequently told to keep their hair in a bun. Taima also "mouthed" the arms and legs of trainers, meaning that she gently bit down on them but caused no harm. Nobody was terribly alarmed by it—the calf wasn't being aggressive, just playful and experimental. It was "exploring" behavior, in SeaWorld's perpetually cheerful argot. But it still needed to go. Senior trainers worked intensively with Taima to stop the mouthing. It was not a complete success. She still pulled the outer layer of socks from Jeff's feet while he was working with her in the water. She was also known to bump trainers and other whales in the pool. Taima was still the most likely whale at Shamu to go "off behavior."

Losing control of a killer whale was not always as serious as it sounded—usually it meant the animal swam off by herself, ignored calls to return to stage, mouthed trainers or their clothing, or something of that order. But loss of control could also potentially lead to danger. Trainers worked diligently to reduce the number of unpredictable interactions. SeaWorld was able to maintain control over the killer whales in the vast majority of interactions conducted each year in the various parks.

This impressive and reassuring record didn't hold true for Taima. She was too crafty, too self-absorbed to be that predictable. She was a handful and broke from control more freely than her tankmates. Despite that, or perhaps even because of it, Jeff secretly considered Taima his favorite whale—a freethinker, she was; a personality with whom he could relate.

The best way to maintain control over trained animals is to remain hyperobservant of everything in the environment that might interfere—other trainers, other animals, distractions in the crowd, even the weather in an outdoor stadium. The trainer must also constantly monitor the animal for any "precursor" of going off behavior, especially if it could lead to aggression. Aggressive precursors might include visual cues such as a widening of the eyes, an open mouth, or jaw-popping; or precursors might be spitting food at a trainer or ignoring hand signals, water slaps, or calls to return to stage.

Recognizing a precursor in time did not guarantee that the trainer could defuse aggressive behavior. Sometimes, social stress or changes in the environment produced completely unpredictable behavior—some of it potentially aggressive. Some whales were more prone to such stress-induced loss of control, such as Tilikum. When frustrated by stimuli around him, Tilly could suddenly display a host of aggressive and dangerous behaviors, such as mouthing the concrete stage, vocalizing in threatening tones, banging on gates with his head, even lunging up from the water at his control trainer during practice or exercise sessions. Tilikum was also inconsistent with "separations"—it was difficult to make him leave a pool when he did not want to go.

Tilikum also became stressed after spending too much time with the females. Ironically, he became agitated after prolonged separation from them, especially if visual access to his tankmates was blocked. He was highly averse to change in his environment, but he also disliked repetition during learning sessions. When bored during a session, he would repeatedly give incorrect responses, to express his displeasure, much as Hyak had done with Paul Spong in Vancouver when he refused to respond to the same piece of music twice.

Tilikum reacted better to some trainers than others, and his bond with John Jett was evident. John loved and trusted the huge whale. He put his hands on the big bull daily. He actually grew to feel sorry for the guy. "He's just a big, misunderstood puppy dog," John once remarked to Jeff. "He's very subdominant. He gets picked on all the time, and he has nowhere to run."

Mostly, John thought, Tilikum was bored in the extreme. John felt this killer whale needed him, needed his companionship. John thought he had made a difference in Tilly's life.

Even Katina had her bad moments. She had once mouthed a trainer's waist and on other occasions bumped her head into a hip, a torso, or a hand. Once, she pushed a trainer around the pool inappropriately.

Katina had other issues. She was highly protective of new calves, especially while doing water work, and would often try to separate a youngster from a trainer by swimming between them. Katina at times also "displaced" (rammed from the side like a pushing foul in basketball) other whales to demonstrate her dominance. She routinely displaced Tilikum when she was left with him for extended periods.

Sometimes Katina showed outright chutzpah, especially around newcomers, engaging in fuck-you moments of her own. Instead of working with someone, she might refuse to make eye contact, slink beneath the surface, play with her food, or refuse to open her mouth for fish. Katina was so bossy that, when she decided *she* didn't want to cooperate, she could actually force the *other* whales into disobedience, as well. Those who still followed their trainers' signals, despite her lead, received a harsh displacement from the queen.

But even Katina had to defend her status at times. In 1994, her daughter Kalina, the original Baby Shamu, returned to the pools of Orlando after spending four and a half years on tour in Ohio, San Diego, and then Texas. Kalina was nine years old. She had left one calf behind in Texas and was pregnant for a second time.

When Kalina returned to Florida, she began fighting with her own mother for dominance.[6] Katina put down the rebellion, but the insubordination was unheard of in whale society—at least in the wild. Jeff located a copy of Kalina's official Animal Profile and discovered that Baby Shamu had developed an entire repertoire of behavioral issues while on tour across the country.

Among the things that might upset the young star, according to her profile, were "major environmental and social changes, unclear/confusing situations, divided attention," and (rather ironically given her travels and many truncated relationships) "long term separation."

Kalina had several "aggressive tendencies" as well. "When excited or confused, she may slide over, push or bow over her trainer in the water," the document warned. "[She] will aggressively and physically displace less dominant whales when frustrated, confused or sees an imbalance in attention." While playing with toys or trainers, Kalina had also "shown extreme excitement to borderline 'aggression.' Aggression involves anything from slight bumping or sliding over her trainer to a complete bow over her trainer." She also "opened her mouth on trainers" on several occasions.

Despite these many incidents and reports, the trainers at Shamu Stadium still felt out of harm's way. As long as they remained consistent, confident, and void of wishy-washy reinforcements, they were taught, as long as they kept a few strategic steps ahead of the animal, they would be successful—and safe. It had still not crossed their minds that working in the water with killer whales might be dangerous.

After all, whenever something did happen, and someone got hurt, most people at SeaWorld assumed that the trainer, not the animal, was to blame.

One day John was practicing a new behavior with Katina in preparation for a show. He had been approved for water work, though not the hotdogging aerial feats. Jeff and their mutual friend Mark Simmons did much of that around Shamu Stadium. Mark, strong, lean, and well built, and also a gifted artist, was probably the most talented and experienced orca hotdogger at SeaWorld and one of its biggest stars. Jeff loved doing combo sequences with his friend. Mark was expert at synchronizing the stand-ons and rocket hops the two would perform in unison. It always dazzled the crowd. John, meanwhile, was allowed to do some of the less intensive behaviors, such as a "foot push," which ended in a spectacular "hydro-slide" across the main stage.

On this practice run, Katina began shunting John around the water by his feet with her rostrum. He rode the force of her flukes on his belly like the prow of a ski boat, a wake of white water dividing at his chest. The audiences loved this behavior. There was something exciting, but also comical about seeing a grown man being pushed around the water by fifty-five hundred pounds of deadly beast. It was part Texas rodeo, part *Gilligan's Island*.

John and Katina did a full lap around the tank. When they approached the stage, Katina already knew what to do. She gave John an extra push and he sailed toward the smooth, wet platform—chest first, arms outstretched. The idea was to slide across the smooth, wet concrete on his belly as if gliding on ice, from one end of the stage to the other, fly off the far end, and slip back into the water like a penguin.

But something went wrong. It might have been the speed, the angle, or the distance from the stage when John launched from Katina's nose, but he heard a loud *whack* as his hip smashed into the edge of the concrete stage. John had to hobble around in pain for the rest of the day. As he explained to Jeff later that evening, "I decided it was better to suck it up and not say anything, than risk being sidelined from the show and labeled a whiner."

"No big deal," he said to the other trainers, trying to laugh off the incident as a rookie's mistake. Mark Simmons helped him limp offstage. "It was probably just my own miscalculation," John told him. Fortunately it was a Friday: John could take the weekend to recuperate. But in the years to come, he would suffer from chronic back inflammation and undergo major surgery to fuse three vertebrae with titanium rods, screws, and spacers. He could never prove it, but John suspected the cause was that collision with the stage.

John was eventually able to shake off his pain—and his doubts about worker safety at SeaWorld. But his days of denial and rationalization about the welfare of the animals were coming to an end.

"You know, every day I go in to work, it becomes more painful for me to see these animals in this environment," John confided in his buddy Jeff over beers at a local pub. "But I keep telling myself that maybe my presence is going to make their lives better. And I really do try to make their lives better, especially Tilikum. I work as hard as I can for that poor guy."

17

Blood in the Water

The more that Naomi read through the files at HSUS, the angrier she became. Though she never regretted the hours she'd spent watching killer whale shows and going backstage to spend time with Hyak in Vancouver, she was shocked to learn how challenging life could be for captive orcas, and how risky—and in her mind stupid—it was for humans to get in the water with them.

Naomi had no idea that killer whales died so often, and so young, as they did in captivity. The carnage ran counter to everything that SeaWorld and other aquatic theme parks had been telling the public. At first blush, the industry's contention that animals lived longer in captivity than in the wild made a certain degree of sense. Some animals *did* live longer in zoos than in their natural environments, including zebras, giraffes, and several species of bears, primates, snakes, and birds.

For top predators such as orcas, one might reasonably think that a controlled environment would increase their survivability. At SeaWorld, killer whales were shielded against pollution, boat propellers, and hunger caused by declining stocks of prey, while given world-class veterinary care, constant human attention, and all that restaurant-quality fish. Supposedly liberated from the threats of illness, famine, and physical injury, popular thinking *expected* them to live longer.

But as Naomi gradually combed through the papers at HSUS, she realized that SeaWorld's gauzy prescription for contentment did not take into account the stress of captivity, which seemed to increase the risk of physical and mental disorders that can produce immune dysfunction, infections, reproductive problems, and potentially lethal aggression among the whales.

Since its passage in 1972, the Marine Mammal Protection Act (MMPA) required all public display facilities to keep a running log of every marine mammal it held, including date of birth or capture—and date and cause of death. Known as the Marine Mammal Inventory Report (MMIR), it remains a helpful tool for animal welfare advocates, even though it lists no details on the animal's demise other than the cause of death. However, these details were in the necropsy reports submitted to NMFS and publicly accessible. Newspaper articles in the files fleshed out more details, though media accounts were sparse in earlier years.

Since the beginning of the orca display industry, dozens of captive killer whale deaths were reported at aquariums around the world, mostly in the United States, Canada, Iceland, and Japan. Twenty of the deaths were at SeaWorld, not counting stillbirths and miscarriages.[1]

The first killer whale death in the inventory was a female named Kandu, back in June 1971. Kandu was one of the Southern Resident K-pod members nabbed in Yukon Harbor, Washington, by Ted Griffin in 1967. She had been towed back to Seattle with four other young whales, including Kilroy, Ramu, and "Walter," who would later be renamed Skana by the Vancouver Aquarium—the animal Paul Spong worked with.

There was little information on Kandu's death in the federal inventory report, other than the cause—pneumonia and liver necrosis—and her approximate age, which was six. Kandu had survived for just four years in captivity.[2]

Just two months later, also in San Diego, death claimed the world-famous Shamu.[3] The original Shamu was initially captured as a mate for Griffin's pet, Namu, but was determined to be too young for that role. She was then sold to SeaWorld. Shamu died of a condition known as pyometra, which usually afflicts unspayed cats and dogs, not whales. When the hormone progesterone comes in contact with the uterus during estrus (ovulation), it can cause rapid cell proliferation of the uterine lining. Bacteria from the vagina and/or vulva then rise into the uterus, leading to acute infection and, in this case, septicemia (blood poisoning).

The great Shamu—the iconic star whose trademarked name would live on forever—passed away on August 29, 1971, after six years in captivity. She was only about nine years old—as opposed to the fifty-to-eighty-year life expectancy of wild female orcas.

Shamu's death was followed by that of Frankie, a male from L pod who succumbed in San Diego in January 1974, of influenza and pneumonia, after only five months in captivity (at about age twelve);[4] and in December 1974 by that of Canuck, a J-pod male probably less than four years old. He died of a systemic fungal infection after just two and half years in captivity.[5] Canuck had the dubious distinction of being the first killer whale to die in Orlando, Florida—some forty miles from the sea.

In 1975, Kandu III died of nephritis and uremia—swollen kidneys and kidney failure, causing urea and other waste products to remain in the blood, instead of being excreted through urine.[6] The condition can be caused by toxins, infection, and autoimmune diseases. Kandu was approximately seven years old and had survived at SeaWorld for four years. She was a member of L pod in the Southern Resident community, captured with two other young whales, Kandu II and Kona, in 1971 near Penn Cove, Washington.

Kona was the next to go, in September 1977. She died of septicemia at about twelve years of age after six years in captivity.[7] (The other whale caught with Kandu III and Kona—Kandu II—died at Marineland Ontario in Canada in October 1979 after eight years in captivity, at about age eleven.)[8] Kona was followed by the female Sandy, who passed away one month later of cerebral hemorrhage (bleeding of the brain), at about age eleven. Sandy had been stranded and rescued on the Washington coast in 1973 and taken to Seattle Aquarium, where she almost died when three boys shoved a metal rod into her blowhole.[9]

The list went on. This was beginning to look like a massacre: SeaWorld was a slow-motion death machine for killer whales. Kilroy, a male from K pod, died in San Diego in 1978 of gangrenous pneumonia, at about thirteen years of age, eleven and a half of them in captivity.[10] In 1981, also in San Diego, SeaWorld lost its first Icelandic killer whale, Canuck II, who died from chronic kidney disease at about age six, after four years in captivity.[11] The next year, Ramu passed away at SeaWorld Florida. The K pod member died in January 1982 of cardiopulmonary decomposition after a remarkable fifteen years in captivity.[12] At the approximate age of eighteen, he was the oldest orca so far to die at SeaWorld.

In January of 1986 SeaWorld's second killer whale born in captivity arrived in San Diego. It was Kenau's calf. SeaWorld naturally called the newcomer Baby Shamu 2. But the infant died of heart failure just eleven days later.[13]

Three months after that, San Diego lost its great breeding bull, the Southern Resident Winston, who had been captured in the notorious Penn Cove, Washington, roundup of 1970, when his dead relatives were found with anchors, chains, and slit bellies. Winston died of heart failure in April 1986 at approximately nineteen years of age.[14] He had survived in captivity for fifteen and a half years. "He was showing signs of age, as any animal would," Lanny Cornell, SeaWorld's zoological director, told the media.

The following year, in October 1987, the Icelandic whale Kona II died in Orlando of a lung abscess, at about age twelve.[15]

Over the next five years, SeaWorld would sustain another seven killer whale deaths, plus two stillborn calves and at least one miscarriage. Naomi, however, knew it was uncommon for whales in Johnstone Strait to die in their teens or twenties.

The company had owned just two Transients in its killer whale collection, and both of them died in 1990. The original Nootka passed away in San Diego on March 13 of pyogranulomatous pneumonia, in which the lungs become a mass of inflamed tissue, often associated with ulcerated infections. She was about twenty years old and had been in captivity for nearly her entire life.[16]

Just six months later, in September, the moody bull Kanduke died in Orlando, reportedly of bacterial pneumonia. Kanduke couldn't have been more than eighteen.

Eight months after Kanduke's demise, in May 1991, the female Kahana was found dead in a back pool at SeaWorld San Antonio.[17] She had smashed her head against the concrete wall and hemorrhaged to death. SeaWorld later revealed that Kahana had a brain tumor, which park officials said drove her to such bizarre behavior. But critics questioned whether aggression by the two other whales in the pool had anything to do with the bloody incident. Kahana had been in captivity for twelve and a half years and was estimated to be about fourteen.

Early the next year a female Icelandic orca named Samoa, approximately twelve, began showing bizarre and repetitive behaviors. Guests at the park reported watching in horror as she repeatedly hurled herself out of the water and crashed down onto the concrete surface of a slide-out area. (SeaWorld staff claimed to have never observed this.)[18] Samoa was pregnant and was in labor when she died in March 1992. Her calf also perished.[19] Samoa was the first killer whale to die at SeaWorld while giving birth, but certainly not the last.

There were more animal files jammed with information, including documents on three of the highest-profile orca deaths ever at SeaWorld.

First was Orky II, the great Northern Resident male, probably from the A5 pod, whom a teenage Naomi had seen performing at Marineland in LA. He died at SeaWorld San Diego in September 1988—just three days after visitors witnessed the live birth of his calf, Orkid, by the female Kandu. The birth happened during a performance.

Orky II, who was about thirty, had been wasting away. Despite veterinary intervention, he had lost four thousand pounds, or one-third his weight, in just a few months. One morning he refused his breakfast and instead chose to swim in tight circles in the middle of the pool. He vomited, lost balance, and listed to one side, made agonizing vocalizations as he tried to breathe, then plummeted to the bottom and died.[20]

Unconfirmed reports said that Orky's trainer was afraid of him and used food deprivation as punishment against the whale when he showed signs of aggression. After he started losing weight, Orky was given massive amounts of food, but could still not gain weight. SeaWorld's necropsy report said he died from kidney failure, which would not have been helped by the increased food he received at the end of his life. The organ damage was already done.

Orky made national headlines. "The death leaves the amusement park chain with only one proven breeding male among its 12 surviving whales, dealing a blow to the breeding program it hopes will supply killer whales for shows at the park," the *Los Angeles Times* noted. Orky had been so critical to that breeding program that SeaWorld had reportedly paid $23.4 million to buy Marineland of the Pacific in 1987. SeaWorld's owner, publisher Harcourt Brace Jovanovich, which bought the chain of parks in 1976, promised not to move the whales from LA. But three weeks later, Orky and Corky were on a truck bed packed with ice water heading south on Interstate 405, the San Diego Freeway. HBJ shuttered Marineland soon after.

While they were at Marineland, Orky and Corky had produced seven offspring, but two were stillborn and the others died shortly after birth. The apparent reason was "at least partly because Corky, isolated from other females, didn't know how to treat a calf," the *Los Angeles Times* said.

Then the most shocking orca death took place on August 21, 1989, in front of thousands of horrified fans, again in San Diego.[21] It involved Corky II and Kandu V, an Icelandic female about fourteen years of age. (In 1987 witnesses reported that Kandu violently collided into Corky, leaving a three-foot-gash along Corky's stomach.)

Kandu had been resting in a back pool with her one-year-old calf, Orkid, along with Corky. Corky had shown undue interest in the calf, something that agitated Kandu intensely. Though younger and smaller than the twenty-five-year-old Corky, Kandu had exerted dominance over her from the beginning. On this day, she engaged in a "normal, socially induced act of aggression to assert her dominance over Corky," according to Dr. James F. McBain, Kandu's veterinarian at SeaWorld.[22]

Kandu slammed her head into Corky, severing a major artery in Kandu's upper jaw. Blood stained the back pool and a ten-foot geyser of crimson spouted from Kandu's blowhole. Over the next forty-five minutes Kandu bled to death as SeaWorld staff and the audience looked on in helpless distress.

Captivity opponents blamed SeaWorld for the disaster. Again they argued how stressful and unnatural it was for orcas to be confined to a tank with other whales with whom they shared nothing in common. Captive whales fought so violently because they had no place to run.

The activist group Greenpeace was the harshest critic of Kandu's death. Benjamin D. Deeble, an ocean ecologist for the group, contended that Corky and Kandu had been aggressive toward each other from the beginning and should have been separated years ago. This type of attack went well beyond the ritualistic displays of aggression found in the wild. Kandu's death was a sign of things gone wrong at SeaWorld, he said.

"Even a four-million-gallon enclosure is a tiny pool for killer whales, because these animals in the wild swim over a hundred miles a day," Deeble told *The New York Times* in a major article on the incident. "An orca in captivity is like an eagle in a parakeet cage." Greenpeace had long opposed captivity for orcas because they "almost invariably" died within eight years of capture, he said, adding that females lived up to eighty years in nature and males could reach fifty.

SeaWorld had rejected the allegations.

"The problem was that Kandu fatally damaged herself in the process of doing this ritualized biting to Corky, which she has always done periodically," SeaWorld vet James McBain told *The New York Times*. "The fracture was literally a freak of the position and point of impact—everything came together to produce a rare outcome."

As for the longevity issue, SeaWorld officials trotted out the "we still don't know" line. "There's a big dispute about that, and the truth be told, we won't know until killer whales are reared in captivity," spokeswoman Corrine Brindley told the *Times*. "There are studies that suggest killers live about thirty years in the wild and in controlled environments."

Plenty of evidence showed that SeaWorld's thirty-year estimate was way off the mark, but park officials simply chose to ignore it—at least publicly.

Kandu's death was enormous news in San Diego, where the city's two hometown papers had different takes on the captivity debate that inevitably followed the tragedy.

The San Diego Union's lead editorial on August 25, 1989, declared Kandu's death a freak accident, even though "the effects of unnatural confinement cannot lightly be dismissed. After all, in the spaciousness of the natural ocean environment, the challenged female, Corky, might have had room to swim away and avoid the encounter."

But that kind of criticism had to be balanced by "the unquestionable scientific and educational benefits" conferred by keeping orcas in captivity, the *Union* insisted. "Educating the public through entertainment does a greater service to marine mammals than scientific research alone." Moreover, Kandu's death would help because it "added to our store of knowledge. As long as the animals were treated well and studied for the benefit of the entire species," the editorial declared, "there is no need to curtail popular programs such as those pioneered by SeaWorld."

Over at the *Tribune*, the editors had agreed that zoos and aquariums were vital "in a world that grows ever more threatening to creatures in the wild," thanks to their education, research, and conservation programs. But there was a "thin line between education and entertainment, between legitimate research and exploitation," the *Tribune* cautioned. "What we should enjoy, perhaps, is not their performance, but the mere fact of their existence. That, we believe, is wonder enough."

Not just the media was beginning to sour on Shamu. Even some of SeaWorld's leading supporters were getting sick of all the whale deaths.

When Kenau died during a stillbirth in Orlando on August 6, 1991, it sparked a wave of outrage among some fans. One file contained a letter mailed to HSUS at the time from Yvonne Alldredge, a SeaWorld "Founder Club" member from San Antonio: "I write this with heavy heart and yet the determination at this time to express my sadness and yes, anger over the tragic death of Kenau." Alldredge said she had tried to justify the "fish bowl conditions" at SeaWorld because, otherwise, so many people would never get to see a killer whale in person.

But, she asked, "When is enough, enough? Is it after all the orcas in captivity are indeed dead?" Kenau's second calf, Kayla, had been taken away when the infant was only two and "I feel Kenau died from stress related symptoms of losing her daughter. Perhaps her big heart either broke, or perhaps she wasn't about to give SeaWorld the gift of another child. . . . As Dian Fossey so passionately replied upon learning of the death of a captive baby gorilla: 'Good for her.' "

Were orcas better off dead than held at SeaWorld? Naomi wasn't sure, but the capacity for cruelty in the human species was certainly striking.

As she reviewed the files over time, Naomi noticed a number of articles, reports, and other documents on killer whales' aggression against their trainers. She could tell it was going to be another eye-opening topic. "Aggression expressed by killer whales toward their trainers is a matter of grave concern," wrote respected veterinarian Jay C. Sweeney in the *CRC Handbook of Marine Mammal Medicine*. Such aggression, he said, had resulted in "butting, biting, grabbing, dunking, and holding trainers on the bottom of pools and preventing their escape," and many situations caused "potentially life-threatening incidents."

Since the 1960s, numerous incidents had occurred with orca trainers. Many more of them, activists had charged, were covered up, except when brave trainers or members of the public spoke to the press. SeaWorld and other park venues seemed to play down the acts, calling them bizarre accidents, just as when Kandu mortally attacked Corky.

Many early incidents were poorly documented, some based on hearsay. Others appeared to be cases of young whales playing a bit too roughly, or genuine accidents caused by miscues during a performance.

The first fully documented act of aggression by a SeaWorld orca happened in San Diego on April 20, 1971. The original Shamu, just four months before she died, was in a publicity stunt gone terribly wrong. A young secretary named Annette Eckis had been convinced by SeaWorld's PR department to put on a bikini and ride around on the back of Shamu for local newspapers and television stations.[23]

The tank was small, its diameter perhaps three times wider than Shamu was long, akin to a backyard aboveground pool. The nineteen-year-old gingerly climbed from the edge of the pool onto the back of the whale, who began slowly moving around the perimeter. The young woman reached behind with her arm to grab hold of Shamu's dorsal fin. As they swam around in tight circles, Annette began to enjoy the ride. Shamu soon grew uneasy and began thrashing about violently, churning the water into a four-foot chop. She shook off Annette, who struggled to tread the churning salt water.

Then Shamu grabbed her legs. Annette was inside the whale's mouth, sitting on the bottom row of teeth with the top row biting down on her. She grabbed Shamu's rostrum for balance and screamed for help as the orca spun furiously around her tiny tank.

When rescue divers tried to enter the water, Shamu blocked them and responded by biting down harder on Annette's legs. Every so often, she headed for the bottom, bringing the terrified secretary underwater with her. Annette thought she was going to drown. (The ordeal, recorded in horrific detail by the news crews, now appears on YouTube, capturing the screams and look of pure terror on Annette's face as Shamu brutally assaulted her.)

Workers hoisted a pole over the water and pulled Annette to the edge. She was panting and out of breath, but she grabbed the wall and held it with the little strength she had left. It looked as if she were going to make it out.

But Shamu had other plans.

The whale surged back up from underneath the water and clamped down once again on Annette's bloody thighs. She shrieked in fear. Eventually rescuers pried Shamu's jaws open with a pole to free the legs and yanked Annette from the tank. The lacerations and puncture wounds that covered Annette from the waist down required two hundred stitches.

"There was a lot of blood, but there was no pain. The water was cold and I was in shock for quite a while," Annette said later. Only after the attack did she learn that Shamu had never been ridden by a female before, and never by anyone not in a wet suit.

Naomi had some sympathy for the victim, but none for SeaWorld. It was a foolish thing to do, especially to get publicity. But back then, perhaps no one actually realized that.

The next attack had come in 1974 at the Windsor Safari Park in England. Four-year-old Winston (called Ramu at the time) lunged out of the water at his trainer, Doug Cartlidge, and tried to pull him into the water from the platform.[24]

In 1978, Orky II took twenty-seven-year-old trainer Jill Stratton to the bottom of the pool at Marineland of the Pacific and pinned her there for four agonizing minutes, before finally releasing her. Stratton, who nearly drowned, was hospitalized for three days. Upon release, she told reporters that she was "not angry" with the whale.[25]

Trainers at Marineland stopped doing water work with Orky II, who was eventually sent down to SeaWorld San Diego along with Corky II. The introduction was chaotic for the newcomers, and for the other orcas already at SeaWorld. Orky had suddenly been commanded to perform with several trainers in the water at once. He'd never done that at Marineland.

Some trainers had grumbled that management was rushing them. They alleged that SeaWorld was in a rush to "break in" Orky and get trainers in the water with him as soon as possible. Even Robert K. Gault Jr., who was then SeaWorld's president, admitted to reporters that management "may have overemphasized" the importance of entertainment over safety, adding, "We did not have enough experienced trainers."[26]

In 1984, SeaWorld San Diego had three incidents, two of them involving Kandu V. On February 24, the Icelandic female grabbed trainer Joanne Hay in her jaws and shoved the woman against a wall, holding her there for several moments.[27] That November, forty-five-hundred-pound Kandu seized the legs of trainer Georgia Jones during a show but did not bite down.[28] On August 12, also during a show, two unidentified whales clamped down on the legs of trainer Bud Krames and pinned him against the Plexiglas retaining wall. He suffered only bruises.[29]

There was another worrisome rash of incidents in 1986, two of them at Marineland Ontario in Canada. One male trainer tumbled from the back of a young male, named Kandu VII, during a performance. The whale then dragged him around the pool. Meanwhile, a four-year-old female named Nootka V smacked a trainer in the head with her pectoral fin, also in front of an audience. One former staff person alleged that Nootka V would routinely shoot out of the water aggressively to whack people standing close to the edge.[30]

That same year, on November 16, 1986, in San Diego, Kandu V pressed her rostrum against a trainer, Mark Beeler, in front of hundreds of tourists, pushing him against a wall for a moment.[31]

In 1987, Naomi would learn, things at SeaWorld San Diego spiraled out of control.

The troubles had begun in March. Jonathan Smith, a twenty-one-year-old business major at Point Loma Nazarene College, was doing water work with several orcas in front of thousands of fans when, without warning, an unnamed killer whale pulled him into its mouth, dove thirty-two feet to the pool's bottom, held him there for two and a half minutes, smashed him against the floor of the tank, then carried him bruised and bleeding to the surface, unceremoniously spitting him out. Smith, a consummate performer, waved to the crowd, trying to reassure them that all was well.

Then another whale slammed into him. Smith continued to pretend to be unharmed. The whales pushed and dragged him back to the bottom of the pool. Now the crowd was no longer fooled: This was a real emergency. Other trainers scurried to the water's edge, banging pails and brandishing food in an effort to distract the whales. Smith eventually managed to climb out of the pool, exhausted and battered. His torso was lacerated on the front, back, and sides, his kidney was ruptured, and his liver had a six-inch laceration. He spent nine days recovering in the hospital.[32]

Smith was a rookie at Shamu Stadium. He'd been transferred from the seal and otter show only a few weeks before. Even so, nobody could explain the attack. Some experts speculated that Smith's seal-like wet suit set off a hunting instinct of sorts. It was typical seal-killing behavior, but no one could provide a definitive explanation for what triggered it. Chief trainer David Butcher, however, played down the attack as mere roughhousing. "These guys were playing and got a little carried away and bumped into Jon," he told the *Los Angeles Times*.

In June, SeaWorld trainer Joanne Webber, twenty-eight, was rehearsing with Kandu V when the whale let all three tons of her body land directly on top of Webber in the water. She was pulled from the pool with a broken neck and suffered permanent loss of head movement.[33] Court papers from the ensuing lawsuit revealed that Webber said Kandu often exhibited "extreme characteristics of aggression when frustrated. She does occasionally bite and aggressively rake other whales."

A little later that summer, trainer Chris Barlow was rammed in the stomach during a performance, sending him to the hospital,[34] and an orca bit Mark McHugh on the hand as he was feeding the animal.[35]

Naomi was astounded to learn that no less than twelve SeaWorld San Diego trainers were injured by killer whales during a four-month period between August and November 1987.[36] It seemed as though the captive whales had decided, all at once, to start biting back. But one more incident was still to happen before the dangerous year limped to a close.

On November 21, twenty-six-year old trainer John Sillick was riding on a female named Nootka I when Orky II slammed down on top of him. It took six operations over fourteen months to put him back together again, with the help of three pounds of pins, plates, and screws, including a plate in his pelvis. All the vertebrae in his neck had to be permanently fused.[37]

Some experts speculated that Orky may have been responding to another trainer's (perceived) signal to breach. Court documents were sealed, but the judge's remarks were not. They showed that Orky may have suffered from "visual limitations" that had not been disclosed to the trainer.[38]

Either way, Harcourt Brace Jovanovich had seen enough. Even though it took an extreme accident like John Sillick's, SeaWorld finally made some fundamental changes.

Company chairman William Jovanovich went to San Diego to take command. He ordered trainers to get off the whales, and to stay out of the water entirely. It would be "dry work" only from that point on, he decreed. He also sacked chief trainer David Butcher, SeaWorld San Diego president Jan Schultz, and longtime zoological director and veterinarian Lanny Cornell. Thad Lacinak was installed as the new VP and corporate animal curator.[39]

But six months later, trainers were back in the pools doing water work with the orcas. Why the policy change? Some critics noted that SeaWorld's twenty-fifth anniversary was fast approaching and the park needed a dazzling new show for Shamu Stadium.

Then the lawsuits were filed. Jonathan Smith's lawyer charged that SeaWorld and HBJ "negligently and carelessly owned, maintained, trained, inspected, controlled, supervised, located, transported and placed" the killer whales, exposing Smith to serious injury. Smith's superiors had concealed the "dangerous propensity of killer whales" and instead "induced" him into the water. The lawsuit accused SeaWorld of telling Smith it was safe to swim with the killer whales, even with little or no training, all the while "knowing well that killer whales have a dangerous propensity for attacking, ramming, dragging and smashing people." Management had assured Smith that the killer whales were "gentle." He was suing the defendants for fraud, battery, and emotional distress as well.[40]

Sillick and Webber also filed suits. All three were settled out of court, with gag orders imposed as part of the settlement agreements. Years later, Naomi regretted that society could never learn or benefit from the thousands of pages of prepared evidence those files must have contained.

Following the spate of dangerous events in San Diego in 1987, the media again began taking note of all the blood in the water, including a *New York Times* article that said the injuries had left the company with "some profound questions":

> *Why did [the whales] turn on two of their trainers, nearly killing them? Are these highly intelligent ocean creatures happy spending their lives in a big tank doing stunts for human audiences? Does the human species have the right to confine them for entertainment and profit, or even research? What, in short, is the proper relationship between the two most highly evolved species in the planet's two biological realms, land and sea?*

The article said some experts considered orcas to be so "emotionally sensitive" that, "like humans, they get ulcers when unhappy and depressed."

18

The Case Against

The second half of 1993 was a high-pressure period in Naomi Rose's life. In early June she was still unpacking at her new Maryland apartment and adapting to her career as an animal advocate/scientist. These were uncharted waters for the recently minted doctor. She had a lot to learn in woefully little time—not only about captive killer whales, but also about the byzantine ways of Washington and its satellite community of nongovernmental organizations, or NGOs in Beltway-speak.

Naomi was by no means only working on orca issues. As the marine mammal scientist at HSUS, she was drawn into a wide range of issues and controversies ranging from the Japanese drive fisheries (which slaughtered hundreds of whales and dolphins a year) to the sport hunting of polar bears, from "swim with the dolphin" programs (which critics charged were unhealthy for dolphins and risky for people) to noisy human activity in the ocean and its impact on marine mammals.

Meanwhile, HSUS president Paul Irwin and Naomi's boss, John Grandy, vice president for wildlife and habitat protection, were keen to ride the publicity bandwagon that was sure to follow the release of the movie *Free Willy* as it opened that summer in theaters across the country. The pending premiere, they told her, was an unprecedented opportunity for HSUS to highlight the darker side of confining whales to aquariums and theme parks. HSUS had hired Naomi, she knew, in part to help the society launch its new anti-captivity campaign for marine mammals.

Free Willy was a quintessential Hollywood heart-warmer about a troubled teen who bonds with a killer whale, played by the orca Keiko (*KAY-koh*). Willy is languishing in a substandard aquarium, where the boy teaches the beast to perform tricks. But progress is slow and Willy's owner decides to kill the whale and collect on the $1 million insurance policy. That's when the race to free Willy begins. The film, despite having no major stars, was already getting great buzz from the entertainment press as a family classic.

Naomi was asked to prepare a report that Paul Irwin could release to the media on the opening weekend of *Free Willy*. She was to gather published data on the well-being of captive cetaceans versus those in the wild and condense it into a handout for the press.

The movie premiere was scheduled for Friday, July 16. Naomi had six weeks.

To Naomi, the case against killer whales in captivity was a slam dunk. She could now name twenty orcas that had died at SeaWorld—and many more at other facilities. In addition to that, she knew, more than a dozen whales had attacked a trainer or another whale. Those facts alone were irrefutable evidence that captivity was unsuitable for these animals.

Still, even more anti-captivity ammunition was stockpiled in the HSUS armory—in addition to the published literature—to aim at the public display industry. The more Naomi learned about captivity, the more it infuriated her. She had definitely made the right career choice: These whales needed advocacy; they needed a voice.

Naomi began wondering how much captivity was contributing to abnormal behavior in killer whales. That was difficult to quantify scientifically. First, what was "normal" when speaking about intelligent and highly adaptable social animals, including humans? After all, it was once considered "normal" for Egyptian royals to marry their siblings and for Europeans to enslave Africans. Today, many people love eating beef and consider it part of everyday life, while many Hindus find that practice barbaric.

Whether a whale's behavior was normal or not could be somewhat subjective. But it was a *fact* that no killer whale had ever been reported to have killed a human in the wild, or even seriously attacked a human in the wild, and no killer whale had ever been known to be killed in a fight with another whale. All three of those things had happened in captivity. Based on that alone, Naomi was ready to say that it was not okay to keep them locked up in tanks.

Was the excessive aggression a sign of impaired mental health? Not necessarily. Naomi believed there was something abnormal about the close proximity of life in a pool, and that putting strange animals together in a confined space with nowhere to run—and then throwing people into the mix—could clearly be dangerous.

On the other hand, it was impossible to rule out mental illness in some of the captives. From what Naomi had observed and read, certain whales had grown neurotic after many years in a tank. Tilikum was spending much of his time in Orlando floating alone in his pool, practically motionless. "When you see a solitary whale like that, endlessly floating around, not swimming but just logging at the surface for hours at a time, when he would normally be traveling for miles with his family," Naomi told colleagues, "it seems perfectly reasonable to compare that to someone sitting alone in a darkened room in a deep state of depression."

Naomi knew she was committing science's cardinal sin of anthropomorphizing—the attribution of human traits to nonhuman animals or objects. But in her opinion, the analogy was appropriate: "They're intelligent, social animals, and so are we. The only gauge we have, the only measuring stick we have, is ours. I don't know how a dog thinks, and I don't know how an orca thinks. But I do know how they *behave*. And I can see how they behave when they're under certain circumstances, and I can draw some conclusions."

Naomi was familiar with several primate studies on social deprivation and mental health. One paper on isolation in infant macaques was particularly disturbing.[1] Each baby monkey was kept isolated, save for a little cloth surrogate mother, made of socks. The infants would curl up pathetically with the only "mother" they knew. It was not normal behavior, even though it was understandable. It hardly painted a picture of sound mental health. Naomi wondered, did male orcas such as Tilikum suffer similar trauma from being isolated from their own mothers?

Another topic she thought about including in her case against captivity was the collapsed dorsal fin seen in 100 percent of the captive adult males. Based on her knowledge of biology and what she had learned about the species in the wild, Naomi considered a number of potential causes of the deformity. They included the stress of captivity, dietary changes, restricted physical activity, and simple gravity acting on the collagen-like tissue in the fin that had been heated and softened by the abnormal time spent at the surface, especially in sunny, subtropical locales such as Florida, Texas, and Southern California.

In the wild, collapsed fins were extremely rare, usually the result of an identifiable cause such as gunshot wounds or collisions with boats. In 1989, following the *Exxon Valdez* oil-spill disaster, two male killer whales exposed to the toxic slick quickly experienced a folding in their fins, which, within two years, had flattened onto their backs entirely. Marine mammal scientists on the scene assumed the animals were ailing. Both animals died shortly thereafter, suggesting that fin collapse could also be a sign of chronic health problems.[2]

Naomi came across even more details on the pitfalls of orca display in a paper commissioned by the UK-based Whale and Dolphin Conservation Society (WDCS) and written by Erich Hoyt, author of the book *Orca: The Whale Called Killer.* Published in 1992, the anti-captivity manifesto was titled *The*

Performing Orca—Why the Show Must Stop: An In-Depth Review of the Captive Orca Industry.

Hoyt opened with the violent death of Kandu V—who had shattered her jaw against Corky II in San Diego during a fierce fight. He described the grueling string of occurrences and injuries in San Diego in 1987, as well as the brutal death of Keltie Byrne in 1991.

"Are these unfortunate, unconnected accidents or telling incidents that ought to be investigated?" Hoyt asked. "Could these be examples of predatory behavior—normal in the wild but dangerous in a confined area and when applied to humans? Are these real signs that orcas are unsuited for life in captivity?"

In one of his chapters, "Health of Captive Orcas," Hoyt reported that of the 127 wild orcas captured and held for public display, only 35 were still alive as of April 1992. The most recent data (from the 1990 paper by Mike Bigg and colleagues) showed wild female orcas with an average life span of 50.2 years and males with 29.2 years. "Only one male in captivity has even reached this average," Hoyt observed, referring to Orky II. At the time, no other captive orca had come close to surviving that long. Moreover, out of twenty-seven known pregnancies in captivity, only nine successful births had been recorded.

Captivity also shattered the tight social bonds that supported orca well-being and survival in the wild. Killer whales did not live alone in nature. But in six of seventeen aquariums with killer whales, they were confined without companions of their own species. Calves were frequently taken from their mothers before the age of five, Hoyt said, "severing the maternal bond." Whales from different pods and different oceans were routinely mixed together, especially for breeding.

All the while, the featureless, acoustically sterile tanks in which the captive whales lived "increased stress levels brought about by overcrowding, poor nutrition, noise, excessive light, transportation and so on," he said, which could lead to impaired immunity and greater susceptibility to illness. Killer whales were curious enough to not "be driven neurotic in a year," Hoyt wrote, quoting Graeme Ellis. "But it's difficult because the novelty wears off."

Hoyt also alleged that some trainers deprived disobedient whales of their full day's ration of food. "A former SeaWorld trainer who requested anonymity told me that whales or dolphins that would not perform were sometimes denied food during or immediately after the shows," he said. Uncooperative orcas received just two-thirds of their daily allotment, plus vitamins, the source alleged, adding, "The whales would start performing when they realized they weren't going to get fed."

Hoyt then attacked the size, design, and water quality of modern orca tanks, saying they failed the industry goal of keeping animals "in conditions that as much as possible replicate those for the species in the wild." For large and social animals such as killer whales, such confinements "would ideally still have comparable areas as in the wild for play, rest and socializing." Killer whale pools, even the sophisticated and complex enclosures at SeaWorld, were six thousand to nine thousand times smaller than the most conservative range for Resident killer whales in the ocean.

Hoyt's full frontal assault offered Naomi some ideas for her own anti-captivity opening salvo. But the response to Hoyt's report by the industry hinted at the pushback she herself would feel once her own report went public. Erich Hoyt had struck a nerve at SeaWorld. The company responded to the author with a barrage of counterassaults.

In October of 1992, SeaWorld released a white paper charging that Hoyt had "overlooked, minimized and distorted information" on captive killer whales. His report was not the work of a scientist, but rather that of an advocate who "relies heavily and selectively on those who support his view."[3]

Many of Hoyt's sources went unidentified and were unverifiable, SeaWorld said. Hoyt had alleged that numerous accidents at SeaWorld were covered up, without defining what "numerous" meant. And if they had been covered up, SeaWorld demanded, then "how does he know about them?" The company also decried Hoyt's use of the word "accidents" in quotation marks as being a "subtle way of implying that marine parks are less than honest."

As for orca longevity, "Mr. Hoyt would have the reader believe that the matter has been settled," SeaWorld contended. "In fact, we do not conclusively know the average and maximum life spans of killer whales in the wild. The best scientific thinking, however, is cited by [the] American Society of Mammalogists. They indicate that the life span of killer whales is estimated to be 25 years but could be as long as 35 to 40 years."

Hoyt's figures of a fifty-year average life span for females were only "theoretical estimates" made from a small population (of Resident orcas) which "may or may not be representative of all killer whales." The estimates were "interesting" but no substitute for "factual, objective, scientific data." More time was needed to complete conclusive studies on the matter.

SeaWorld also denied that food deprivation was ever permitted. All animal training within its system was based solely on the positive-reinforcement model. As for the size of its pools, SeaWorld said it provided killer whales with more cubic feet of water than any other operation, and far in excess of US government standards. Although its tanks were not the equivalent of the space provided by nature, Erich Hoyt had ignored "the fact that these animals thrive at SeaWorld."

Killer whales in captivity exhibited nothing but "normal behavior," and they reproduced successfully, indicating "that they are comfortable in their surroundings." Animals bred at SeaWorld were at no greater risk for health problems than animals bred in the ocean. "In fact, they probably are at less risk at SeaWorld."

Hoyt was likewise off base in attacking SeaWorld for its alleged paucity of scientific research on killer whales. His paper had ignored "basic research on animal physiology and blood chemistry" conducted at the parks. SeaWorld staff had authored or coauthored more than three hundred abstracts, book chapters, technical reports, and peer-reviewed papers. SeaWorld also provided access to its animals for outside researchers to conduct investigations.

Finally, SeaWorld mocked Erich Hoyt for declaring that the days of captive orcas "are and should be numbered" by saying that the 10.5 million people who visited a SeaWorld park each year "obviously do not share this belief. They have seen and understood the magnificence of these animals and many have requested information on how to become more involved in saving habitats for ocean-dwelling species. For that reason alone, we feel we are accomplishing our purposes and remain committed to pursuing them."

Naomi was exasperated. Arguing over the use of quotation marks around the word "accidents" was petty, she thought; stating that wild killer whales lived for just twenty-five years was preposterous; and suggesting that all orcas "thrived" at SeaWorld, exhibited normal behavior, and were perfectly healthy ran counter to the evidence. Furthermore, of the three hundred papers worked on by SeaWorld staff, only a handful, at best, had anything to do with killer whales or were published in peer-reviewed journals.

Naomi wondered how many visitors had actually requested information on becoming more involved in saving habitats for marine species. She saw no evidence that SeaWorld guests went out and got personally involved. Nor was it clear how such efforts might specifically protect killer whale habitats.

SeaWorld's PR push continued. Naomi found another article, published by United Press International, about how content killer whales were in captivity. "Trainers at SeaWorld say they know the killer whales at the park are happy because of the animals' sex lives," the article said. SeaWorld had the world's most successful breeding program, with an 88 percent survival rate, as opposed to the "best estimate" of 44 percent in the wild, said park official Bill Hughes, without providing evidence. "If animals aren't happy, they won't reproduce," he told UPI.

Chief trainer Ted Turner upped the ante. The animals at SeaWorld were happy "because I've devoted my life to making them happy," he said. "I see more of them than I do my own children." He said he would not get in the water with the animals if they were unhappy.

Now Naomi was fuming. This belief that "the animals are having sex so they must be happy" was demonstrably ridiculous. While some species were so strongly affected by confinement that they did not reproduce readily or at all, the sex drive was one of the most powerful in nature. Many animals—including humans—had sex and reproduced even under the most stressful conditions. Libido was especially powerful in members of the dolphin family. All dolphins, whether free roaming or captive, were sexually charged, seeking sexual contact even when reproduction was not involved. Humans were the same way. Prison inmates had sex with each other, but that didn't mean they were "happy." Even inmates at internment camps throughout history still managed to reproduce.

These controversies—poor mental health, violent aggression, the question of whale "happiness" in an artificial environment—fascinated Naomi. But little discussion about them appeared in the scientific literature. And Naomi's investigation was focused on mortality rates, using the US Marine Mammal Inventory Report (MMIR) and its rich lode of data on captures, births, and age at and causes of death. Her task was to compile data on the four species of small whales most commonly found in public display: belugas, pilot whales, false killer whales, and orcas.

Naomi had no idea she was about to wade into a treacherous pool of statistical quicksand.

The debate over whale longevity in captivity-versus-nature had long been contentious and complicated, especially when it came to killer whales. The industry, for example, said it was impossible to determine killer whale life expectancy in captivity until all animals currently on display had lived out their full lives.

Naomi was unaware of these pitfalls. She did not know what she was doing. She felt lost. But she had no idea how far into the reeds she had wandered.

To make her estimates on mortality, Naomi took her cues from past HSUS statements on survival rates that counted the total number of animals from each species that was taken captive and the number of animals that had died. She then calculated a simple percentage, compared that with data from wild whales, and called the result the relative mortality rate.

The finished product was called *Small Whale Species: The Case Against Captivity*.

Among the wild orca population in British Columbia, Naomi wrote, a total of 248 animals were alive in 1973 or had been born since then. In the subsequent twenty years, 45 of their members, or 18 percent of the total, had died. But among captive killer whales in a comparable time period, 30 out of 62 whales, or 48 percent of the total, had died. SeaWorld had the highest number of whales, 37 in total, 20 of whom (54 percent) were now dead.

"In other words," Naomi concluded, "killer whales are more than 2.5 times as likely to die in captivity as in the wild." Most whales were captured at about the age of three, and for orcas, the average life span after capture was about six or seven years, "reducing the life expectancy of these animals to just one-quarter that of wild orcas."

Only one animal out of thirty achieved the average life span found among free-ranging whales: Orky II. None had even come close to the maximum—at least not so far. Several females, meanwhile, had died after giving birth to just one or two calves.

The three other whale species were also faring poorly. Among belugas, seventy-seven had been held in captivity since 1961, and thirty-two had died—or 42 percent of the total. The average age at death was ten. Among false killer whales, sixty-seven had been held in captivity since 1964, and thirty-eight of them, or 57 percent, had died. The average age at death was seven. Pilot whales had the grimmest record: thirty-eight had been captured and thirty-five were dead, or 92 percent of the total. The average age at death was six.

Given these facts, "it should be clear that the small whales commonly held in U.S. facilities are not suitable animals for captivity," Naomi said. "Their mortality rates are aberrantly high." The whales' size, space requirements and social needs "all prohibit captive facilities from providing adequate conditions. Most tanks are at best seven times the length of the animals." Given their wide range, a killer whale or beluga would have to swim over five hundred times around its tank to complete a normal day's travel.

"The world of a whale is three dimensional," Naomi concluded. "A tank is for all intents and purposes two-dimensional. Whales require the wide and deep ocean in order to live a healthy, long life. Concrete tanks guarantee an impoverished life cut short and an inhumane existence at every level."

Once the document *Small Whale Species: The Case Against Captivity* was approved by Naomi's superiors, the HSUS publicity team prepared to drop it on the public. A press conference was planned to present the paper to the media on July 16, opening day for *Free Willy.* The venue was strategic: the Radisson Hotel on International Drive, in Orlando, a short walk from SeaWorld's main gates.

Four days before the premiere, Paul Irwin faxed a letter to SeaWorld's vice president for zoological operations, Brad Andrews, with copies forwarded to Anheuser-Busch and all aquariums and theme parks that kept killer whales in captivity. Drafted by Naomi, the communiqué laid out the Humane Society's case, based on what it called "a major investigation into the mortality of captivity occurring amongst captive whales."

Irwin solemnly presented the mortality figures that Naomi had compiled. Whales were confined to small pools, he said. Denied their natural free range, they were left to float idly at the top of their tanks. In nature, they might spend 10–20 percent of their time at the surface. But this trend was reversed in captivity, where they "must spend over 50 percent of their time at the surface, resulting in skin problems and collapsed dorsal fins." Tank water was usually manufactured. And because the whales were cut off from their families and social structures, "many in captivity are solitary," Irwin stated. "Neurotic stereotyped behaviors abound."

The consequences of captivity on whales were "appalling," he said, including "aberrant, neurotic, and destructive behaviors, diseases, abortions, high mortality, and a drastically lowered life expectancy." The HSUS position was resolute: "Whales cannot be and should not be maintained in captivity for public display."

Irwin said, "The time is now for change and commitment. With increasing knowledge and understanding, the public no longer accepts that wild cetaceans need to be captured, traumatized, and isolated for public display and profit." The industry could no longer raise "a chorus of objections" anytime an outside group questioned the ethics of whales in captivity. The day was gone when aquariums and theme parks could get away with repeating the mantra that captive whales were essential for education, for gaining public support, and for species conservation.

"Captive whales teach nothing of the majesty and habits of wild whales or their natural habitat," Irwin wrote. "At best, keeping these spectacular social creatures weighing up to several tons in small pools teaches only misconceptions based on human domination, not conservation stewardship."

Irwin's letter drew an immediate and unexpected response from SeaWorld: The company invited the HSUS president on a backstage VIP tour of SeaWorld while he was in Orlando for the press conference. He gladly accepted and asked John and Naomi to accompany him.

SeaWorld was clearly bracing for a fight. Until recently, the only organized opposition to captivity had come from a small cluster of politically liberal groups that lacked the influence to provoke a meaningful national debate. Industry had been swatting at flies, Naomi thought. Now, with HSUS in the fray, it would be swatting at a large hornet.

"I think we've made them nervous," she told her colleagues. "This VIP invitation was an attempt to head us off. SeaWorld wants to see if they can work things out with us, lay our concerns to rest, and have us call off the campaign."

Everyone agreed: *That* wasn't going to happen.

On the morning of the big press conference, Friday, July 16, 1993, the American Association of Zoological Parks and Aquariums (AAZPA) issued its own preemptive statement to "denounce the alarmist tactics adopted by animal activist groups" to use the *Free Willy* opening to "further their own agenda." That agenda was now clear, Executive Director Sydney Butler warned. "It is to deny the public the thrill of seeing whales and dolphins and learning about them. It is to free all captive whales and dolphins to an uncertain future in the open ocean."[4]

Future generations of children would be "denied the vital, up-close experience of the young boy in the film" because whale-watching trips on the ocean were the province of "only a privileged few." HSUS and other groups were simply trying to turn a nice film about a young boy and a whale "into a heavy, message-laden film about animals in captivity."

Naomi rolled her eyes when she read that assertion. *Free Willy was* a film about the pitfalls of animals in captivity. That was the intention of the producers in the first place.

Meanwhile, Butler employed the oft-repeated industry line about the high-quality food and medical care given to animals in captivity. He suggested that confinement was superior to life in the wild, which he derided for being "not an idyllic playground." The ocean was "not a heaven," according to the AAZPA, and marine parks were "not hells." It was simply "irresponsible and reckless to the cause of conservation to make universal declarations based upon a Hollywood portrayal."

This was going to be quite a fight. Naomi was ready for it.

On Friday morning, in a nondescript meeting room at the Radisson Hotel, Paul, John, and Naomi held their news conference and presented their case against captivity. A dozen reporters showed up, including a writer from the Associated Press. This story was going national.

Paul Irwin welcomed the media and announced, "We have chosen today to release our conclusions from this study because of the opening of the courageous and meaningful movie *Free Willy,* which addresses the world of captive whales." The society had selected Orlando for the venue "because this is the home of SeaWorld, the best-known and most highly advertised keeper of killer whales in the world."

When Naomi's turn came to speak, she held nothing back: "I have studied marine mammals for many years, especially the behavior of killer whales in the wild. Comparing this to the behavior of these animals in captivity, I can tell you that visitors learn almost nothing about whales by watching them in a small pool. I say this because the lives of captive whales are so different from the lives of wild whales."

Naomi had never seen a lone animal "aimlessly floating in place" in the wild, she said, but it occurs frequently in captivity. "In six years of observing killer whales in the wild, I only saw one head-first leap into the water, which is called a porpoise leap," she continued, "yet this leap is a routine trick performed in captivity to dazzle the crowd."

Then she went for the emotional jugular with a compelling narrative of orca life in the "non-idyllic" wild: "I have seen a grandmother share a fresh-caught salmon with a granddaughter. I have seen three brothers play together for hours. I have seen older brothers babysit younger siblings. On one calm and foggy morning, I encountered fifty whales lined up side by side, all resting in family groups, all breathing together—a sight I will never forget."

In contrast, observing unrelated whales in a pool "can teach nothing about events like these. It is very stressful for an animal as sensitive and intelligent as a whale to be captured and kept in a small tank forever." Whales in captivity typically died from disease and accidents, not old age.

Finally, Naomi repeated HSUS's call for the release of eligible captive cetaceans. "Although no whale has yet been released back into the wild, the Humane Society of the US believes it is time for marine parks and aquaria to try." Rehabilitation and release might not be possible for orcas held for long periods, she admitted. But longtime captive dolphins had successfully been returned, "and the risk for whales is worth it, since life for them in captivity is so poor. Besides, it can't be said that it is impossible until it is tried."

The time had come for keepers of captive whales to "do what is best for the animals rather than doing what is best for the bottom line," Naomi concluded.

Rounding out the event was a recorded statement from John Sillick, the SeaWorld San Diego trainer who was seriously injured by Orky II. Sillick was under a gag order about his legal case, but that did not prohibit him from speaking out against captivity for "Shamu."

"I want to support the Humane Society of the United States in its call to stop catching whales from the wild to display in marine parks and aquaria," the former trainer said. "These whales are wild animals. It's important to remember that they're wild. The perception is that they seem to be obeying their trainers' commands. The truth of the matter is no trainer has complete control over a whale."

Wild whales had no place at marine parks and aquariums because "they can be dangerous to the humans who work with them. I speak from firsthand experience," Sillick said. "I can walk today, but I still feel the pain of the accident."

After a brief celebratory lunch, Paul, John, and Naomi arrived at SeaWorld for their VIP tour. It was a blistering summer day. The trio was met by Brad Andrews and the park's top executives. All comported themselves cordially and respectfully—making no mention of the press conference that morning—as they toured the grounds in a golf cart. They took in the show at Whale and Dolphin Stadium, which Naomi now found cheesy. She thought the bottlenose and false killer whales were being treated like clowns at a children's circus.

It left a bad taste in her mouth.

After attending the show "Shamu: New Visions," they were led back to examine the medical pool and other tanks backstage.

Naomi noticed Tilikum by himself, his massive black shape floating at the surface of a back enclosure. The females were in the front pool, swimming and breathing in unison in endless circles. The animals were not in the shade, and the temperature—91 degrees—was not exactly Icelandic. Naomi watched Tilikum "logging" motionlessly under the relentless sun. It made her angry to think how majestic these animals were up in Johnstone Strait, compared with this pathetic shadow of a former whale, languishing here in the humid soup of summer.

The HSUS party was led to the back corner of the Shamu complex, where the park was building yet another, somewhat larger rectangular tank, called G Pool. Brad Andrews explained that SeaWorld was adding a "Dine with Shamu" feature, where guests could have lunch or dinner at water's edge as one of the killer whales was brought in for an intimate and low-key exhibition.

Workers were installing large rocks and a few evergreens near the pool to make it resemble something of a northern-tier habitat. The tank was still empty, its harsh and hard-edged sides exposed to the Florida heat. The juxtaposition was striking: the ocher-colored boulders and scenic shrubbery were placed there above the water surface solely for the esthetic enjoyment of paying guests, and to make them "feel" as if SeaWorld's animals lived in a world quite reminiscent of their own natural environment.

But underneath the surface, the killer whales were afforded no such considerations. Their flat concrete world—laid bare now by the construction—looked grim, featureless, and utterly devoid of anything that made the natural world enriching and inviting to sentient mammals. To Naomi, it was a chamber of dreariness; but the tank was much easier to clean this way.

"Hey, Dr. Rose, can I ask you something?" The lead curator, Frank Murru, had spotted Naomi staring at Tilikum in a back pool.

"Sure, go ahead," she said with a guarded smile.

"It's not really true what they say about the males, is it?"

"What do you mean?"

"The male killer whales. They don't *really* stay with their mothers their whole lives, do they? I mean, that's just a myth. Right?"

Naomi was surprised. Here was an expert at one of the leading aquatic parks in the world, and he was actually questioning one of the basics of the natural history of the animals under his charge. She stood in place and stared at Murru for a moment.

"Well?" he said finally.

"Yes. They stay with their mothers, Frank. At least Resident whales do."

Murru gave Naomi a hard look. Then he turned and walked away. She wondered if he even knew what a Resident whale was.

When the tour ended, Irwin thanked Brad Andrews and his team for the VIP treatment. However, he added, SeaWorld was not going to dissuade them from dropping their anti-captivity campaign. Things were just getting started. Then everyone shook hands.

It would be the last cordial exchange between the two parties.

Coverage of the press conference was considerable. The AP story was picked up in newspapers across the country. Many in the media treated the showdown like a clash of Titans, though Naomi knew that HSUS and its little captive-whale campaign were going to be thoroughly outgunned, outspent, and outlawyered by the other side.

"The Humane Society, the nation's largest animal-rights group, and SeaWorld, the nation's largest owner of marine mammals, squared off in Orlando as moviegoers lined up for the premiere of *Free Willy,* a film about a young boy who liberates a killer whale from an abusive aquarium," the *Orlando Sentinel* reported.

HSUS had taken off the gloves "in a departure from its less-strident past," the *Sentinel* said. The group had always opposed catching or importing wild whales, "but until now, it had never advocated their release, a position of more radical animal-rights organizations." The paper said that HSUS was now calling for the release of animals such as Katina, the beloved queen of SeaWorld Florida, "and scores of other whales kept in captivity by theme parks and aquariums."

SeaWorld pushed back hard. They were ready for combat, despite the niceties of Paul Irwin's tour.

Curator Frank Murru led the assault. Naomi's conclusions were "based on sloppy science and a selective review of available research," he told the *Sentinel.* "Their agenda is to close down zoos and aquariums around the world," he said, then added, rather awkwardly, "In my opinion, maybe dogs and cats aren't sexy enough." As for Katina, he announced that the seventeen-year-old was pregnant again and expecting a calf that September. In other words, she wasn't going anywhere.

Naomi figured that SeaWorld had been holding on to this joyous piece of news about the pregnancy for many months, waiting to spring it on the world at just the right moment, when publicity was not going their way. The thought irked her. Katina was a wild animal held in confinement, not some PR chip to be played in a high-stakes marketing game.

Murru did not argue with the mortality figures Naomi had culled from the MMIR, but he insisted that her conclusions about longevity were way off base. SeaWorld's animals enjoyed the same full life expectancy as wild whales, which Murru claimed was not terribly long. "We know [wild orcas] live to be twenty-five to thirty years old," he told the *Sentinel.* "A thirty-plus-year-old killer whale would be an exceptionally long-lived killer whale."

Jim McBain, director of veterinary medicine at SeaWorld parks, truncated the life span of wild killer whales even more. He told the paper that other studies had revealed that females lived until their late twenties, on average, and males lived until their mid-twenties.

Naomi was ready to explode when she read that utterly outrageous claim. SeaWorld knew practically nothing about killer whales in the wild, she thought—or else they *did* know and were deliberately misleading people.

Why, she could personally identify dozens of Resident killer whales who were well past the "exceptionally long-lived" milestone of thirty years, let alone their mid to late twenties.

As for SeaWorld's contention that Naomi's report was "based on sloppy science" and relied on a "selective review of available research," that was befuddling. Her source for captivity data was a US government publication, which in turn got its information from the industry. If anything, Murru was trashing science that his own company had helped compile.

The figures on wild whale mortality were taken from the seminal 1990 International Whaling Commission study that Mike Bigg—along with Peter Olesiuk, Graeme Ellis, John Ford, and Ken Balcomb—had just completed when he died. It was not only the most recently published data available, it was widely considered by cetacean scientists to be the gold standard of killer whale censuses. She stood by her statement that 18 percent of the killer whale population in British Columbia had died between 1973 and 1992. That 48 percent of the captive population had died in a comparable period was irrefutable.

It angered Naomi to be labeled "sloppy"—she was anything but *that.* But she quickly brushed off the insult. She knew more industry barbs would be flung her way.

Naomi returned to Maryland satisfied but exhausted. Orlando had been fruitful, but there would be no resting on any laurels. First, there was the pending reauthorization of the Marine Mammal Protection Act, and many fights ahead over the attempt to enact a permanent federal ban on the capture or importation of any marine mammal from the wild.

Naomi would also have to expand the new case-against-captivity campaign, to ramp up public opinion against the industry during the summer of *Free Willy.*

The movie turned out to be a sleeper: Opening weekend box office put *Free Willy* in fifth place, behind megahits such as *Jurassic Park* and *The Firm,* but ahead of another summer favorite, *Sleepless in Seattle*. The whale movie moved into fourth place the following weekend and remained popular all summer, finishing out the season with a respectable domestic gross of $70 million (about $118 million in 2012 dollars). It was a good movie. Even Naomi enjoyed watching it, and she was a tough critic of Hollywood's slick style.

A few weeks after returning from Florida, while relaxing at home one evening, Naomi received an unexpected phone call. It was Dave Bain, the PhD student from Santa Cruz who, along with Janice Waite, had befriended Naomi and invited her to be their field assistant on West Cracroft Island. Naomi still recalled Dave fondly as the tall and somewhat nerdy guy who was recording and trying to decipher the mysterious language of the orcas in Johnstone Strait. After getting his doctorate, Dave had gone on to conduct research at the National Marine Mammal Laboratory, the Canadian Department of Fisheries and Oceans, and the University of California campuses at Santa Cruz and Davis.

Naomi was happy to hear from her friend and former mentor. But she could tell right away from his voice that this was no social call.

Dave was now doing work for the Marine World Foundation—an offshoot of the Marine World/Africa USA amusement park in the Bay Area community of Vallejo, California. Marine World had asked him to review and critique Naomi's anti-captivity paper, plus the letter that Paul Irwin had mailed out with the report. Dave, as usual, was blunt, brief, and matter-of-fact.

"Your report is riddled with errors," he told Naomi. "You made a lot of careless mistakes."

Naomi shifted uncomfortably. She was tempted to dismiss Dave as she had done the other pro-captivity advocates. After all, he now worked for the industry; Dave had gone the *other way.* But Naomi could not ignore what he was saying—she knew the ring of truth when she heard it, no matter how unsettled the delivery made her. Dave was an excellent mathematician and one of the most forthright people Naomi knew.

"What do you mean, Dave?" she said. "The data I used was from Mike Bigg and the MMIR. I don't understand."

Naomi listened stoically as Dave calmly ripped apart her work. First of all, it was inappropriate to use longevity data for Pacific Resident whales when most of the orcas in captivity were of Icelandic stock. It was likely that Resident killer whales lived longer than other ecotypes, so the comparison was invalid on its face.

Even worse, the data on wild whales contained nothing about infant deaths: Mike Bigg and colleagues did not include new orcas in their study until they had reached one year of age. Naomi, however, had included *all* live births in captivity, even those animals that died in the first year. It was an unfair comparison.

Finally, it was a statistical sleight of hand to calculate longevity figures using only data on overall mortality. Just because a greater percentage of whales had died in captivity than in the wild over a comparable period did not mean that whales in nature lived longer. Dave explained this to Naomi but it went over her head. She was never great at advanced statistics.

Dave said he was disappointed by what Naomi had done. He was honestly trying to help her out, give her a heads-up that she had made a fool of herself. She had stained her scientific integrity in her quest to promote animal welfare. The scientific community would not soon forget her blunders.

Before hanging up, Dave told Naomi he was helping Marine World draft a letter to Paul Irwin with a point-by-point rebuttal to the HSUS report, letter, and press communiqués on captivity. It was going to be long and tough.

Naomi told Dave she would look into his critiques, but advised him she would never give up her advocacy work. She was glad to be out of academia. She could never have gone to work for the display industry, the way he had. She was right where she wanted to be.

She thanked her old friend and hung up. More than anything, she was embarrassed. She was mortified—and angry at herself—for tripping into such rookie pitfalls right off the bat at work. She held no animosity toward Dave. He was defending public display because he sincerely believed in the value of researching the whales at Marine World. His work was important, and he was going to defend his ability to do it. Naomi did not mind his intervention. She was actually somewhat grateful.

It was a good thing Dave called, Naomi thought to herself. He kept me from making an even bigger ass of myself.

But the next day, the episode was not sitting well with her. Naomi decided to respond to Dave in writing. She sent him a long e-mail, in which she upbraided Dave for clinging to his claim of scientific objectivity while collecting a paycheck from the industry, which is excerpted here:

> *Dave: I am writing this letter as a personal friend rather than as a professional colleague or, for that matter, an adversary. I will understand, however, if you feel obliged to show it to your employers. It is, of course, in regards to our conversation last night. I would like first of all to thank you for taking the time to call me with your concerns. . . . I am still a bit uncertain as to your actual motivation in so doing, but I suspect it had something do with a misguided notion on your part of "rescuing" me from my "wrongheadedness." I hope I put such notions to rest. I made a very conscious decision to become an advocate for a cause I believe in deeply; I feel I paid my dues in the academic world of science and am now free to use my knowledge and experience to contribute to and aid a non-profit organization whose [ideals] I have always supported. You, of course, have done something similar; you have chosen to use your knowledge and experience to further the cause of a for-profit organization whose work supports you.*
>
> *However, I would like to point something out. Although you are quite right in saying I have compromised my objectivity by becoming an animal-welfare advocate (and I did so gladly), you are in the same position. If you believe that you still wear the pure mantle of objectivity securely around your shoulders, please think again. I am not the only one who considers that those who work for a for-profit enterprise such as Marine World are mouthpieces for that enterprise (hardly objective) when the topic of conversation is the very thing that keeps that enterprise alive, in this case captivity for marine mammals. The only thing*

you have that I do not (and, incidentally, even this is debatable) is a doorway back to academia, which is an advantage only if that is where one wants to go. . . .

It is unfortunate that there is a statistical argument against the set of numbers presented in our letter, but frankly, as you yourself pointed out, statistics in many cases are just smoke-and-mirror shows. With some hand waving, one can make any case one wants to, using the same set of numbers. . . . I must admit I felt your implicit accusation that I lacked integrity because I was willing to use some smoke to further my cause to be somewhat hypocritical. Although you do not directly operate the smoke machine, your employers are past masters at operating that device. I take great pride in my integrity, Dave; I will endeavor to play this game we play with truth and fact, but let us not kid ourselves. This is a game. I suspect in the integrity sweepstakes, you and I are neck and neck. If anything, I'm a bit in the lead, because we both know that what really motivates Marine World is not a desire to educate the public, but money. The HSUS, as a non-profit organization, is hardly in the league of marine parks when it comes to money.

That is all I have to say, really. As I said on the phone, I intend to forge ahead. I have much to offer to this cause and I am having the time of my life, because I fit here and I am following my conscience. . . . As I said on the phone, I did not deliberately wave my hands over the numbers the HSUS used; I believe those numbers tell a true story. You now have your chance to tell Marine World's story.

Then Naomi waited for the letter from Marine World to arrive. It was, as Dave had warned, long and tough. The five-page broadside was signed by Marine World/Africa USA president Michael Demetrios, who said he would be citing information supplied by Dr. David Bain, "a colleague and personal friend of your in-house scientist Dr. Naomi Rose." Dave had spent more time observing wild and captive orcas than any other biologist in the country, he contended. He had the "expertise and integrity required to comment fairly on the accuracy of Dr. Rose's work."

Demetrios began with the longevity and survivorship numbers for Northern Resident whales, and the HSUS's methods for comparing them with animals in captivity. Again, as Dave had warned, Demetrios attacked Naomi for choosing one specific population of orcas "which are not representative of most populations in the wild." The Northern Resident community was "doubling every generation," he wrote, "with calving rates that are higher and mortality rates that are lower than most populations in the wild. Making universal statements about longevity for all killer whales based on this unrepresentative 'best case' scenario is unscientific and misleading."

Then he laid into Naomi's inclusion of captive newborns in her mortality data but not infants who died in the wild. "The number you missed is not inconsequential—the wild neonatal mortality rate was calculated by Mike Bigg at over 40 percent for the first six months of life. That amounts to about 100 deaths of northern Resident calves over a 20-year period. To overlook 100 deaths in a survivorship study of the wild population severely undermines confidence in your credibility, methods and intentions," Demetrios complained. If Naomi had included infant deaths in the wild as well, he claimed, "then the mortality rates of that wild population and the captive population become roughly the same: about 2 percent per year. The average age at death for the wild population becomes about 7 to 10 years of age, about the same as you quoted for captive whales."

Then came the assault on Naomi's methodology that Dave had warned her about. "Most so-called 'longevity' statistics are absolutely meaningless for drawing conclusions about the well-being and life expectancy of wild versus captive killer whales. They are good, however, for manipulating perceptions for propaganda reasons, which you do, either knowingly or unknowingly, throughout your material."

All scientists who had analyzed data from the MMIR had stressed how inappropriate it was to estimate longevity "using measures that depend heavily on factors other than life expectancy." Those researchers, Demetrios added, "specifically say that numbers for average age are irrelevant until all the animals in the study group are dead, yet time and time again, that is what anti-captivity activists do."

Naomi's statistics on wild whales were "equally misleading." The HSUS report had claimed that females lived fifty years on average, with a maximum life expectancy of eighty, "with the implication that something is wrong if captive killer whales don't reach these same ages. What you don't say is the fact that about 8.5 percent of wild females don't reach the age of 50 and about 99 percent don't reach the age of 80. (Those percentages are for animals that have survived infancy.)"

Naomi was also wrong to have criticized the industry's captive breeding program. Here, she got her numbers wrong, too. There had been ten successful killer whale births in captivity, not six, as she had put in the HSUS report's introduction.

"We have clearly reached a major milestone in the care of these magnificent animals: sustained captive breeding. It is one of the greatest success stories in the history of zoological park animal care," Demetrios said. "Rather than offering congratulations to the many caring and dedicated training, veterinary and husbandry staff professionals who worked hard to reach this milestone, or even acknowledging its reality, the HSUS and Dr. Rose have told the media and the public that the effort is a failure. It is difficult to comprehend how a biologist could take such unsupported statements about the killer whale breeding program, apparently sacrificing accuracy to further a philosophical position."

Captivity, Demetrios said, does not mean a "loss of freedom" for marine mammals, but rather the "absence of the hour-by-hour, day-by-day, season-by-season struggle for survival."

Now he had crossed the line, Naomi fumed. It was one thing to critique her math, it was quite another to repeat the tired ocean-as-dark-and-scary-place mantra. "Their life here is different from in the wild," he wrote, then added, without any proof whatsoever, "But it is not worse, and in many ways it is better."

Finally, the letter blasted HSUS for trying to divert badly needed funds for research and habitat conservation into a naïve and extremely costly quest to free a few whales. That was money that would no longer be spent on "a couple of 'sound-watch' boats in Puget Sound to intercept and advise boaters on the proper ways to whale watch in order to stop harassment, or to fund a study of the effects of pollution on orca calf mortality in that same area, or to clean up and restore a Pacific Coast salmon stream, or to monitor the alarming rise in heavy metals and other toxins found in wild killer whale tissue."

In other words, HSUS was doing nothing to make the ocean less dark and scary for killer whales. Naomi wondered what Marine World had done—and spent money on—in that regard as well, especially when it came to conserving killer whale habitats in the wild. She suspected it amounted to little.

Naomi was bruised but hardly down and out. Fortunately, her superiors supported her efforts 100 percent. Besides, they could hardly reprimand her for using statistical calculations taken right from the HSUS's own playbook.

HSUS decided to let the whole thing blow over. Naomi vowed to take up the longevity argument again one day, but only with statistical analyses that were conducted by those with expertise in the field and reviewed by peers. Naomi was a behavioral biologist; statistics were not her forte. And right now, she had a lot of other work to do.

19

Free Willy

Long before Tilikum maimed and killed Dawn Brancheau in 2010, the world's most famous killer whale was unquestionably Keiko, the fifty-eight-hundred-pound male and international star of the Warner Bros. movie *Free Willy*. Keiko and Tilikum had remarkably similar stories. The records were sparse, but it appeared that Keiko had been captured sometime around 1979, a few years before Tilikum, along Iceland's rugged eastern coast. It's conceivable, though we will never know, that the two whales belonged to the same pod.

Keiko had been captured by a local fisherman aboard the *Gudrun*, the same vessel that had taken the female orcas Kenau and Gudrun a few years earlier. Like the infant Tilikum, young Keiko was abruptly snatched from his family. Men who captured Icelandic killer whales employed two favorite techniques for distracting the animals long enough to encircle them in purse seines. Both Tilikum and Keiko were surely captured in one of these ways.

The first method was to trail herring fishermen and wait for them to close their nets and haul the fish aboard. The orcas knew that a generous portion of live herring would spill from the net as it was lifted from the sea. An experienced killer whale hunter, with precise timing and a good shot of luck, could deploy his own net and surround the unsuspecting whales just as the herring boat steamed away.

The other technique was to purchase seven hundred to nine hundred pounds of fresh herring and spread it out on the water like chum as a pod approached. When the whales gathered for the meal, the seine nets would be tightened around them.

The infant Keiko, just like Tilikum, had been lifted in a canvas sling, placed on a boat, floated to another part of Iceland, then trucked to another location—either Saedyrasafnid Aquarium, where Tilikum went, or possibly a private zoo in southern Iceland. There was no documentation of the movements. Anti-captivity activists called this practice "hiding" or "laundering" whales that were caught in excess of capture permits.

The youngster was named Kago.

Ultimately, like Tilikum, he was sold to a theme park in Canada—in this case Marineland Ontario, where he was shipped in 1982. There, the calf was put in a tank with five older and more aggressive killer whales. They began beating him up relentlessly—slapping him with their flukes, bumping and ramming his flanks, raking him with their teeth, much like the domination that Tilikum had suffered at SeaLand.[1]

The new whale would not learn show behaviors and demonstrated little interest in taking part in the performances. "He was kind of a weak animal," recalled Angus Matthews, who was director of Marineland at the time. "His stimulation and attention span was kind of waning. He didn't deal with stress well at all." Kago was so stressed-out that Marineland ended up stowing him in a small tank inside a warehouse, where he spent most of his time while in Canada: alone and indoors without fresh air or sunlight. His living quarters were so wretched that grotesque lesions began to sprout on his skin.

In 1985, an amusement park in the far southwestern section of Mexico City—a Six Flags–style attraction called El Nuevo Reino Aventura (New Adventure Kingdom)—bought the orca for a reported $350,000.

The calf arrived amid much fanfare and desperately needed a name change: *Kago* in Spanish sounds like "I shit." The whale was rechristened Keiko, the female Japanese name for "blessed child."

But Keiko wasn't all that fortunate. He was kept in an extremely small pool along with a rotating number of bottlenose dolphins. Built for dolphin shows, the tank was ninety feet long, forty-three feet wide, and less than twenty feet deep. The water, warmed by the tropical sun to temperatures reaching eighty degrees, was unfiltered Mexico City tap water mixed with chlorine and giant sacks of table salt. He was not in prime health. Keiko's skin infection, most likely from a papillomavirus, doctors said, was getting worse.

But Keiko, warts and all, became staggeringly famous in Mexico, especially beloved among children. He learned new behaviors and performed five shows a day. He started reproducing the sounds of the dolphins that shared his tank, especially a bottlenose named Ritchie. The two grew close and often played together. Keiko even began imitating the murmur of the pool pump and the wails of police sirens screeching outside the park.

He was by all accounts a sweet, gentle, and even compassionate little whale. One day a caretaker's baby son fell into the pool. Keiko raised the child up to the surface and gingerly deposited him on the walkway before any humans noted the danger (the touching scene would be included in the movie *Free Willy*).

Keiko grew fast. By 1991 the owners of Reino Aventura realized he was quickly outgrowing his tank. His immune health was deteriorating and it looked as if Keiko might die if not relocated to a better situation. The owners even offered to sell him to SeaWorld. Naomi found some of the internal memos—delivered anonymously to HSUS headquarters back in 1991 and put in the files—in which SeaWorld officials asked for funds to buy Keiko.

SeaWorld wanted two males: Keiko and a second bull born in captivity at Marineland Ontario. "The acquisition of two male killer whales will improve the SeaWorld long-term breeding program and increase the genetic diversity of SeaWorld's whale population," said the unsigned memo, which was addressed to Busch Entertainment Corporation, part of the Anheuser-Busch beverage conglomerate, which bought SeaWorld in 1989. "The whales are trained and should be adaptable for use in SeaWorld shows."

The memo warned that members of the public would be allowed to make written comments on the import permit and that SeaWorld would be required to respond. A public hearing might also be called and "protests from some animal activist groups can be expected." The author recommended approving the purchase of the Reino Aventura whale for $1.2 million, plus an additional $500,000 for transportation, permit fees, legal costs, insurance, and so forth. The Canadian whale had a price tag of $850,000, plus $500,000 in costs.

SeaWorld Orlando had just lost the male Transient Kanduke and desperately needed fresh sperm. That's when Tilikum killed Keltie Byrne and was put up for sale by SeaLand. When Tilikum became available, SeaWorld abandoned its plans to buy Keiko. Better to buy a proven stud, even one with a murder rap, Naomi imagined was the thinking, than to buy one with a potentially contagious skin disease.

Then Warner Bros. came calling. The studio had spent nearly a year searching for an orca to play the role of Willy, and an aquatic park to film in. Their first choice had been SeaWorld, followed by the Miami Seaquarium, but both places turned them down after learning that Willy is freed at the end of the movie. SeaWorld officials called the script irresponsible and reportedly told Warner Bros. they would agree to a deal only if the title character was sent to a superior captive facility at the end of the film.

The producers discovered the hapless Keiko alone and confined to a tiny tank in a second-rate facility. He was the perfect whale, and Reino Aventura the perfect location. As luck would have it, the park was about to close for major renovations, making it easier for Warner Bros. to film there, while also giving Keiko something to *do* in the interim. Reino Aventura's owners agreed to the filming. Naomi assumed that the owners knew the fictional park would be portrayed harshly. But they also must have known that Keiko would have to be sent somewhere else sooner rather than later. Perhaps they thought that making him famous would help him find a better home.

Keiko proved to be an adept and cooperative actor, and the camera loved him. He reenacted the baby-rescue scene like a true professional. Everyone believed that he enjoyed the attention, the lights and the crew, and the retaking of shots. When it was over and workers began to dismantle the set, Keiko seemed to sense there would be no more film acting. On the last day of shooting, he breached from the water repeatedly, something he had never before done without being given a signal. Keiko doused the crew, perhaps sensing that the excitement was about to wrap up.

The whale became an overnight sensation and international movie star (though many shots were done with an animatronic body double). *Free Willy* was a surprise hit, not only domestically but in Europe and parts of Asia as well.

Not surprisingly, SeaWorld hated *Free Willy*. "Its curators are steamed at the film's depiction of an animal theme park as an inhumane cesspool," film critic Richard Corliss wrote in *Time* magazine. Jim Antrim, SeaWorld San Diego's general curator, complained to Corliss that the movie was not a "fair portrayal" of the reality of public display. "The trainer seems to be feeding the animal an inferior type of fish and often walks by the animal in an uninterested manner," he said in what Naomi thought was an odd critique. SeaWorld also scoffed at the "miraculously rapid" bond that developed between human and whale in the film, and the "malicious" depiction of the park's owner and staff—and the visitors who bang on the glass walls so loudly that it sends Willy into a desperate frenzy.

In late summer, when US reporters went to Reino Aventura to have a look at the famous whale, their stories about his poor health and substandard housing shocked the American public. Though he seemed spritely in the movie, Keiko had grown listless and stopped eating properly. His immune system was weak, he suffered from ulcers and digestive distress, and his muscles had lost tone. His skull and rib cage were showing. Keiko was sluggish and unable to hold his breath for more than three minutes. The high altitude and Mexico City's noxious and notorious air pollution were not helping matters. Keiko would have to be moved if he was to survive.

Even as the movie was opening, a plan had already been developed to relocate Keiko from Mexico City to a facility where he could be nursed back to health. There was even talk of rehabilitating the whale and releasing him to his native waters of Iceland.

Warner Bros. and *Free Willy*'s executive producers, the husband and wife team Richard Donner and Lauren Shuler Donner, of *Lethal Weapon* fame, took the first steps toward getting Keiko out of Mexico. The Donners were firmly in the anti-captivity camp. "If I had my druthers," Richard Donner told *Time* magazine, "these places wouldn't even exist."

The Donners contacted Kenneth Balcomb, founder of the Center for Whale Research on San Juan Island and one of the fathers of modern orca photo-identification techniques, along with Mike Bigg and his Canadian colleagues. Ken had since emerged as one of the leading authorities on wild orcas.

Ken went to work on a formal plan. He proposed that Keiko be flown to the Bahamas, where Ken had found a sea pen location at a private research center that was willing to help reintroduce Keiko into the ocean. The plan included a comprehensive "rehabilitation and reintroduction" package that would slowly lead Keiko out into the sea regularly behind a ninety-two-foot research vessel Ken had secured for the project. Ken wrote that the US Navy had successfully taken captive orcas out into the ocean for extended periods, periodically calling them back to the ship with underwater tones. Keiko would also learn to forage for live fish and learn other skills needed to survive in the wild.

Eventually, the team would slowly head north in the ship, with Keiko following behind, all the way to Iceland. Ken figured the thirty-three-hundred-mile journey would take about a month and half, assuming that Keiko traveled at sustained speeds of about three miles per hour. Once he was in Iceland, the town of Eskifjördur had already agreed to "adopt" Keiko if needed and to construct a fence across their fjord, with a gate so that Keiko could come and go.

Every attempt would be made to locate Keiko's Icelandic family—including DNA samples and acoustic comparisons of distinct pod calls—and reunite him with his pod. Another possibility was that Keiko would become a "rogue" or solitary whale. This was acceptable as well, Ken said. "The major point is that many more options and enrichments to his life will be available to him than could ever be made available in captivity," Ken wrote. "If perpetual maintenance is to be his fate, we will ensure that it is in a natural healthy seawater environment with as much enrichment as possible."

Scientists and vets working for the industry pooh-poohed the plan as unrealistic and dangerous for Keiko. But that didn't stop the momentum to free him.

Ken had been contacted by an experienced whale and dolphin activist named Ben White. Ben was working freelance with several groups including the Animal Welfare Institute, In Defense of Animals, and The Corky Project, whose goal was to free Corky from SeaWorld and return her to her Northern Resident pod, the A5s. Ben suggested assembling a scientific working group to develop a protocol for the rehabilitation of captive whales in general, and another group of environmentalists and animal protection groups to form a coalition to support Ken's Mexico-to-Bahamas-to-Iceland proposal. Ken loved the idea. When he and Ben approached Naomi for her support, she jumped at the chance. HSUS would act as coordinator for all communications with Reino Aventura.

Even as all of this was happening, another group, Earth Island Institute, had formed a partnership with Warner Bros. to fight the international whaling industry, especially in Norway, which was undermining a 1986 international ban on commercial whaling. Earth Island Institute, based in San Francisco and run by Dave Phillips, was a grassroots outfit founded in 1986 to promote conservation, preservation, and restoration of the earth's ecosystems.

At the end of *Free Willy,* theatergoers were shown a special announcement asking for help to stop the hunting of whales. "You can personally help save the whales of the world by calling 1-800-4-WHALES," it said. The number was run by Earth Island. Within weeks, more than three hundred thousand people had called. Most of them wanted to know about the fate of Keiko. "Please save Willy," many of them said.

By that time, Naomi had struck up a friendly relationship with Earth Island Institute's Mark Berman, an indefatigable activist known to send out dozens of faxes a day to his network of anti-captivity allies. People called him the Faxinator.

Mark told Naomi that most of the calls to the toll-free number were coming from people wanting to know when the "real" Willy would be set free.

It seemed pretty obvious what to do next: HSUS, Earth Island, the Center for Whale Research, the Fund for Animals, and other groups would join forces to persuade Reino Aventura to let Keiko be rehabilitated and reintroduced into his native waters. Naomi would see what kind of funding HSUS might be able to contribute. Mark Berman would see if the toll-free number could be used to raise funds from the public.

Meanwhile, SeaWorld was trying to brush off the implications of the popular movie and its anti-captivity message. Naomi was interviewed by WJR radio in Detroit, along with SeaWorld's Jim McBain, though the format did not allow them to interact with each other. McBain said he hoped the movie would encourage people to come see orcas at "good, clean facilities" such as SeaWorld and stressed that the park in *Free Willy* was completely fabricated. The film was having zero negative impact on ticket sales, he noted.

McBain insisted that it was unrealistic to expect that a captive killer whale could ever be released. They were "too dependent" on people; they would die if returned to the wild. To even mention release was tantamount to denying the value of what the animals offered as "ambassadors for their species." Naomi loathed that term when used to describe animals in captivity. Human ambassadors drive hard bargains to negotiate treaties, trade deals, and other tangible items that benefit their homeland. And, of course, ambassadors come and go as they please.

McBain insisted that SeaWorld's ambassadors "help people care" about whales in the wild and let them know what a nasty place the ocean had become, with all its pollution, boat traffic, and drift nets.

Naomi was undeterred by the usual rhetoric. She, Ken Balcomb, and Ben White pushed ahead to convince Reino Aventura to let them have the whale. Toward the end of the summer, they achieved significant progress: At the prompting of Mark Berman at Earth Island Institute, the park's director general, Oscar Porter, agreed to a meeting in Mexico City, on August 31, 1993. Ken invited Busch Entertainment to send a representative to the meeting and told Warner Bros. about the release plan for Keiko.

Ken, Naomi, and Ben, along with Kate O'Connell, US representative from the Whale and Dolphin Conservation Society, who spoke fluent Spanish, flew down to Mexico City and headed out to Reino Aventura to meet with Oscar Porter and his team.

First they went to see Keiko. Naomi was dismayed to see his undersize confinement. If anything, it seemed even more confining than it had in the movie. His tail was hitting the bottom of the shallow pool, which was not deep enough for him to breach completely out of the water. It only took three pumps of his flukes to propel him from one edge to the other. The bases of his pectoral fins were covered in thick layers of black, bumpy warts. It was depressing to even look at him.

The meeting itself was a formal affair with tea and cookies, attended by Mario Aguilar, Mexico's minister of environmental affairs. JoBeth McDaniel, a writer for *Life* magazine who was doing a story on Keiko, was also invited.

Ken explained past protocols for the release of captive cetaceans, including killer whales from the US Navy, and whales that had been held in sea pens for a year or more before being set free. The Americans explained that Keiko's proposed release program would meet every concern that Reino Aventura had expressed.

Porter and colleagues looked impressed. They liked the idea, they said. They voiced their agreement that Ken should direct a project to return Keiko to his native waters in the North Atlantic. Porter further agreed not to negotiate with any other group, including SeaWorld, for six months while Ken drew up an even more comprehensive protocol for Keiko. Porter promised to draft a letter of intention to spell out the agreement by September 15.[2]

Then, halfway through the meeting, the Mexicans were called away for an urgent phone call. They returned to the conference with entirely different expressions on their faces. Naomi shot Ken a look that said, "What the hell happened?"

"Thank you very much. We appreciate your coming all this way. We'll take what you said under advisement," Porter informed the confused Americans. His body language suggested that the meeting was over and it was time for the groups to go back to the States. Tea and cookies were canceled. Naomi was never able to confirm it, but she was certain that the phone call had come from SeaWorld. On the flight back to Washington, she realized she had been out of the country for a total of nineteen hours.

The next day, Vancouver Aquarium director John Nightingale sent a fax to Oscar Porter in Mexico informing him, "All institutions holding killer whales in North America have pledged to work together to solve the problem of providing better conditions for Keiko immediately." That communication was followed up by another fax from the Alliance of Marine Mammal Parks and Aquariums to "reconfirm the commitment already made to ensure the health and well-being of Keiko." Days later, Warner Bros. told Ken Balcomb they would not support his release protocol.

Then, on September 9, the Alliance issued a news release announcing its offer of "a comprehensive rescue program" for Keiko, stating that soon "a home will be provided for long term recovery."

Naomi had waited long enough for a response from Porter. On September 13, after consulting with HSUS attorneys, she faxed him a letter demanding to know the status of both the "verbal commitment" and his promise to sign a written agreement by the fifteenth. "Events concerning Keiko's future are proceeding quickly and our coalition would find it easier to respond to circumstances if we were able to clearly indicate that an agreement has been reached," she wrote. HSUS wanted to issue a press release, in two days' time, on September 15, she said, to announce the mutual agreement to rehabilitate and release Keiko.

She meant business. Naomi urged Porter to get back to her promptly "as we do not wish to proceed without your valuable input and cooperation." She added, "These are exciting times for everyone who takes a special interest in the welfare of marine mammals."

The next day, Naomi arrived at work to find a *USA Today* article in which she was quoted, and an urgent fax from Mexico City.

"Movie Star Trapped in Whale of a Hassle," the *USA Today* headline blared.

"The popularity of the film *Free Willy* is fading along with summer memories, but the real Willy—a killer whale called Keiko—remains in hot water in a Mexico City marine park," the article said, adding that animal activists were feuding with the film's producers over Keiko's fate. Meanwhile, Mexican officials were eager to see Keiko sent to a more appropriate habitat, "but that's where all the agreement ends."

But Naomi had upped the ante in the high-stakes game. "Negotiations with officials are very, very encouraging," she told the newspaper.

Her encouragement was stomped on by a Warner Bros. executive, Rob

Friedman, however. His company would never get involved with a proposal that would be "more injurious to this whale than his current situation," Friedman said.

Naomi could not believe what she was reading. How could a long-term, comprehensive program to slowly reintroduce Keiko into the ocean possibly be worse that his eighty-degree, table-salt pool in Mexico? Warner Bros. had been grossly misinformed, she thought.

The studio would continue to "monitor" Keiko in Mexico, Friedman said, and "try to help him by getting coolers for his tank and looking for further opportunities." Yeah, right, Naomi scoffed, "opportunities" like spending the rest of his life at SeaWorld.

Friedman said that Keiko's lesions rendered his release impossible because he could infect wild whales. But Dr. Lanny Cornell—the former SeaWorld veterinarian who had been let go following the rash of orca attacks in San Diego in 1987—said the warts were the result of suboptimum water quality and temperature. In his estimation, they were not contagious to other whales.

The Mexico City fax, meanwhile, had come from Oscar Porter. Reino Aventura was not yet ready to sign a letter of intent. The fax confirmed her suspicion that SeaWorld and the Alliance had intervened while they were in Mexico City.

Porter said he tried to explain that the Alliance was in "the best position to provide us with all their support to optimize Keiko's living and medical conditions, as well as to help get ready for the first stage of the preparation program for Keiko's departure, that is to start cooling the water. We are requesting them to be more specific on this help."

Once the agreement letter was ready, however, he would send it to Naomi and Ken.

"We have always stated our main interest is Keiko's health and well-being, and we gladly accept any positive help offered for this purpose. I have verified that the Alliance is doing this without asking any promises from us. We expect the same . . . from your organization."

Naomi wrote back immediately, expressing patience and conciliation. "We are grateful that the Alliance appears ready and willing to cooperate," she wrote, adding that the park would soon be receiving letters of support from US and international groups "praising your decision two weeks ago to commit to a release program for Keiko." She conceded that no release of Keiko would be authorized without the blessing of an international panel of experts who would "be chosen by our coalition and the Alliance."

The days went by and no reply came from Mexico City. Ken wrote to Oscar Porter again, offering to help with logistics and even financing to improve conditions at Reino Aventura. If the Alliance of Marine Mammal Parks and

Aquariums was going to help with water cooling and filtration, perhaps Ken's group could arrange to fly in fresh Icelandic herring to help accommodate Keiko in his transition to native prey.

Porter gave Ken the brush-off. He said that recent medical reports had revealed a compromised immune system in Keiko. Until that was brought under control through chilled water, proper salinity, and a new filtration system provided by the Alliance—in another six months to a year—Keiko was not going anywhere. Once his health was deemed robust enough to travel, "we will decide where to relocate Keiko to continue with his rehabilitation program," Porter wrote. "In due time, we will keep you updated."

Reino Aventura had declined Ken's offer of assistance, but Porter invited him to join a Keiko Working Group the park was forming.

Naomi's heart sank when Ken faxed her a copy of the letter. What did he mean "in due time"? And what was this working group? That was not in the original commitment. Clearly, SeaWorld and the Alliance had intervened well, she thought.

She was right. After the Free Keiko group had departed, the Mexicans were visited by executives from the Alliance, who made an offer of chilling and filtering the tank, and adjusting the salinity. They would also provide Keiko with a complete medical analysis and treatment for his skin condition. Only then would the park owners consider finding a location in the United States for his further rehabilitation "before reaching any decision pertaining to the reintroduction matter."

Porter nonetheless encouraged Ken to continue his research in Iceland on genetics and orca calls to try to locate Keiko's family, which Reino Aventura would be interested in reviewing "at some time in the future." But at the same time, he noted that Ken's coalition was raising money to finance the Keiko project and ordered him to stop doing so.

Ken was not amused. He had his attorney write to the Alliance, and to Busch Entertainment Corp., threatening legal action for "engaging in efforts, including possible defamatory conduct, to induce Reino Aventura to break its contract with the Center for Whale Research and, instead, allow you to obtain Keiko and prevent his release to the wild."

Ken, Naomi, and the others were undeterred. In late October, Ken traveled to Iceland to collect DNA samples and orca recordings for comparison with Keiko's own. Money was starting to come in for the project, including $5,000 raised by kids at one school in Tampa.

Meanwhile, Earth Island had convinced Warner Bros. to allow the conservation group to include Keiko information (including the release plan and information critical of captivity) in packets sent out to people calling the 800 number. The studio also agreed to a 900 number when the movie opened in the UK, to be managed by the Whale and Dolphin Conservation Society. All proceeds would go toward Keiko's release effort.

It was a huge victory to have so much support from Warner Bros. SeaWorld and the Alliance could pick off HSUS and other animal groups, Naomi knew, but taking on Warner Bros. was another matter entirely. This was now a battle of wits between two huge entertainment conglomerates: Hollywood glitz versus St. Louis beer. Naomi had her money on Tinseltown.

Keiko's release gained another boost in November, when *Life* magazine published its cover story, written by JoBeth McDaniel. Headlined "Won't Somebody Please Save This Whale?" it described the horrible conditions for Keiko at Reino Aventura and detailed Ken's release plan.

"Clearly bored and frustrated, he gnaws at the concrete walls of his prison. His teeth are worn down, and blood seeps from ragged abrasions on his chin," McDaniel wrote. "Reared to swim in 40-degree arctic seas, Keiko is now required to give three strenuous performances a day in 80-degree water." Keiko "lives in a cesspool," McDaniel went on. "His tank has a weak filtration system that can't handle the mass of excrement he releases; he must swim for hours in his own wastes before they are flushed away."

Photos showed Keiko's bloody lip from scratching the poolside. So many readers called the magazine wanting to help Keiko it had to open a dedicated phone line. *Life* received more calls about Keiko than any other story in the preceding decade.

Naomi was feeling confident. Donations were expected to total between $1 and $2 million by January or February of the new year. Michael Jackson, the pop star, had expressed an interest in the Free Keiko endeavor and planned to visit the whale during an upcoming "Dangerous" concert tour that would bring him to the Mexican capital. "I really think we should pursue that interest here at HSUS," she said. "If Michael Jackson wants to fork over a million or so to help send Keiko back home," then he should be fully encouraged, she said. (Naomi did not realize that Jackson wanted to bring Keiko home to his private Neverland Ranch outside Santa Barbara, California.)

"I don't think Oscar Porter will be able to hold out against this tide of money and feeling for much longer," Naomi predicted. "And when the movie opens in Mexico in December, he will be under enormous pressure to give in." Only time would tell.

The media played up the battle between industry and activists. Ken Balcomb was happy to fuel the flames. "We suspected SeaWorld would do anything to block Keiko's release," he told the *Los Angeles Times*. "They stand to lose millions of dollars if a movie-star killer whale is released into the wild successfully and the public realizes there is no reason that whales need to be kept in captivity."

Despite Reino Aventura's claim that it had not ruled out releasing Keiko back to nature, Ken said that would never happen "as long as SeaWorld is in control." He said the company did not want "a bunch of enviro-crazies like us maybe getting the public more interested in whales in the wild than whales as circus performers." Indeed, the Mexican park was now denying it had entered into an agreement with the activists, even though they had videotaped the entire meeting at Reino Aventura.

SeaWorld's Jim McBain jumped into the international mêlée, saying he had been involved with Keiko's care since 1991. "I don't think Ken Balcomb and the others had even heard of Keiko in 1991," he told the *Times*. "As romantic a notion as it may sound, taking an animal like Keiko and doing the *Born Free* thing just isn't realistic."

Naomi also spoke to the media about the industry's interference in the Keiko affair. "He could die tomorrow," she told a reporter from Gannett News Service. "It was definitely an ambush so they could look like the knight in shining armor, rescuing Keiko. If they cared so much about him, why didn't they do this three years ago?"

The Keiko saga was far from over; it was just getting started.

20

Protection

Up until the early 1970s, marine mammals were having a hard time in the United States and its territorial waters. Pollution and coastal development had threatened entire populations; polar bear, seal, and sea otter communities had been decimated by hunting and trapping; whaling operations were still in business on both coasts; tens of thousands of dolphins were dying in tuna nets; and aquariums, marine parks, and research teams had culled hundreds of mammals from the ocean without federal permits or rules governing their care and transportation.

Onshore, things were little better. Marine mammals were bought and sold like commodities, to be displayed at zoos, theme parks, aquariums, and roadside attractions that were often second-rate, and even at trade shows such as the boat show that had bought Skana in 1967 and tried to force-feed her raw hamburger. Uncounted numbers of animals were held in substandard environments: tiny and sometimes filthy tanks and pens, often overcrowded with sick and neurotic creatures.

Meanwhile, huge gaps existed in knowledge about most marine mammals, their natural histories, and their ecological importance.

Congress took note. Cetaceans, pinnipeds, manatees, sea otters, even polar bears—all mammals that were essentially tied to the marine environment—deserved federal protection, many lawmakers now believed. Their growing appreciation of these species as living beings—creatures of beauty and an international resource to be treasured and studied, rather than hunted, harassed, and abused—led to the landmark Marine Mammal Protection Act (MMPA) of 1972.

Under the act, it became a felony for anyone to "take" marine mammals from US waters, or for any US citizen to take these protected species from the high seas. Congress broadly defined "taking" as the commission or even *attempted* commission of "any act of hunting, capture, and/or harassment of any marine mammal." The MMPA also prohibited the importation of marine mammals or marine mammal products, among many other provisions.

The intent of the bill was to protect mammal species in the oceans, but Congress also specified that a small and regulated number of mammals could be taken for the purposes of public display, scientific research, or conservation, or incidentally during certain important activities such as commercial fishing or oil and gas exploration and extraction.

Under the act, marine mammals from nondepleted stocks could be captured or imported into the United States for public display, with certain restrictions. The bill authorized NMFS to issue permits only to those facilities that the secretary of commerce deemed to have "acceptable" programs of conservation or public education. Display venues also had to meet the minimum requirements for care and maintenance for animals in their custody, as set forth by the Animal Welfare Act of 1966.

Clearly, Congress felt that captivity for these species was justified as long as it benefited the public and the animals, through research, conservation, and education—and as long as marine mammals were treated according to federal standards on animal well-being.

The MMPA was written to be reauthorized roughly every four years, which generally led to a number of amendments being fought over by various parties, leading to deletions and additions to the law. When Naomi began working at the Humane Society of the United States in May of 1993, negotiations on the next reauthorization were well under way. She was fully expected to get herself up to speed and conversant on the legislation, its history, and the intent on both sides of various issues, including captivity, to propose changes this time around.

It was a daunting education. Naomi went back to the HSUS files for a quick primer on the legislation, and how it had been conceived, implemented, and amended over the years.

On the captivity issue, she found a mess of federal departments, agencies, commissions, and services entangled in a typically bureaucratic jumble. Sorting through what Congress had originally intended for the law was difficult, and competing and overlapping government entities had interpreted that intent differently.

To begin with, the management of marine mammals in the wild was divided between NMFS—an agency of the Commerce Department's National Oceanic and Atmospheric Administration (NOAA)—and the US Fish and Wildlife Service of the Department of the Interior. NMFS took jurisdiction over whales, dolphins, porpoises, and all pinnipeds except (for some odd reason) walruses, while Fish and Wildlife handled walruses, sea otters, polar bears, manatees, and a related species known as the dugong.

This division of labor was codified in the MMPA, including provisions for oversight on the care and maintenance of marine mammals in captivity.

Then it got even more complicated. NMFS did not have the staffing or infrastructure to go out and inspect conditions in hundreds of locations around the country where whales, dolphins, seals, and sea lions were put on display. Many of the people at NMFS were dedicated field scientists and highly talented marine biologists. They knew a tremendous amount about marine mammals, but their agency was not set up to inspect captive facilities and issue animal-welfare violations.

That task had traditionally fallen under the jurisdiction of yet another federal agency: the Animal and Plant Health Inspection Service (APHIS) of the US Department of Agriculture. APHIS had hundreds of inspectors spread out in regional offices across the country, mostly in farm states, with thorough knowledge of the Animal Welfare Act and its provisions. In 1979, the two agencies entered into a formal agreement—APHIS would establish minimum standards for holding and displaying marine mammals and would conduct inspections of public display facilities to ensure compliance with the Animal Welfare Act, while NMFS would permit the holding and display of marine mammals in those facilities, with the ability to add specific conditions to any display permit.

This complicated and unclear relationship, Naomi quickly learned, didn't seem to serve the animals well. NMFS staff knew all about marine mammals, but they had no capacity to inspect facilities holding cetaceans. APHIS staff knew all about facility inspections, but they knew virtually nothing about marine mammals. Moreover, APHIS had long been viewed by animal welfare activists as being in the hip pocket of Big Business. Corporate profits and rising exports seemed to be its driving force far more than protecting the well-being of animals.

The main focus of the reauthorization in 1993 was the need for a new regulatory regime for the commercial fishing industry—the "fish fix." Commercial fisheries were allowed to take (that is, injure or kill) marine mammals incidental to their operations, but the law remained in need of a mechanism to control this accidental carnage and reduce it.

Free Willy could not have come at a better time for HSUS and other anti-captivity groups. By June 16, Naomi was working to put together the Marine Mammal Protection Coalition, more than twenty NGOs committed to strengthening the MMPA. Naomi became the organizing secretary of the coalition. On the captivity issue, the movie was their rallying point.

Even Senator John Kerry, the Massachusetts Democrat, was moved by

Willy's message. Kerry chaired a hearing on the public display provisions of the act in July 1993, just two weeks after the Humane Society press conference in Orlando. He referred to the movie in his opening remarks. Naomi was there, seated behind her boss, John Grandy, who was at the witness table.

"The people that have seen the movie have had a chance to do what many say the public display of marine mammals is supposed to do, which is bring people closer to an understanding of our relationship to them and an understanding of nature and ecosystems," Kerry said. But it also raised questions about the quality of life for captive marine mammals, he said.

The rising popularity of viewing marine mammals and interactive programs such as "swim with the dolphins" had brought "more and more Americans into contact with marine mammals," Kerry said. That had gone a long way toward fostering public education and a growing respect for the animals, "but those activities also raise questions about treatment."

At the hearing, Grandy laid out several of HSUS's key proposals for amending the public display provisions of the MMPA. The ideas were controversial and vehemently opposed by industry. But HSUS considered this an opportunity to ask for as much as it could, no matter how pie-in-the-sky the request. Among the amendments John proposed were a permanent ban on the taking of any marine mammal for the purpose of display; a prohibition on direct contact with marine mammals through interactive programs such as swimming with dolphins; and the transfer of all oversight responsibilities for captive marine mammals away from APHIS and over to NMFS.

"APHIS has not demonstrated in the past that it can adequately ensure the humane treatment and welfare of marine mammals on public display," John said in testimony prepared by Naomi. NMFS "should have sole responsibility for marine mammals in captivity."

As 1993 drew to a close, Naomi worked diligently with HSUS legislative staff, congressional allies on the Hill, and her coalition of marine mammal and animal welfare organizations, which engaged in a massive letter-writing campaign to Congress.

The anti-cap group came up with a six-point wish list they wanted from Congress for public display amendments. Naomi knew they would get few if any of them, but it was worth making the case. In addition to a ban on capture and swim-with programs, and the transfer of oversight to NMFS, the coalition was now seeking a prohibition on marine mammal exports, a ban on the intentional feeding of mammals in the ocean, and a law against the taking of whales and dolphins from the notorious "drive fisheries."

By any standard, drive fisheries are brutal, bloody, and intensely inhumane. They are also the source of many of the dolphins and small whales held in captivity today, particularly in Asia. Found in places such as Taiji, Japan, drive hunts are done primarily to eliminate toothed whales that compete with humans for fish, and to provide their meat to local consumers. The HSUS website provides this description of the awful scene in Japan:

> *Fishermen take out several small motorized boats to locate a pod of bottlenose dolphins, Risso's dolphins, or false killer whales (and possibly such other species as pilot whales). Once the fishermen locate a pod, they begin herding the animals toward shore, using the noise of the boats' engines and the banging of pipes underwater. There are some reports that they also use underwater explosives. The fishermen will then either drive the animals right onto the shore or trap them in a bay. Either way, shallow water is necessary, because fishermen slaughter the dolphins by getting into the water and moving through the pod, stabbing animals to death. . . . Animals destined for slaughter may be hauled out onto land with cranes, often still alive. The cruelty is enormous.*

The Japanese drives had a secondary revenue stream: The youngest, most attractive animals were, and still are, set aside and spared the knife, to be presented instead to buyers from aquariums, theme parks, and swim-with programs around the world.

As Congress kept wrangling over amendments from various factions, Naomi continued to learn more about killer whales in captivity. She wondered what SeaWorld had been telling its visitors about killer whales.

SeaWorld's Corporate Zoological Department had produced a pair of pamphlets specifically on orcas, which Naomi located in her HSUS files. Both were riddled with inaccuracies and misstatements. Naomi was so astounded by the liberties SeaWorld was taking she mailed a four-page letter of complaint to chief animal vet Jim McBain in Orlando.

"These brochures apparently are being handed out to the general public when they visit your parks," Naomi began. "There are several problems with the information presented in [them] and I wanted to discuss them with you."

The first publication, "The Facts About SeaWorld's Killer Whales," for example, referred to the vitamin and mineral supplements in the whales' diet "as though these were a benefit of captivity," she said. "It reads, 'supplements are even added to ensure nutritional requirements are being met.' It disturbs me that SeaWorld would mislead the public in such a way. Wild orcas obviously do not require vitamin supplements to live healthy lives and the reason they are necessary in captivity may be because some nutritional needs are not met by the restricted diet of frozen fish that captive whales are provided."

The brochure also claimed that "much of what the scientific community knows about killer whales comes as a result of knowledge gained at SeaWorld." Such a statement was "out of context at best and untrue at worst," Naomi charged. Yes, some knowledge about physiology had been gleaned largely from captive animals, "but little if any of the information regarding behavior, population demographics, or social structure has come from them, as it is widely accepted that these characteristics are greatly altered in the captive environment. These days, the vast majority of knowledge about these topics was coming from studies of wild populations."

Naomi continued, "By far the most disturbing statement in this brochure is that 'killer whales have a maximum life span of 25 to 35 years.' Jim, you know this is not true. There are two dozen individual whales in the Pacific Northwest who indisputably demonstrate the inaccuracy of this statement; first observed as adults in 1973, they are still alive and healthy today, with several of the females still reproducing."

Next, she attacked the claim that the social setting for the park's killer whales was "carefully monitored to ensure it remains healthy." But none of the groupings at SeaWorld reflected any natural social structures known in the wild, Naomi stated. "I am at a loss to understand how you and your colleagues have determined that SeaWorld social groupings are 'healthy.' They mix stocks (Atlantic and Pacific) and races (transient and resident); in my professional opinion, it is in fact highly unlikely that these social groupings are healthy. SeaWorld's social groupings are artificial in every sense of the word."

The second brochure was called "The Real Story on Killer Whales." According to Naomi, it implied that fishermen would return to shooting orcas "if these animals are no longer held in captivity as 'ambassadors.' This is absurd; the public has amply demonstrated that the whale is now institutionalized as deserving of our protection."

This brochure correctly stated that humans do not yet completely understand killer whale behavior. But then it added, "We are able to tell that our whales are happy by the way they act." The two positions, Naomi said, were "contradictory" and "poor science." Animals have strong survival instincts, she wrote. They can adjust to their circumstances. Intelligent animals use whatever means available to them to ward off tedium. Therefore it was "self-serving for SeaWorld to say that its whales are happy. The state of mind of captive animals certainly cannot be discerned through the behaviors that are mentioned in the brochure."

It was abundantly hypocritical for SeaWorld to maintain that the behavior of its wild-caught killer whales was natural while also claiming they couldn't survive a return to the wild. SeaWorld implied that years in human care had transformed them into helpless creatures unable to return to their natural environment. "Maintaining these two positions simultaneously is a serious flaw in logic."

Naomi ended with this final jab: "Jim, I take great exception to the following sentence: 'To subject one of these magnificent animals to an untried experiment [rehabilitation and release] would be inhumane.' This is enraging.

Thirty years ago, SeaWorld subjected several of these magnificent animals to an untried experiment—captivity. And it was extremely inhumane, because several animals promptly died, painfully and completely unnecessarily, as a result of the flawed protocols of this experimental captivity program. It is hypocritical of SeaWorld to criticize an experimental release program with such a statement."

Naomi never got a response to her letter.

The MMPA reauthorization had its final hearing on public display in the US House, in February of 1994. Naomi attended along with her boss, John Grandy, to present their six proposed amendments. They had no idea they were walking into an eleventh-hour ambush. The industry and its powerful allies on Capitol Hill had quietly devised a raft of changes to the law behind closed doors and presented them here for the first time. Naomi felt sick to her stomach. Voting on the final bill was only weeks or at most months away. Now, SeaWorld and its formidable coalition of zoos and aquariums had finally shown their hand on display issues.

Among the amendments proposed on behalf of public display facilities:

- Gut NMFS's authority over marine mammals once they have been taken from the ocean: Industry wanted APHIS to have full control once any animal "made it through the door."
- Allow the export of captive marine mammals without a permit.
- Eliminate the requirement that educational programs must be acceptable to the secretary of commerce.
- Exempt animals taken for research or species enhancement from inclusion in the Marine Mammal Inventory Report.
- Drop the requirement to submit necropsy reports to NMFS.

In the end, the Alliance and the Association of Zoos and Aquariums (AZA; formerly the AAZPA) got almost everything they wanted on the captivity provisions of the bill; HSUS and its allies *lost* almost everything (they did manage to win one slight victory: The MMIR would now have to be updated within thirty days of an animal's change in status, rather than annually). It was a kick in the gut for marine mammal activists everywhere. Naomi had never felt so defeated before in her life. She and the coalition had been outfoxed by their opposition. Naomi was beginning to appreciate just how powerful and influential the industry was. She had chosen a formidable foe. Its pockets were deep, its attorneys were skilled, and its lobbyists were everywhere.

The display industry had sauntered away from Capitol Hill with a massive and well-orchestrated victory. In exchange, industry had agreed to provisions that Naomi considered to be mere window dressing. From now on, marine mammal facilities would have to meet three rudimentary criteria to obtain a permit (from NMFS) to capture or import a marine mammal for public display purposes.

Industry lobbyists themselves had drafted the new criteria. Naomi figured that executives at the trade association—despite their influence—knew that asking for total freedom to do anything they pleased was unlikely to fly. Congress members, even close allies, would likely push back, at least somewhat, on that idea. So they included these three simple provisions:

- The facility must be open to the public with no barriers to access except for an admission fee.
- The facility must have a license and adhere to all best standards and practices set forth under the Animal Welfare Act.
- The facility must offer conservation efforts and education programs that meet "professional standards."

It was a pretty low bar for the industry to meet. Keeping regular hours for the public was certainly no problem (only a small number of facilities were private by that time). The AWA standards were nothing new. Besides, those standards had been established with heavy influence from—yes—the industry. Who knew more about captive marine mammals than the captive marine mammal industry? Members of Congress were hardly in a position to dictate such things as pool size, water quality and salinity, nutrition, number of animals per enclosure, enrichment programs, and so forth. Congress (and APHIS) felt they had no choice but to rely on the experts. Those experts were mostly at SeaWorld.

The same was true for educational standards. It was no longer up to the secretary of commerce to decide what was acceptable; it was now up to industry itself. In both cases—animal welfare and public education—it was the height of self-regulation and it repelled Naomi to no end.

Another crushing blow for Naomi and the anti-captivity people was that, under the reauthorization bill, facilities that did not display captured or imported animals were completely exempt from the rules and protections of the MMPA. Any aquarium could now open up with domestic, captive-bred animals and not have to be open to the public or have an education/conservation program.

Naomi also worried about what would happen now that NMFS's marine biologists would no longer have any jurisdiction over the health and well-being of cetaceans in captivity. What would happen, she wondered, when the only experts judging the mental and physical health of captive marine mammals were inspectors from the USDA?

It seemed like a recipe for disaster.

21

Kiss of Death

Jeff Ventre was growing more disillusioned every week with his training job at Shamu Stadium. He kept up his lengthy conversations with Astrid van Ginneken, the Dutch woman who split her vacation time each year between SeaWorld Orlando—home to her beloved orca Gudrun—and the Orca Survey project at the Center for Whale Research on San Juan Island. Whenever they got together, Astrid would fill Jeff in on all the data that Ken Balcomb and his team were gathering from the Southern Resident community, often bringing news about recent births, deaths, and other developments in the eighty whales or so belonging to J, K, and L pods.

Astrid continued to urge Jeff to make the trip up to the San Juan Islands to see the wild whales for himself. He would get a much better perspective of the orcas among their pods, she assured him.

Jeff was finally accepting that SeaWorld had crafted a tightly orchestrated façade of happiness and old-fashioned fun, then blended that with the pretensions of hard science and public education. This carefully constructed image that SeaWorld sold to a gullible, entertainment-hungry public was highly profitable, Jeff now understood. But it was a fairy tale.

And Jeff had helped preach the party line. SeaWorld had been spoon-feeding its staff a big box of bullshit, he now reckoned with a modicum of shame for having been deceived. It was all jovial pabulum cooked up soft and dished out to tourists about how "happy" the whales were, how their fish was restaurant quality, how safe it was to work with them.

Jeff was told to say that killer whales live about twenty-five to thirty years, whether in the wild or at SeaWorld. But Astrid told him that some females in Washington State were well into their seventies. One matriarch, the undisputed chief of J pod, was believed to be in her eighties. Her ID number was J2, but everyone knew her as Granny.

If asked about dorsal fin collapse, Jeff had been taught to dismiss it as something that happened both in the wild and captivity. But Astrid said that wasn't so. A few male orcas in the Southern Resident community had slightly bent, curved, or wavy fins, she told Jeff, but none had the kind of complete collapse seen with Tilikum, Keiko, and all adult males living in tanks. At SeaWorld, even most of the adult females had some form of dorsal fin bending, which was rare in wild populations.

Now, in the fall of 1994, Jeff had been at SeaWorld for nearly seven years, and it felt like the time to leave was approaching. He might have been a leading star in the killer whale pool, but it was starting to feel like a lowbrow circus act. Jeff had even made plans to start classes in January at Seminole State College in the northern suburb of Sanford, to begin preparing for the next stage of his life. Jeff wanted to go to medical school and become a physician, but he needed more premed credits to be accepted. In addition to hotdogging the Shamu show, he was now studying for exams in algebra, calculus, biology, and physics. It was a tough load, but Jeff was not going to spend the rest of his life at SeaWorld.

Jeff had seen some of his best friends at SeaWorld quit, including Carol Ray and Sam Berg (who had left SeaWorld in 1993 on good terms) and watched as another stream of people came in. John Jett and Mark Simmons, of course, were now his closest buddies and confidants. Jeff had also grown fond of Kelly Flaherty Clark, an orca trainer rising through the ranks. She was Jeff's kind of person: smart and educated. He liked Kelly's husband, Stew, as well.

Kelly's circle of friends at Shamu included a young woman named Lindsay Rubicam, and a relative rookie to the stadium, another young, blond woman named Dawn Loverde (*Low-VAIR-dee*). Jeff hung out with Lindsay and Dawn at their place in the Metrowest district of Orlando, and he and Dawn together represented the animal training department at the annual SeaWorld golf tournament. Jeff liked Dawn. She was unwaveringly kind to everyone, fun to be around, shared his keen love of music, and was a terrific animal trainer. Dawn's roommate, Lindsay, and Jeff would become close.

Dawn had always adored animals. As a kid growing up in Indiana, she would take in stray cats, dogs, birds, and other creatures. When she was eleven, her family came to Orlando for a vacation. She was so excited at the prospect of going to SeaWorld she could barely sleep. When the big day arrived, she walked into Shamu Stadium, grabbed her mother's hand, and said with certainty, "This is what I'm going to do. I wanna be a Shamu trainer."[1]

Dawn had received degrees in psychology and biology from the University of South Carolina, where she was active on the dance team and a member of

Alpha Delta Pi sorority. After college, she started working at SeaWorld in Sea Lion and Otter Stadium. Within two years she was an apprentice trainer at Shamu Stadium. While at work, Dawn met a tall, husky man with dark hair and a goatee, Scott Brancheau. He was a performer in the park's highly acrobatic waterski show, and the two of them fell in love. In 1996, they would marry.

Jeff experienced friction with higher-ups. One of the least popular superiors among the top brass was the chief training curator, Thad Lacinak. With his short auburn hair, bristly mustache, and muscular frame, Thad looked like a marine sergeant—and he could bark like one, too. Thad had a temper. Jeff, John, and Sam sometimes called him Thug behind his back.

Thad, like Jeff, had grown up in the Orlando area, where he spent a lot of time fishing and swimming in the many lakes that dot the landscape. Also like Jeff, Thad had been inspired in his youth by the TV program *Flipper*. He dreamed of one day training animals.

Thad had quit college in 1973 to begin working at SeaWorld as an apprentice trainer and rose through the ranks to become vice president and corporate curator of animal training for the entire SeaWorld park system, with more than a hundred people working under him. He spent much of the year in San Antonio. Thad was a key figure in developing and implementing SeaWorld's celebrated training method (which was divided into discrete sessions such as husbandry, exercise, learning, etc.) and various animal "enrichment" programs.

Enrichment is the provision of stimulating environments for captive animals that let them engage in typical behaviors and allow them to exercise some choices over their highly controlled and unnatural environment. For orcas, that might mean introducing new objects, sounds, or other stimuli into the pool. Thad was also instrumental in developing new husbandry programs and helped create many of the crowd-pleasing behaviors the animals performed.

Despite his gruff drill-sergeant demeanor backstage, Thad was also a public face for SeaWorld and was often featured in local and national news stories about the park. He was fond of telling reporters that the tank at Shamu Stadium was like a "mini-ocean" and a "big playground" for killer whales.[2]

Thad had little patience for animal-advocacy types, including the staff of HSUS and their new marine mammal scientist, Dr. Rose.

"All these so-called environmentalists who say the whales are not happy here—what proof do they have?" Thad once told a local reporter. "Everything I see indicates that they don't sit in the pool thinking about the ocean." Captive killer whales did not miss the ocean because "they seem to enjoy what we do," Thad said. "The animals recognize you by face. They come swimming over and act like they want to be with you."[3] In other words, an animal such as Tilikum was having too much fun to ever miss his mother and family back in Iceland.

Jeff's work at SeaWorld was getting harder to stomach. He hated having to spin lies for the public; he felt chronically and pathetically underpaid. He felt that SeaWorld was little more than a glorified circus. On many occasions, Jeff had seen killer whales ignore callback stimuli such as water slaps, stage slaps, and underwater tones. Callbacks were the fundamental means of controlling whales during an emergent, potentially hazardous situation. All trainers had experienced this intentional disregard of their commands; it was part of their workaday lives.

When a killer whale was unhappy, Jeff observed, it might well let its trainer know. There might be an instant expression of displeasure—a rough "love tap" for example, a bruising bump or a quick dunk under the water. Or the retribution might come later. The whales could potentially remember past slights. The vast majority of such incidents went unreported.

Jeff was not all that reassured by the safety procedures that SeaWorld had developed for Shamu Stadium over time. A brand-new tank had recently been completed—the large, rectangular G Pool deep in the backstage area—where special events, VIP performances, and the new "Dine with Shamu" events were staged. During the "Dine" show, guests paid extra to eat an all-American buffet lunch or dinner at tables set up along the pool's edge as one of the killer whales swam around quietly. After the meal, they were treated to a low-key, twenty-minute demonstration, without all the leaping, splashing, and hotdogging found at A Pool.

When G Pool opened, Shamu trainers were handed photocopies of two-page, handwritten notes on safety procedures in the new tank. It spelled out what should happen if one of the killer whales became aggressive. It didn't reassure Jeff that the trainer would get out alive. In summary, the "Emergency Action Rescue Procedures 6-Point Flow Chart" called for:

1. Spot trainer directs Education staff to sound trainer alarm and call 911. As soon as possible, a Senior Trainer or Supervisor will direct all rescue personnel.
2. Spot trainer attempts quick control response with whale(s) from behind rocks, with emergency food.
3. Arriving trainers: Bring additional food for control trainer and A) Shepherd's hook to poolside, B) Life ring/ pony bottle [small oxygen tank] to poolside, C) E/F gate operation & control trainer spotter also brings water rescue equipment.
4. With whale(s) controlled at stage, extricate victim from water using a shepherd's hook. If victim is unreachable, move whale(s) to E/F Pool. Once whales are separated & gated securely, trainers implement water rescue procedures.

Are they *kidding*? Jeff said to himself when he read that. Out of six points, they were going to wait until point 4 before extricating the trainer? The trainer was probably dead by point 3, he mused darkly. Still, the victim wasn't necessarily out of the water yet.

5. If whale(s) refuse quick control response and/or separation to E/F pool: 1) Spot trainer continues control attempt; 2) Assisting trainers shift or maintain poolside positions with shepherd's hook and life ring for optimal rescue opportunity.

Optimal rescue opportunity? This was beginning to sound like a Monty Python sketch, theater-of-the-absurd, gruesome-death edition. Finally, point 6 listed all of the things that should have got under way at the same time as point 1, begging the question, why was it listed last?

6. Activities occurring during rescue procedures:
   1. 1–2 extra trainers suit up with full scuba and be prepared to enter water as soon as whale(s) are gated away.
   2. One extra trainer to stay with whales once they are gated away.
   3. Education staff communicates information to the public.
   4. Operations staff control public for safety.
   5. Paramedic and security arrive for assistance.

One day Jeff was shooting video of the Shamu show, which Chuck Tompkins wanted to send to the training department in San Diego, to share techniques and other information. It had been a stupendous show: the animal behaviors, musical cues, and trainer lines all came together. Chuck, who could sometimes come across as authoritarian, seemed pleased as he watched the taping.

As Jeff recalled, right toward the end when Tilikum was under control at the stage, he pumped his tail and surged up out of the water, lunging at trainer Liz Morris, who was standing at the edge of the stage. He was huge next to her, maybe fifty times bigger. Morris took three steps back to prevent him from grabbing her. It was a surreal moment. Tilikum slowly sank back down into the water. Morris adeptly gave the renegade animal a neutral response—standing still for three full seconds—and sent him on a side breach before finishing the show. Jeff thought that Chuck seemed more upset about the video than the potentially dangerous event that had just transpired—on video and in front of thousands of visitors.

As for Nootka IV, she had been capricious and volatile since the day she arrived in Orlando. She was known to lunge ferociously from the pool at people standing near the deck. The main job for Nootka, who was banned from water work and not used in shows because of her unstable past, was getting pregnant.

Nootka got to work right away. She arrived in Orlando in January of 1993. By February, Tilikum had impregnated her once again (their first calf, a male, was born while Nootka was still at SeaLand, but only survived a month).

On August 18, 1994, Nootka went into labor. The calf arrived dead.[4] The stillbirth took its toll on Nootka's health, and she struggled to recover. Less than a month later, she appeared sluggish in the water and her condition began to worsen. Vets prepared the forty-six-hundred-pound female for surgery, but she died during pre-op examination. SeaWorld VP and curator Frank Murru said blood samples showed some type of infection. He was not sure if the animal's death was related to the stillbirth.

A columnist for the *Boca Raton Beacon,* noting Nootka's loss of two calves in a row, asked if SeaWorld was perhaps "overlooking the complications of a broken heart." Nootka IV was approximately thirteen years old.

Two months after Nootka's death, her former tankmate from SeaLand, Haida II, who was now at SeaWorld San Antonio, gave birth to a seven-foot-long, four-hundred-pound female calf after less than two hours of labor—record speed for a SeaWorld killer whale.[5] The father was Kotar, the bull who was exiled from Orlando after biting and severely injuring Kanduke's penis back in 1987. This was the tenth calf born at a SeaWorld park, and the third in Texas. Glenn Young, the park's general curator, called the new arrival "another tremendous opportunity to learn more about this fascinating species of marine mammal."[6]

For the first three weeks or so the calf seemed to be doing fine, swimming and nursing with her mother, and even playing around with her half brother, Kyuquot, whom Haida had given birth to on Christmas Eve 1991 at SeaLand. ("Ky" was Tilikum's first calf.) But in late December, the infant female began showing signs of breathing troubles. She died soon after that, of pneumonia.[7] Unfortunately, the world got to learn little about this mammal, despite the rosy predictions.

SeaWorld claimed that Haida was "unaffected" by the loss of her calf. She was eating normally and otherwise showing no signs of stress. As for the father, Kotar, it was not known whether male killer whales recognized their own offspring. Indeed, Kotar seemed more interested in playing with the metal gates that separated the killer whale pools than paying attention to a baby whale. Jeff remembered this habit from when Kotar was in Orlando. He was obsessed with mouthing and tugging on the metal bars.

Three months after the death of his calf, on April 1, 1995, a gate that Kotar was playing with closed onto his head. It crushed his skull and he quickly bled to death.[8] SeaWorld listed the cause as "acute hemorrhagic pneumonia" in the Marine Mammal Inventory Report.

Jeff was unsettled by the many deaths at SeaWorld. These animals were being exploited for profit, and they were paying with their lives. But that wasn't the only exploitation going on at work, in his opinion. Jeff was now a senior trainer, one of the leading performers in the killer whale show and a seven-year veteran. But he was still only making about $15 an hour. Jeff had helped SeaWorld put millions of dollars into the pockets of Anheuser-Busch executives and stockholders every year. He was rewarded with about $31,000 a year in return.

In 1995, Jeff was ordered to perform as a stunt double for a television series being filmed on the Orlando campus. He did not get paid extra for it—it was part of his job—but it was kind of fun to do. The show, called *Out of the Blue,* was a teenage sitcom featuring young Latin American actors speaking in both English and Spanish. Many of the characters worked at SeaWorld, including Charlie, the orca trainer, who was played by the actor Carlos Conde. The camera crew would film close-ups of Conde at the stadium, getting ready to do a show. Then, when it came time for "Charlie" to get in the water and do stand-ons and rocket hops with a whale, Jeff would step in to play the role. The producers said he had the most "Latin-looking features" of all the male Shamu trainers.

During breaks in between shots, Jeff would sometimes sit down and shoot the breeze with Carlos. Inevitably talk turned to salaries. The young Screen Actors Guild member, who arrived for two hours to shoot a brief scene, earned $650. He got $250 just for showing up, plus $200 per hour of filming. In those same two hours, Jeff had earned $30, even though he was doing just as much "performing" as Carlos. Jeff was deeply bothered by the inequity. He began discussing it with friends at work, and even with one of his supervisors.

"I'm the one flying through the air, busting my ass, shoveling fish, and risking injury and death for fifteen bucks an hour and Carlos shows up and makes six hundred and fifty dollars?" Jeff told one of his bosses, who said nothing and looked glum. "This whole situation has got me thinking how poorly paid we really are."

Jeff casually mentioned to friends at work that it would be nice to have some type of union representing trainers. But everyone knew that would never happen, including Jeff. SeaWorld viewed its trainers as expendable parts, he thought, and politically found collective bargaining an anathema.

Within weeks after grumbling about trainer salaries and safety hazards to his supervisor, Jeff was ordered to take a drug test. He passed, but the incident stuck in his mind. Jeff had never been subjected to drug testing before—not even when he got hired. Something just didn't add up right.

Then there were the Taima incidents.

Taima's intelligence was noteworthy. She was quick to learn new behaviors and how to interact with trainers, but she would also find clever ways to misbehave and amuse herself. Taima was rambunctious. Like her father, Kanduke, and like Tilikum, she was difficult to control at times, freethinking and experimental. In Orlando, she was Tilikum's best friend. Only senior people worked with her and only senior trainers got into the water with her.

Jeff's first misdemeanor with Taima came in October 1995. Taima had regularly been "gaming" during the foot-push segment of the show. When presented with the signal for a foot push, she would occasionally refuse. Sometimes she would begin the maneuver and break off before it was completed. Sometimes she would use the anatomical equivalent of her lips (the soft tissue around her jaws) to separate the outer sock from the inner sock on Jeff's foot. Other times she would remind Jeff of her presence by nibbling on his toes.

On October 27, Taima broke from control in the middle of a foot push. She rolled away from his feet with her mouth open. Jeff thought she had spotted some low-hovering seagulls and was going off to goof around with them. When an orca breaks from control, things can escalate into a dangerous situation. Jeff's trained response was not to react to Taima in this situation. He delivered a neutral response—three seconds of doing nothing—then regained control and finished the segment without incident.

But two of his managers didn't agree with Jeff's judgment in the pool. Thad Lacinak and Chuck Tompkins pulled Jeff into a closed-door meeting at Shamu Stadium. They were not pleased. "First of all, you can't trust the animals, you can't get too comfortable with them, but you did," Thad told him. "You should have exited the pool the second Taima broke from control." Jeff was going to be written up in the incident log, they told him. The slipup resulted in a three-day suspension without pay.

Jeff was taken aback by the reprimand. Nothing had gone wrong. Nonetheless, Jeff had clearly fallen out of favor with Thad and Chuck. First came the "random" and surprising drug test, and now this bogus charge about Taima. Jeff figured that management was probably unhappy with his grumblings about low trainer wages and his talk of organized labor. He had also continued his friendship with Astrid, had an open invitation to visit the Center for Whale Research, and had subversively questioned SeaWorld claim's about the health, longevity, and mental stability of the killer whales.

On the other hand, Jeff was still a star performer and valuable staff member. His file was brimming with positive reviews from the public, and from his superiors. On April 3, 1995, for example, a guest had called in just to say he was "very happy with the trainer 'Jeff' and said that, in the 50 or so shows he had seen, this was the best and that 'Jeff' was excellent, and wanted to call and let us know of his pleasure," according to the Guest Services Report that was forwarded to Chuck Tompkins.

"GREAT job Jeff—Chuck," Tompkins wrote on the copy sent to Jeff.

And on May 10, Shamu Stadium had received a 99 percent rating from surveys completed by guests who watched Jeff and Mark Simmons perform together in the hotdogging segment. It was the highest show rating to date at any Anheuser-Busch theme park, Jeff and Mark were told. The two men were a big reason why.

This time, Thad delivered the accolades. "Congrats on getting 99 percent for the categories on the shows and attraction ratings on the May surveys," he wrote to Jeff. "Your contribution has helped trigger a tremendous amount of cooperative energy toward the goals at Shamu. You're so close to perfection—100 percent in the ratings—and I know you can do it. Go for it!"

And in late August, Jeff led a guided VIP tour at Shamu Stadium, earning him a rave review from the guest services staffer who was there that day. "He gave us a tour of the backstage areas—he took time to answer the many questions of these curious guests and their children," the memo said of Jeff. "He was patient with the children and impressed the adults. I believe that his presentation was a highlight to our guests' day. Please extend to him my thanks for a job well done."

Jeff figured everything would blow over. Then one day in December he was performing with Taima when, at the end of the performance, instead of the normal closing routine, Jeff impulsively stuck his entire head inside her mouth and planted a kiss on her big, pink tongue.

The endearing scene was caught by one of the camera operators and projected onto the JumboTron. Jeff had no idea it would end up in a demo tape the camera operator was putting together for her portfolio. Little did he know his playful act would come back to haunt him.

A couple of weeks later, Jeff learned he was going to appear in a special promotional video that SeaWorld was producing. The company wanted to film a trainer doing a stand-on—when he rose straight out of the water perched on a whale's rostrum. But because of the desired camera angle—up close and shot from below—Jeff would need to perform the behavior near the Plexiglas wall.

Stand-ons were done in the middle of the pool because the whales did not always rise from the water perfectly perpendicular. Sometimes they leaned back a little bit; sometimes they leaned forward—sending the trainer flying in one direction or the other. There needed to be plenty of water for the trainer to land in. Doing a stand-on while hugging the wall was extremely difficult and perilous. It took Jeff several takes until the directors were happy with the shot.

Thad showed up for the filming. Jeff could see him as he worked from the water. But Thad wouldn't make eye contact. Jeff thought it strange. Something about it was ominous, like jury members who won't look a defendant in the eye. Jeff completed the series of stand-ons. The crew got what they wanted. He got out of the pool and dried off.

Once again, he was called into a closed-door meeting with Thad and Chuck. The men were sour-faced. They showed Jeff video footage of his big, fat kiss with Taima.

Thad and Chuck told Jeff he had acted unsafely and recklessly—again. He had violated the park's "no tongue-tactile" rule. This time, he would be subject to a full-week suspension without pay, plus a formal investigation.

Jeff was outraged. Yes, he knew about the tongue-tactile rule, and he knew he had got carried away that night. But lots of people did tongue-tactile during shows—it just wasn't that big a deal. Besides, management had no problem telling him to do all those dangerous stand-ons near the wall for the promotional video. He was obviously a safe and talented trainer and a crowd favorite.

A week later, when Jeff returned to work, he was fired. Animal training curator Robin Friday drove him around SeaWorld in a golf cart as he collected his things, signed a bunch of forms, and said good-bye to many people he had known for a long time.

"This is total bullshit," Jeff told John that night. "I don't think anyone has ever been disciplined for tongue-tactile."

With modest savings and no income, Jeff applied for unemployment benefits: a few hundred dollars every two weeks. The checks came in the mail for several months, until SeaWorld contested his benefits. At the unemployment arbitration hearing, Chuck Tompkins and the human resources director argued that Jeff had been an unsafe trainer. He was fired with cause; it was his own fault, Tompkins said. The arbitrator knew nothing about killer whale training. Jeff lost and was ordered to pay several thousand dollars back.

It was not the end of Jeff's sparring with his former employer.

22

Catch and Release

Over the years, more than a hundred dolphins—and a smaller number of other cetaceans—have been released back into the wild after spending varying periods in captivity. In the earliest days of reintroduction, animals were taken from public display tanks and essentially dropped into bays and inlets in waters nearby. Many of these dolphin rejects were no longer of use to their owners. They were not monitored after release and no survival data exist on these earlier cases.

For anti-captivity forces fighting for Keiko's freedom, demonstrating that successful release could be accomplished was crucial. When Ken Balcomb prepared his scientific proposal to Reino Aventura for the rehabilitation and eventual release of Keiko, he included a long list of dolphins and whales that had been held in captivity but were later set free or escaped.

Much of the research was contributed by Ken's half brother, Howard Garrett. Howard went on to create the Orca Network, a nonprofit whale conservation group, which he now ran with his wife, Susan Berta, out of their home on Whidbey Island, not far from Penn Cove, site of the infamous 1970 roundup of the entire Southern Resident orca community. Howard had met Susan in 1995 when she was the volunteer coordinator for Beach Watchers, a local environmental program.

Most of the animals on the release list were held temporarily, though several dolphins were kept for a number of years—sixteen of them spent more time in captivity than in the wild before being reintroduced into the ocean and reliably reported to be alive some time afterward.

As for the killer whales, Ken wrote, "All reintroductions can be considered successful." He then listed those freed animals. Again, the majority of the orcas taken and released were rounded up in the early cowboy days in Washington and British Columbia. Most were held in netted pens for days or weeks, and in some cases months. A few were held for a year or longer. None had been held for anywhere near as long as Keiko.

Several mainstream scientists, especially those affiliated with the display industry, ridiculed Ken's assertion that these cases proved Keiko could be set free. They mocked his proposal to "swim" the animal from the Bahamas to Iceland and reminded the media that the Icelandic government—beginning with the sale of Tilikum back in 1991—had issued several statements repeating its ban on captive killer whales returning to its territorial waters.

Ken was still unbowed.

In his proposal to Reino Aventura, he conceded that a "major point of contention" in the red-hot debate over cetacean returns was whether the animals could readapt to catching live prey after being "fed piecemeal in prolonged captivity." Another battle raged over the risk of formerly captive animals introducing pathogens into wild communities and whether freed cetaceans would have sufficient immunity to fend off natural diseases in the ocean. Yet a third dispute was being fought over whether released cetaceans could reintegrate with their family and pod or "be condemned to a life of loneliness."

Ken countered that all three were possible, especially following proper treatment, training, and acoustical and DNA research on Icelandic orca pods to locate Keiko's relatives. Besides, it wasn't as if the whale center had drafted a plan to "dump Keiko in the ocean and see what happens," as critics suggested, he said. But he warned his opponents, "We will continue to make it clear that his fate shall not be biased toward further captivity. We want to bring him to a condition of health in which he can comfortably live in the ocean, and then let him decide what to do."

Many critics were already making dire predictions about Keiko's certain demise should his release be attempted. Some said he would die within a month. But if postcaptive release was so "lethal, dangerous and irresponsible," Ken asked in his proposal, "then why has it been done so many times by organizations that are generally considered responsible?" None of the released cetaceans were known to spread diseases to wild animals. When Keiko was brought to Mexico from Canada, he did not sicken the dolphins at Reino Aventura, not even with his unsightly skin condition.

In addition to Keiko, at least two dozen other killer whales being held in captivity—about half the total—could be considered candidates for reintroduction, Ken stated, adding that most wild animals of all species return to nature when given the chance. "We think that probably any captive killer whale would go back to the wild, even to an uncertain fate," he said. "In this particular case, we don't know what Keiko wants, but we propose that at least his needs be guaranteed and that he have some control over the options for his future."

SeaWorld officials dismissed the Center for Whale Research's quixotic dream of freeing the real "Willy." They made it clear that their intervention in Mexico to improve Keiko's water quality and temperature meant that *they* would help determine the whale's fate, and not that troublemaker on San Juan Island, Ken Balcomb, or anyone else in the opposing camp.

After all, SeaWorld could rightly claim that it had stepped in and saved Keiko's life.

In January of 1994, SeaWorld announced that its efforts to stabilize Keiko's living conditions had cost more than $200,000 in equipment and services. The improvements were "part of a responsible and comprehensive plan to help Keiko," a company statement claimed.

But part of the plan to "help" Keiko meant shipping him to another marine park, where he would "benefit from medical treatment, a more suitable facility and the companionship of other whales." After Keiko had spent six months to a year in his newly chilled, fifty-five-degree pool in Mexico, he "could be evaluated to see if he can be moved to another marine park."

But he would not be released. The character of Willy may have been "returned to the sea and reunited with his family," SeaWorld said, but "we do not believe that scenario is in Keiko's best interest." SeaWorld experts and "independent scientists around the world" had studied the feasibility of releasing long-held captive animals and concluded that "such a proposal is ill-advised and would most likely result in the whale's death."

Ken was not interested in SeaWorld's opinion, and he fundamentally disputed the conclusion of the independent scientists the company had consulted. Instead of being cowed by their bluster, he got on a plane and spent two weeks in Iceland to start the daunting task of identifying Keiko's family via DNA samples and pod vocalizations.

That same month, the Keiko saga took yet another unexpected detour. Ken was contacted by Craig McCaw, a high-tech-entrepreneur billionaire from Washington State. He and his wife, Wendy, had read Keiko's *Life* magazine profile and were inspired to help. McCaw and his brother had just sold their cell phone megabusiness to AT&T for a staggering $11.5 billion. Now McCaw wanted to discuss financial support for Keiko's release plan. Ken went down to Kirkland, the affluent suburb northeast of Seattle, to meet with the billionaire.

Ken secured a commitment of $2.5 million.

The case for Keiko's freedom got another lucky break on March 7, 1994, when Brad Andrews told the *Chicago Tribune* that SeaWorld had ruled out bringing the whale to one of its parks. Perhaps the Miami Seaquarium might be interested, Andrews ventured, or maybe Hong Kong's Ocean Park aquarium. But Keiko would never come to SeaWorld. Industry observers speculated that Keiko's warts had scared off SeaWorld—even if the condition wasn't contagious, the lesions were not something people wanted to see in a show. Reino Aventura was stuck with a high-profile whale that it was not able to care for. Keiko needed saving.

Naomi was following the drama intently. In May of 1994, Ken returned to Mexico City to present his finalized plan for Keiko's release, now backed with the $2.5 million commitment from Craig and Wendy McCaw. Oscar Porter, the park's director, rejected the proposal. For now at least, Keiko would stay in Mexico City. Sadly for the whale, his best friend, the bottlenose Ritchie, had to be separated from Keiko by a glass partition in the pool because the dolphin could not survive the cold water. Keiko spent much of his time making dolphin vocalizations and staring at Ritchie through the glass.

Ken was appalled and disappointed. He, his half brother, Howard, and the staff at the Center for Whale Research had fought hard. They had pulled together a comprehensive proposal and secured impressive amounts of funding. But they couldn't win the release of Keiko.

The idea, however, lived on. In June, Mexico's *Proceso* newspaper reported that Reino Aventura was getting more than a hundred letters a day urging the company to approve Keiko's release to the wild. Letters also began pouring in to US companies, and to the Alliance of Marine Mammal Parks and Aquariums, including one from Nancy Bendsten of Pittsburgh. Her letter was accompanied by a petition demanding an end to whale and dolphin captures for public display, and calling for the release of Keiko. Bendsten added that whale watching on the ocean was a far superior way to observe the animals—for people and orcas alike—than in artificial tanks.

But Alliance executive director Marilee Keefe wrote back to Bendsten, who forwarded a copy to Naomi.

Keefe suggested whale watching in the wild was an elitist activity that only affluent people could enjoy. "Each year, more than 30 million people come to the marine life parks and aquariums that make up the Alliance," Keefe wrote. "Many of these people cannot afford to get to the coast to have the opportunity to go whale watching, and common sense suggests that the impact of closing public display facilities and urging these millions of people to whale watch each year would be dangerous to the animals and their environments."

Naomi found the affordability argument to be preposterous. To begin with, while it did indeed cost money to travel to the coast, once there one could view whales for free from the shoreline at many points around North America. Second, if a family of four wanted to see killer whales at a US theme park, they would still have to travel to California, Florida, Texas, or Ohio, where they would have to lay out well over $100 just to get in, plus parking, meals, snacks, dolphin food, orca plush toys, and Shamu T-shirts.

"The Alliance is as concerned about Keiko's future as are you. Our members have spent thousands of dollars trying to improve his health and environmental conditions." Release would be a "death sentence" for Keiko, who had been in an aquarium for at least twelve years and was "no longer capable of foraging for food in the wild." Keiko also had a skin disease, which Keefe said was "believed to be contagious."

The director went on to take a swipe at groups that "publish misinformation about the public display community," especially when it came to the debate over longevity. Published, peer-reviewed studies, she insisted, "document that marine mammals in aquariums live as long or longer as those in the wild." Keefe closed by asserting, "The irony is, I believe that without our programs caring by people such as yourself and the signers of your petition would wane and disappear. Your caring assures that these animals will be available for your children and grandchildren to admire and love, as does ours."

But neither HSUS nor the Earth Island Institute were giving up on winning the whale's release. Naomi decided to up the pressure, and to direct it to Hollywood. She knew how bad it would look for Warner Bros. to have used Keiko to make millions and millions of dollars worldwide, then to allow the real Willy to languish in a tank. In August she sent a mass e-mail on her CompuServe account to activists and animal protection groups around the country.

"To everyone interested in Keiko the whale," she wrote, "what is needed now is letters and calls to Warner Brothers to persuade them to try the release plan. It may not work, but Keiko will still be better off in a sea pen in Iceland (where, after all, he was born) than just another concrete tank, which is what the Alliance is proposing. The Alliance does not want to see a captive killer whale go free—people might start demanding that all captive killer whales who were caught in the wild be allowed to go home."

The effort paid off. By early fall, Earth Island Institute had entered into talks with the Oregon Coast Aquarium to negotiate the possible transfer of Keiko to a specially designed tank that would be built to facilitate his rehabilitation. On November 21, 1994, Earth Island Institute announced the creation of the Free Willy-Keiko Foundation with $2 million in seed money from Warner Bros. and New Regency Films, the production company.[1] The foundation's mission was to raise money and assist in the "care, treatment and potential future release" of Keiko. Meanwhile, schoolchildren were raising money to free the famous whale, including $31,000 that was collected at just one elementary school in Tampa, Florida.

Two months later, the Free Keiko forces won the big victory they had been waiting for. Reino Aventura agreed to donate Keiko to the Free Willy-Keiko Foundation. The foundation also announced plans to build a new $7.3 million rehabilitation facility at the Oregon Coast Aquarium, with the intention of eventually returning Keiko to the North Atlantic. Most of the money would come from Warner Bros. and the McCaws.[2]

HSUS also took the extraordinary step of contributing $1 million of its own funds to the project. It was a major victory for Naomi, and for Keiko and his supporters. But the Keiko war was for from over.

In 1993 a new Internet e-mail group called MARMAM had been formed through the University of Victoria—funded for some years with a $500 annual grant from HSUS. It quickly became the electronic soapbox for all things pertaining to marine mammal science, policy, and politics.[3]

When it came to captivity, list members ran the gamut from pro-cap animal trainers to people such as Naomi who were fighting to end the need for trainers at all. She had a number of pitched arguments, including over the proposed release of Keiko. One of her most biting—though respectful—critics on MARMAM was a British trainer named John Dinely. On July 8, 1995, for example, he complained that no research had been done "regarding the release of Icelandic orcas back to their area of capture." As a result, "such releases would be an experiment with an unknown outcome. I cannot see how this can be presented as a viable option that will enhance the INDIVIDUAL welfare of the captive animals targeted by various groups."

Naomi took her time to craft a response. She knew that hundreds of people in science, industry, animal activism, and marine environmental protection would be reading it. She offered an eloquent defense to the many critics who were online:

> *Although to some extent I concur with Mr. Dineley that the risks involved in a precedent-setting, experimental return-to-the-wild for a killer whale do not entirely support the goal of enhancing the individual welfare of the experimental subject (in this case, we are speaking of Keiko), I do not think these risks are prohibitive or unreasonable and frankly, the first steps must be taken, risks and all, or no benefit will ever accrue to that first or any other whale.*
>
> *Since I firmly believe (and history and the odds support me) that Keiko will not live much longer in captivity, the risks involved in rehabilitating him and at the least retiring him in a natural sea-water pen in Iceland (to emphasize a point that seems frequently to be lost in this debate, release is not a rigid goal for most of those who support it as a concept) seem balanced and in Keiko's best interests. In the same sense that a terminally ill human submitting to an experimental drug treatment that may cause an even earlier death, but then again may give the person several more years, seems worth the risk to many people, submitting Keiko to an experimental rehabilitation and release program seems worth the risk to those who support such a program for him—yes, he may die (for any number of reasons—and he will die sooner than later anyway), but he also may retire in a sea pen in Iceland and live a*

few more years there, and better yet, he just may successfully return to his pod for the remainder of his life. . . .

I find the argument that an experimental rehab and release project, whose primary goal is enhancement of the individual whale's welfare, should not proceed because there are risks involved to the individual whale to be an unreasonable, Catch-22 kind of argument. I also find it incongruous coming from an industry that didn't consider it a valid argument thirty years ago when captivity for killer whales was an experiment (and arguably still is) and the risks of being brought INTO captivity to the individual welfare of all the animals captured from the wild then were considerable (and in many cases promptly fatal).

It had been another long and trying year for Naomi, but fortunately it seemed to be ending on a positive note. By December 1995 construction on Keiko's new, high-tech tank by the sea in Newport, Oregon, had been completed, with its oscillating water jets, natural-looking rubbing rocks, and underwater viewing windows all artfully installed. The pool was nearly four times larger than his cramped quarters in Mexico and featured clean, cold ocean water pumped in fresh from the Pacific. It was an ideal environment to restore his health, energy, and appetite. Perhaps the seawater would heal his skin. Given the opportunity, he could be trained to catch live fish again. Within a year or two, Keiko might be ready to make his final journey home to Iceland.

On January 7, 1996, residents of Mexico City lined the streets by the thousands to bid a tearful farewell to Keiko as he was driven to the airport. United Parcel Service flew the famous whale from Mexico to the Oregon Coast Aquarium, without charge, aboard a C-130 Hercules transport. Once inside his specially designed traveling container, whale and water weighed in at about forty-two thousand pounds.

Keiko was greeted by equally adoring crowds in his multimillion-dollar home in the picturesque town of Newport. He seemed to take to his new environment immediately. Within just five days, aquarium officials announced joyfully, Keiko's appetite had doubled from a hundred pounds of fish to two hundred pounds. He had also grown more energetic, they reported. Naomi was thrilled to hear the news. No matter what happened, she knew Keiko was already in a much better place than he had ever been since he was snatched from his native waters in Iceland.

But a fairy-tale ending was still far from guaranteed. The government of Iceland, for one, was hardly enthralled by Keiko's taking one step closer to home. Officials in Reykjavík restated their prior decision that Keiko could not be released in Iceland because of fears he would transmit diseases to wild whales.

Critics of the project increased in number and acrimony, even though Keiko was clearly on the road to recovering his health and strength. The MARMAM list was ablaze with division as people fought over the merits of the project. Naomi and HSUS were frequent targets. Most people complained about the huge costs involved, and how that money could have been spent saving thousands of animals, rather than just one. The point was not lost on Naomi.

A pointed critique, for example, was posted by Timothy Desmond, the animal trainer who'd coached Keiko for *Free Willy,* under the subject head "Keiko—a blown opportunity."

Desmond was following the debate over "spending $10.5 million on a sick, but famous killer whale who is a terrible candidate for release by any reasonable measure." The campaign was "a terribly misguided bit of conservation and/or animal welfare activism. The money could and should have been put to much better use."

Desmond had been hired by *Free Willy*'s producers to plan a facility that was originally proposed for Keiko on Cape Cod. Keiko should have gone there instead, he insisted. The Massachusetts site would have doubled as a "critical care facility for the hundreds of marine mammals that strand every year on the northeast coast of North America," while also serving as a regional marine research center.

That plan had enjoyed local and federal support, and from "several universities, aquariums, several moderate animal welfare organizations, and many potential private donors," Desmond continued. "However, the animal welfare activists couldn't stand not having complete control over the political agenda and the money." Instead, they "conducted a coordinated attack on that center and killed it.

"So now, Keiko goes to guess what? A captive display facility—where there are very few cetacean strandings. Many, many marine mammals will die because of the lack of that critical care facility while Earth Island will float the illusion of releasing Keiko as a long-standing fundraising mechanism for their agenda."

Groups such as Earth Island and HSUS had reached the point "where they will start killing more animals than they are saving. In this case, hundreds of nameless whales and dolphins will die on the shores of New England over the next several years because the $10 million to save Keiko was not made available to them at the same time."

Naomi thought it was a stretch to blame those deaths on the Keiko project. Besides, the money had been donated to free Keiko. It's not as if the campaign could simply sign over the millions in donated funds to another project. It was a harsh reality: no Keiko, no cash.

23

Tilly's Willy

One of John Jett's biggest regrets in his life was that on that December day in 1995 when his best friend, Jeff, got fired from SeaWorld, he did not walk out with him in solidarity.

"These are totally trumped-up charges," John agreed with his buddy. "Nobody has ever been fired for a tongue-tactile before, as far as I know." However, John decided to stay at SeaWorld; he had no money and nowhere else to go. But John knew his days at the park would not last long. He had become as cynical as Jeff about killer whales in captivity.

"I wish I could go with you," John said to Jeff. He was disillusioned with the whole operation, with watching the animals getting chased and beaten up, with the teeth drilling, the chronic infections, the small pools. Tilikum's tail flukes dragged along the bottom in all but two of the tanks. Most of the orcas sat in the enclosures for much of the day with nothing to do but get bored. John had learned through Jeff, Astrid, and other sources about wild killer whales and their natural behavior. He was aware of the disconnect between what happens in the wild and what was going on in captivity. John was sickened by the living conditions of the orcas and put off by the urine- and feces-smelling backstage area at Sea Lion and Otter.

"It's just not working out for me," he told Jeff a few weeks after his co-worker was fired. "I can't continue to ignore how I feel about it."

John's unease toward SeaWorld was about to get worse. Chuck Tompkins called him into his office one day to inform John of a new assignment. As Tilikum's team leader, John was to begin teaching Tilly approximations—small, discrete training steps—to present his penis to trainers. After Tilly learned to do that correctly, John and his team were supposed to masturbate the phallus, collect the semen, and freeze it for use in SeaWorld's new artificial insemination (AI) program.

John was shocked and disgusted. "Sorry, Chuck," he said. "I'm not going to put Tilikum through that. Get someone else." For that insubordination, John was banished from Shamu and transferred to Sea Lion and Otter Stadium. But John was already on his way out the door, psychologically speaking.

Working with these smaller animals sometimes seemed more dangerous than working with the killer whales. John was bitten by an otter—a painful laceration on his leg—and often became uneasy around the other animals. Sea lions are smart, cunning, and surprisingly mobile on land. If they get mad, they can dash across the stage and bite somebody. Walruses are more dangerous—almost as quick on land as sea lions, they can easily push people against a wall and crush them with their heft and power.

John did not especially like working with pinnipeds, but at least it was less painful for him than going to Shamu Stadium every day had been. The pinnipeds were extraordinarily intelligent, although John got the sense they either didn't realize or didn't care that they were enclosed behind walls. But John did feel that the killer whales *knew* they were in captivity, even those that were born at SeaWorld. They fully understood their predicament.

"When you get to know an animal like a killer whale," he told a friend, "you know that somebody's home. And when they look at you, and you look at them, and they're like 'Yeah, man, this is pretty fucked-up'—well, it just breaks your heart. Going to work every single day had become agony for me."

Soon after John was kicked out of Shamu Stadium for refusing to masturbate Tilikum, Gudrun went into labor with yet another calf sired by the Icelandic bull.

John and Jeff had previously discussed Gudrun's pregnancy since she had begun eating less food in the final weeks of gestation. They were also concerned about the health of the fetus.

With the completion of G Pool, SeaWorld had figured out a way to use Gudrun to generate new revenue streams. As one of its rare adult orcas whose dorsal fin had not bent over or collapsed, she was perfect to pose with tourists for photographs. Several times a day, SeaWorld would bring in tours of people to wait their turn in line for a picture shot by a professional photographer and sold to them at a premium. The pregnant whale would remain "dry" in the slide-out area as she held her pose for many minutes at a time. The weight on her unborn calf must have been immense.

When Gudrun went into labor, the staff veterinarians could not get a pulse on the unborn calf. It was presumed dead. Since Gudrun was not expelling the calf, they needed to pull it from her. Gudrun was taken into the med pool, which was drained to immobilize her. They put a cable up her vagina and wrapped it around the flukes of the calf.[1] The dead infant was reportedly pulled out manually by the animal care team.

The pain must have been unearthly. Gudrun began to hemorrhage severely. Her dorsal fin collapsed, probably due to dehydration. She refused to eat and ignored all attempts by people to make contact with her. She remained motionless in one spot, unprotected by shade, so staff lovingly lavished her back with zinc oxide. After the bleeding stopped, Gudrun stayed that way for four days as her worried caretakers did all they could to nurse her back to health.

On the fourth day, Gudrun finally moved. She slowly swam over to the gate where her disabled young calf, Nyar, was watching. Nyar had had to be separated from Gudrun after the mother began attacking her daughter. Now, Gudrun gently nudged Nyar's rostrum through the bars, as if to ask for an overdue rapprochement. Gudrun died a few hours later.[2]

John was devastated. Astrid took the news particularly hard. As for the brain-damaged Nyar, her own fate was uncertain. She still listed to one side when she swam. Her food was stuffed with drugs every morning. She could not perform, could never breed, and was of no use to SeaWorld. Still, a special four-person team was assigned to exclusively look after all of Nyar's "special needs." Some people speculated that management was basically waiting for the sad little calf to die.

On April 2, 1996, John got a phone call from his friend Lindsay Rubicam. She was sobbing. John knew what it was about. "Nyar died," Lindsay told him.

Chuck Tompkins told the Associated Press that Nyar had "very obvious physical problems," including poor motor skills and learning disabilities. "In a human, these could be likened to autism or something like that," he said. "We filled the role of her family."

That was it for John Jett.

The next day he presented his letter of resignation to Robin Friday, walked out the main gate, and never went back.

24

Better Days

Naomi Rose's first two years at HSUS had been a wild ride, a trial by fire played out on a steep learning curve. First she had butchered the statistics in her "case against" report by including infant deaths in captivity but not in the wild. She'd also committed the rookie error of comparing the overall percentage of orca deaths in nature with those in captivity, without correcting for age and then treating them like comparable "mortality rates."

Then there was the MMPA drubbing, when the public display industry got almost everything it wanted in the reauthorization and the marine mammal coalition got almost nothing. That was followed by the nail-biting roller-coaster ride of trying to get Keiko away from Mexico and keep him out of another marine park, accompanied by the slings and arrows hurled at her on the MARMAM list for thinking that anyone could free "Willy" in the first place.

With those defeats already notched into her belt, Naomi was thirsting for a victory on the captivity front. Despite the setbacks, she was even more determined to prove to her scientific colleagues, the industry, the media, and the world at large that keeping killer whales in captivity was unethical, indefensible, and hazardous to both animals *and* their trainers.

Then she got a break.

A pair of scientists from NMFS, Robert Small and Douglas DeMaster, were working on an analysis of survival rates among captive marine mammals. They offered to send Naomi an advance copy of their paper. They explained to her why it was incorrect to calculate mortality rates the way she had in her original report. Simply comparing percentages of deaths in each population without accounting for age was misleading.

It was better, instead, to examine the annual survival rate (ASR, or percentage of animals that survived in a population from year to year); one could then compare the values between wild and captive populations. With the ASR method, if one took the total number of killer whales in a population, grouped by age, and calculated the total number of days they survived over any particular period, one could arrive at the annual survival rate for each age cohort, and for the population as a whole.

Naomi jumped at the chance to have an early look at the paper. What she read that day altered the course of her battle against the display industry.

Because many killer whales died in the first year of life, Small and DeMaster analyzed annual survival data for non-calves—in both captivity and the wild—from 1988 to 1992. Statistics on captive animals came from the Marine Mammal Inventory Report and wild survival data came from the landmark paper that Peter Olesiuk, Mike Bigg, and their colleagues had published on Resident orcas of the Pacific Northwest.

"Survival of the wild population . . . based on approximately 250 non-calves, was significantly higher than our estimates for non-calf captive whales," Small and DeMaster wrote.[1] They had found an annual survival rate of 0.938 among killer whales in captivity, meaning that 93.8 percent of the population survived from year to year. Among the wild whales, the ASR was 0.976—97.6 percent of those whales survived each year.

That might not seem like much of a difference, but if looked at inversely, the distinction was glaring. If 93.8 percent of the captive whales survived from year to year, then 6.2 percent of them died. By contrast, just 2.4 percent of the free-ranging whales died from year to year.

The evidence, when laid out this way, could not have been clearer. The annual mortality rate among non-calf captive killer whales was more than *two and a half times higher* (6.2 percent vs. 2.4 percent) than the rate among non-calf whales swimming in the ocean. Statistically speaking, the difference was highly significant, and Small and DeMaster put the figure out there for all to see in their 1995 paper.

DeMaster had published a similar study five years earlier in 1988, with Jeannie Drevenak of the US Marine Mammal Commission.[2] Since that time, annual survival rates for California sea lions and bottlenose dolphins had actually increased in captivity, likely due to improved husbandry practices. But captive killer whales enjoyed no such rise. Industry assurances aside, things were not getting better for killer whales in captivity.

Many advocates might have felt vindication—or even jubilation—to learn that hard science was backing up what they had been saying in public: Naomi's original HSUS report coincidently had reached the same conclusion as Small and DeMaster's peer-reviewed, published paper, even though her methodology was faulty. "Killer whales," she had written, "are more than 2.5 times as likely to die in captivity as in the wild."

Instead of pride, Naomi felt a sense of urgency—and renewed energy—from Small and DeMaster's data. The war she had engaged in was predicated mostly on the question of survival. By SeaWorld's logic, all marine mammals should be thriving in captivity, and living even longer than in the wild, given the outstanding food, protection from predators and pollution, and medical care they enjoyed. But now Naomi had powerful evidence that the opposite was true.

Most important, she thought, if captive killer whales were dying off at 2.5 times the rate as their wild cousins, then how could captivity possibly be justified? (Later, Naomi would realize she had been naïve about the impact that Small and DeMaster's paper would have on scientific and public opinion toward captivity.)

Meanwhile, good news was breaking on the Keiko front as well. He seemed to take to his new high-tech, 2-million-gallon tank almost immediately. The pool included a closed-loop, ionization filtration system that would prevent any contamination of the ocean, while still providing Keiko with real seawater—the first time he had experienced it in fourteen years. He quickly began frolicking in the cold brine and explored the deeper depths of his new 2-million-gallon "halfway house" in Oregon.

Aquarium staff wasted no time in working with Keiko on his aerobic reconditioning, mental stimulation, and husbandry procedures—including regular blood sampling to monitor his health. Keiko began eating more and quickly put on pounds. His vigor returned and his cardiovascular health and muscle tone were soon on the rebound. Best of all, Keiko began playing again—his lethargy had lifted like a dark fog and he was now exerting energy of his own volition, without any commands from trainers.

"Within six months, we began to see a profound change in Keiko's energy level. And in his character in general—it was as though he rediscovered his own personality once he began to feel well," Beverly Hughes, president of the Free Willy-Keiko Foundation, told the media. "He is beginning to act like a killer whale." So far, she added, no one had observed anything that would "knock Keiko out of the running for release back to the wild."

Keiko was a smash hit among people in Oregon, who came by the thousands to see the famous whale. No longer required to perform, he seemed to love spending his free time staring at the humans—and especially the children—who lined up to look at him through the underwater viewing windows. Keiko had also become a consumer rather than an object of mass entertainment. He liked watching TV shows and movies on a donated television set in the staff office, which had a large picture window visible from his pool. Vets at the aquarium prescribed television to help him maintain mental stimulation after all his adoring fans had gone home for the night. Keiko was interested in videos of other orcas, but his favorite seemed to be *Monty Python and the Holy*

Grail, the only movie he watched in its entirety. He also showed interest in parts of *Blazing Saddles* and *The Lion King,* but reportedly turned his back on *Free Willy*.[3]

Officials began limiting Keiko's time in front of the television screen to prevent him from becoming "an aquarium potato," as the Associated Press put it in an April 1996 article, even though aquarium president Phyllis Bell said the odds of that happening were not great. "He loses interest; he doesn't just sit there and watch it the whole time," she said. "He's not like us."

Keiko continued to improve. By the end of 1996, the aquarium and the Free Willy-Keiko Foundation put out a joint press release celebrating Keiko's remarkable recovery in just one year. Not only had he gained a thousand pounds, nearly all of his skin lesions had cleared up.

The year had been a smashing success for the Oregon Coast Aquarium. Attendance blossomed after Keiko arrived: 1.3 million visitors came through the door in 1996, more than double the year before. Clearly Keiko had not lost his magnetism; he was still big business.

By May of 1997, Keiko's rehabilitation staff began releasing live fish into his tank regularly. At first, Keiko seemed to think he was playing a game of fetch. Instead of eating the fish, he would obediently return it to the trainers, even though that was the exact opposite of what they wanted him to do. Some people, such as Howard Garrett, said Keiko was behaving like any orca in nature would, by trying to share his food with others. Food sharing is a well-documented behavior among most killer whale populations.

Before long, however, Keiko caught and ate his first live fish. Encouraged by that, his team brought in two thousand live herring—the native staple for wild Icelandic orcas—to Keiko's tank. Some of the herring he ate, some he presented to his trainers, and the rest he simply ignored.

Keiko's weight continued to increase, and by summer of 1997 he clocked in at 9,620 pounds, an astonishing gain of nineteen hundred pounds (the equivalent of ten or eleven grown men) since arriving in Oregon.[4] Naomi was thrilled with his progress and delighted that the Free Willy-Keiko Foundation staff was now beginning discussions about relocating Keiko to a bay pen in the North Atlantic, sometime in 1998.

Foundation members were growing anxious about getting Keiko out of Oregon. They charged that aquarium staff had failed to change the sand in the filter for Keiko's closed-loop tank. The water was becoming rank and murky; Keiko was swimming in his own waste. By August 1997 he had contracted a respiratory infection and developed various parasites. Fortunately he responded well to treatment. Foundation staff replaced the filters and the tank water cleared up.

But tensions were clearly mounting between the aquarium and the foundation. In September, local media were receiving anonymous communiqués claiming that the aquarium was frantic over Keiko's health. One source implicated the Free Willy-Keiko Foundation for trying to conceal that Keiko had been sick. But the foundation fired right back, accusing aquarium staff of fencing off and padlocking the water filters to prevent them from being changed. Aquarium officials countered that it was the fault of the foundation, which they alleged had postponed the replacement because of financial constraints.[5]

Suddenly, things weren't looking so rosy along the Oregon coast. But the bickering only increased the foundation's determination to get Keiko back into the ocean soon. The next month, foundation officials announced that Keiko was nearly ready for release, something that alarmed the aquarium. Officials there called for an independent evaluation of the whale to see if he truly was suited for reintroduction.

Before long, supporters of the Keiko release program were accusing aquarium management of trying to prevent Keiko from getting the health clearance necessary to let him ever leave Oregon.

Despite the open warfare—and perhaps because of it—media fascination with the ongoing saga never waned. Keiko became the centerpiece of an important, groundbreaking one-hour special on the captive orca industry that was being produced for the PBS investigative-journalism series *Frontline*. The special, called "A Whale of a Business," included interviews with everyone involved in the long, bitter war over orcas in captivity—from SeaWorld's Brad Andrews on the pro side, to Naomi Rose and others on the far opposite end.

It was hardly a flattering portrait of the billion-dollar industry.

"*Frontline* examines the money, power, and politics behind the captive marine mammal industry," PBS announced before the show aired on November 11, 1997. Producer Renata Simone promised to "take a hard look at the industry behind the spectacle." Viewers, she said, would encounter "a war between activists who fervently believe these animals should be free and corporations like SeaWorld where entertainment, image, and sales are the objective." Caught in the middle was "a ten-thousand-pound whale named Keiko."

The show was narrated by reporter Linden MacIntyre. "The consequences of releasing Keiko into the wild could be dire for him, but the prospect of his freedom raises even greater alarms in the billion-dollar industry that brought him here in the first place," he began. "The marine zoo has become the hottest entertainment phenomenon since Disneyland, assembling the same elements of fantasy and kitsch. Nature, stripped of all that's wild and strange, becomes almost human in scale, behavior, and personality."[6]

A bit later, the host confronted SeaWorld's Brad Andrews, pressing him to concede that SeaWorld was little more than a "massive entertainment and amusement institution." Andrews was unfazed and unapologetic. There was nothing to be ashamed of. "Businesses that survive are businesses that make money," he said, stating the obvious. SeaWorld had no choice but to keep its visitors consistently amused. "You really have to impact that learning process on what they might learn about the environment, the animals, what they didn't know, but you have to do it in a very fun and entertaining way."

As *Frontline* pointed out, most of that fun and entertainment—and profit—was built squarely on the back of Shamu. John Hall, a former research director at SeaWorld, said the company had estimated that seventy cents on every dollar of total revenue was due to the presence of killer whales. The animals would be the driving force behind SeaWorld's unbridled success heading long into the future. "If you wanted a strong cash flow to continue, you needed a steady supply of killer whales, especially if you're expanding and building new parks," Hall said. "You need whales to fill them up."

Naomi took on the animal welfare issues posed by the whale business. "There's just no way that a facility can provide for these animals," she said. They had been stripped of their basic natural behaviors. "To put them into a concrete environment where there's simply no variety, no texture, no substance, no depth, to the environment, why even use their echolocation? They know where the four walls are."

The show did not shy away from the feud between the foundation and the Oregon Coast Aquarium. The general impression it left was of a desire, if not an outright effort, on the part of aquarium officials to keep the world-famous whale securely on their premises for his revenue stream.

Keiko had become a marketing phenomenon. "In his two years at the Oregon aquarium, there's been a booming trade in Keiko products, a variety of consumables limited only by the fertile marketing imagination," host MacIntyre said.

All that steady cash was alluring, and the aquarium was falling under its spell, Craig McCaw, the billionaire benefactor, complained to MacIntyre. "It is not in the aquarium's interest for Keiko to be free . . . there's a lot of money at stake," he said flatly. "And we have felt that perhaps the aquarium was not doing everything it could to possibly bring about the release of Keiko to his highest level."

Phyllis Bell, the aquarium director, hardly went out of her way to dispel viewers of that notion. "Well," she answered, "it depends on how—if Keiko's ready to be released or not. We always have supported their goal of releasing Keiko, if it was possible and if that was the best thing for Keiko."

But then she showed her hand. "So if the best thing is not releasing him, then he's welcome to stay here."

The PBS special ended with lingering question marks. Would Keiko ever go free? *Should* he? And who would continue to pay for this ridiculously high-priced undertaking?

"He may one day return to the wild, but the wild may prove to be a lot more difficult to get to than anybody thought, and a lot more expensive," MacIntyre said. But that would not be a problem for Keiko, he said, thanks to his "guardian angel, the billionaire Craig McCaw."

McCaw assured MacIntyre, "If we fall short, I'm sure I'll be able to find the money to make it happen. This will not be stopped for lack of money." (No one, least of all McCaw, foresaw the looming high-tech bubble burst that would soon erase enormous swaths of personal fortunes.)

Money questions aside, SeaWorld stood firm on the proposed release of Keiko. "Keiko is not a good candidate," Dr. Jim McBain, its director of veterinary medicine told *Frontline*. "He's been dependent upon humans for his food, his interaction. He's an animal that's adapted to living in an oceanarium environment and has done so successfully for many years." McBain said it made little sense to "try to somehow train this animal to then go and survive in the wild."

Brad Andrews, meanwhile, insisted that no animal belonging to SeaWorld would ever meet the same fate as Keiko: "We're not going to release any of the animals in our collection because they have been in our collection for long periods of time. We're not going to put them at risk where they can die."

When Naomi watched the program, she wanted to shout at the screen after Andrews spoke. If SeaWorld was truly opposed to putting killer whales "at risk where they can die," she thought, shaking her head, it would never have condemned them to captivity in the first place.

For now, she knew it was time to get this whale back to Iceland, where he belonged.

25

The Salish Sea

After SeaWorld terminated Jeff Ventre—for kissing the hybrid Taima on her tongue—he ramped up his efforts to finish the premed credits needed to get into graduate school. Jeff not only wanted out of animal show business, he wanted to leave Florida.

The ex-trainer was yearning to get to the Pacific Northwest. The year before he was fired, Astrid van Ginneken had offered Jeff a formal invitation to visit the Center for Whale Research and volunteer for the Orca Survey project, Ken Balcomb's long-running documentation effort, which had photo-identified every member of the Southern Resident orca community of Puget Sound and the San Juan Islands.

While still at SeaWorld, Jeff had hardly concealed his excitement at the thought of seeing wild killer whales. He boasted to colleagues at work about going to see the free-ranging orcas. It won him few allies in upper management. Jeff could not think of any member of the staff who had actually *seen* a killer whale in the ocean, except for a few senior officials such as Thad Lacinak, Chuck Tompkins, and SeaWorld Florida president Bill Davis.

After SeaWorld, Jeff moved into a three-bedroom suburban home just north of metro Orlando with John Jett and Mark Simmons. The house was on Peace Pipe Drive, adjacent to thousands of acres of pastures and pine forests waiting to be explored. Jeff, John, and Mark were all back in school. On several occasions, Jeff brought his buddies over to his parents' lakeside home near Oviedo, and Mark once gave Jeff's mom an exquisitely detailed pencil drawing he had done of a red-tailed hawk.

That June of 1996 Jeff headed off for adventure in Washington State, where he would spend the summer. In Seattle he was met at the airport by his old friend and colleague Carol Ray, who had quit SeaWorld back in 1990.

After she left, Carol had embarked on an exotic adventure across the South Pacific and returned to Florida eight months later without a clue as to what she wanted to do next. She took some graduate classes and discovered a love and talent for linguistics, especially speech and language pathology. Near the end of her graduate program, she secured two clinical internships near Seattle and fell in love with the place and its tremendous outdoor opportunities: rock climbing, mountaineering, mountain biking, snowboarding, and backpacking.

Carol and Jeff spent a few days together catching up, reminiscing, and going out on the town for live music, fresh Pacific seafood, and frosty mugs of local microbrewery beer. The highlight was driving over the North Cascades into central Washington and catching a live concert at the Gorge Amphitheatre. Jeff had always loved the Pacific Northwest, and now he remembered why.

Eventually, Carol drove Jeff to the scenic port town of Anacortes, about two hours north of Seattle, where Washington State ferries departed for the San Juan Islands. It being the beginning of the high summer season, the dock area was packed with cars carrying couples, families, dogs, kayaks, and great expectations for fun and natural beauty in one the prettiest corners of the planet.

The San Juans are an archipelago of three main islands and several smaller ones situated in the far northern stretch of Puget Sound. The remnant of a primordial continent unrelated to North America, the chain is separated from the US mainland and the Cascade Range to the east by Rosario Strait. San Juan is the largest and most populated island, followed by Orcas and Lopez Islands. The names derive from early Spanish explorers who arrived in the area in 1791. Orcas Island, ironically, is not named for killer whales, but for Juan Vicente de Güemes Padilla *Horcasitas* y Aguayo, Second Count of Revillagigedo, Spain.

The islands' beauty is legendary. San Juan Island is graced with meadows and farmland divided by thick stands of evergreen woods covering a few looming peaks. Protected by coastal mountains on three sides, its climate is drier and milder than that of nearby Seattle. The rocky coastlines are pocketed with dozens of coves and inlets, some with quaint villages clinging to their shores, with fishing and pleasure craft moored in the harbor. It is all reminiscent of Maine, save for the snow-covered peaks such as Mt. Baker and Mt. Rainier that rise from the horizon.

The islands were originally home to several Native American peoples belonging to an ethnolinguistic group called the Coast Salish (*SAY-lish*). The entire area, stretching north to Johnstone Strait and south to Olympia, Washington, is now officially recognized by the United States and Canada as the Salish Sea.

Jeff's ferry from Anacortes took about two hours, with a stop at Orcas Island, to navigate the sounds and winding channels that led to Friday Harbor, the main settlement on San Juan Island. He was invigorated by the cool Pacific air, the stunning scenery, and the abundance of wildlife: Porpoises, otters, seals, and sea lions populated the waters, while bald eagles and red foxes could be seen amid the trees and fields of the islands. Jeff fell in love with the Salish Sea before the ferry even landed.

From Friday Harbor it was about a twenty-minute drive past goat herds and expensive vacation homes to the northwest corner side of the island, between Smallpox Cove (the European disease wiped out many Salish people) and Smugglers Cove (named for its proximity to Canada and once-unguarded remoteness). There, on about an acre of cliffside property, stood Ken Balcomb's large but down-home scientific headquarters.

The two-story, cedar-shake house doubled as the Center for Whale Research. Out in the yard were trucks, boats, a barbecue area, and vintage navigational flags from around the world. Inside, the rooms were filled with whale-watching paraphernalia: binoculars, spotting scopes, telephoto cameras, and speakers hooked up to hydrophones to announce the arrival of an orca pod. Occupying a sizable chunk of the living room was a massive killer whale skull, a souvenir from Japan that Ken had acquired in the seventies while working as a navy acoustics analyst.

When Jeff walked into the place, he felt as though he'd somehow come home. Astrid greeted him with a huge smile and warm hug. She introduced him to the legendary Kenneth Balcomb, the cetologist with the bushy Kriss Kringle gray beard, about fifty-five years of age. The place was bustling with young people who'd signed up through the Earthwatch Institute—a nonprofit group that engaged volunteers in field research and educational projects pertaining to environmental sustainability. Some were busy developing black-and-white photographs of dorsal fins and saddle patches, while others compared the images with databases of the entire Southern Resident community, about eighty-five whales from J, K, and L pods.

Ken gave Jeff a firm handshake and led the newcomer out onto the wide wooden balcony that overlooked a calm green cove tangled with long strands of bull kelp, a favorite hangout for Southern Residents. In the summer, the entire western shore of San Juan Island is patrolled almost daily by J, K, and L pods, who travel up and down the coast, sometimes quite close to the cliffs, in what locals refer to as the Westside Shuffle.

Jeff was staggered by the view. Beyond the cove was Haro Strait, churning like a river with strings of frothy rip currents as the tidewaters raced toward the Pacific. Ten miles west across the channel, the green hills of Vancouver Island rose above the tidy homes of Victoria. To the south, across Juan de Fuca Strait, the towering stone peaks of Washington's Olympic Peninsula glinted in the June sun, their highest summits still dusted in winter white.

He was going to love it here, Jeff thought.

Ken's younger half brother, Howard Garrett, came out to join them. Howie was tall, lanky, affable, and as passionate about orcas as Ken. Both men loved killer whales and intensely disliked SeaWorld. "Shamu is an acronym for 'shame on you,'" Ken joked with a wicked grin. Jeff took to the half brothers right away. He and Howie quickly became close friends.

Ken regaled Jeff with story after story of wild whales. He explained how gregarious the Southern community was—more prone to greeting ceremonies and giant leaps from the water than their Northern cousins in Johnstone Strait. He talked about individual whales and the personalities that he knew so well after twenty years of tracking them. Ken and the orcas were so well acquainted that the older whales recognized the sound of his boat's motor and would often come pay a visit, especially proud mothers with their newborns.

"When they have a new calf, they actually bring it over to my boat," Ken said. "I think they're teaching their offspring that we are okay, they have nothing to fear from us. But I also think they're showing off the youngsters." Sometimes, the mothers would turn their children over, allowing Ken and his team to sex the calf at an early age (males and females have different black-and-white patterns around their genital slits, and females have mammary slits). "I don't know if they do it on purpose," Ken said with a sly grin, "but it sure is a big help."

The men stood in silence for a while. Jeff took in the water, sky, and mountains, and the bald eagles circling nearby. He could *feel* the majesty of the place as much as he could see it. After a few moments of quiet, Ken pointed to the bull kelp rocking gently in the cove, about twenty yards offshore.

"When I first moved here in the seventies, the whales used to stop by that kelp almost every afternoon in the summer," he said with melancholy nostalgia. "They would come into the cove, maybe a dozen at a time, and just loll around in that seaweed there for as long as they wanted. They'd be out there rubbing against each other and rolling in the water. It was just one big orgy. Nonstop sex."

"So what happened?" Jeff asked. "Don't they do that anymore?"

Howie answered, "Not as much. They don't have *time* for that kind of thing these days. They're too busy looking for salmon."

Howie and Ken explained how the strait once choked with migrating chinook in the summertime. But overfishing, pollution (especially PCBs from the Seattle area), and dams that impeded the adult salmon's return to spawn upriver were pushing the numbers down.

"Chinook are energy bars," Howie explained. "Back when the whales had good, full bellies and lots of fat reserves, and they didn't need to store any more up, they moved on to do other things that they naturally *prefer* to do.

When they got a chance, they'd stop and socialize and engage in a lot of sex play. It's just their normal attitude toward life, it seems."

Ken nodded his head. "If they could do that all the time, they would. But of course, they've got to make a living. And it's become a lot harder for them to do that, which leaves a lot less time for playing around in the kelp." The population was more or less stable, for now, he said. The Southern Residents were finding food somewhere. But it kept people up at night worrying about what would happen if the salmon population collapsed entirely.

"There's an old saying among the Salish people," Ken said. "No fish, no blackfish."

Both Ken and Howard had long and storied histories. They grew up with a third brother, Rick, just north of Sacramento, raised by a single mother who worked night and day, leaving the boys to their own devices. Ken spent most of his time out in the fields of central California, bringing home frogs, snakes, birds, and even abandoned lambs to take care of.

Ken went on to get his bachelor's degree in marine biology from UC Davis and worked at the last whaling station in the United States, in nearby Richmond, which did not close down until the early 1970s. Howie, after graduating from college, was drafted to fight in Vietnam. The left-leaning hippie-type passed a petition around at his draft physical in Oakland saying, "We, the undersigned, are against the war in Vietnam and object to being forced to fight in it." Howie got about a dozen signatures before they ejected him from the building "and straight to Canada," as he put it. After Canada he traveled through Europe and South Asia. When President Nixon offered conditional amnesty to draft evaders in 1972, Howie returned to the States, moved to New Mexico (where he was born), got married, and had a baby. The marriage did not last.

Howie moved to San Juan Island in December of 1980 to help Ken smooth over a tumultuous transition at the whale museum that Ken had founded in Friday Harbor. Howie spent the next two years assisting with field research and writing about new findings in cetology, especially the orca studies he was participating in. After a three-month research expedition to study humpbacks in the Caribbean, Howie ended up in Gloucester, Massachusetts, working for about a decade as a freelance naturalist and wildlife guide for whale-watching tours in New England. In June of 1993, he agreed to move back to Washington State to help Ken on the Orca Survey. Driving on the road west, just twenty miles outside Gloucester, Howie came across a movie theater that was screening *Free Willy*. He decided to stop to see the movie, as a form of inspiration before the long drive out West to see the real whales. (The movie's shots of wild orcas swimming in the ocean were filmed in the San Juan Islands.)

When Howie got to Friday Harbor, he discovered that Ken was already deeply involved with the plan to free Keiko. That's when Howie became an orca activist.

Two years later Howard started his own nonprofit called the Tokitae Foundation. Tokitae, which means "nice day" in the Coast Salish language, was the original name given to Lolita, the female killer whale locked in a small pool at the Miami Seaquarium in Florida. Tokitae was a member of the L25 matriline of L pod. Her mother was thought to be L25, Ocean Sun, who was now over eighty years of age and still going strong, still running her family. Tokitae emitted calls used only by the L25 subpod.

Tokitae was taken on August 8, 1970, during the infamous Penn Cove roundup led by partners Ted Griffin and Don Goldsberry. More than eighty members of the Southern community—possibly the entire population—had been captured, and seven of the youngest whales were sold to marine parks.[1] Tokitae was sent to Miami as a playmate for a young male named Hugo, who was taken from Puget Sound in 1968. Hugo was an anguished cetacean. He would often send high-pitched shrieks echoing across the park grounds and repeatedly banged his head on the tank wall. One time he actually broke a viewing window and sliced off the tip of his rostrum on a glass shard. The piece of flesh had to be stitched back on by a veterinarian.

Once in Miami, Tokitae's Salish name was switched for something a bit more Latin sounding for South Florida: Lolita. She and Hugo were probably closely related, though no one realized that at the time. The two Southern Residents lived and performed together in their cramped pool for ten years, until March 1980, when Hugo slammed his head into the wall for the last time. He was about fifteen. The Seaquarium listed the cause of death as a brain aneurysm. After he died, Lolita kept performing the show all alone.

By 1987, Lolita was the only survivor out of an estimated fifty-eight killer whales taken captive from Puget Sound or killed during captures. The same year, while attending the biennial conference of the Society for Marine Mammalogy, in Miami, Ken Balcomb approached the Seaquarium with a scientific proposal: Let Lolita listen to audio recordings from her pod, "just to see, and record, what might happen," he said. When the facility refused, Ken offered to lease the whale for the experiment, but received a similar rejection.

In 1992, after Hurricane Andrew flooded the Seaquarium, electrocuting six of its sea lions, Ken offered to buy Lolita outright. He had picked out a clover-shaped cove on the west side of San Juan Island, Kanaka Bay, used in 1976 to hold two orcas for two months prior to release as stipulated by a court order. It still had metal hooks in the rocks where nets were stretched to corral the orcas. It was the perfect location for a retirement pen, Ken told Seaquarium officials. When they once again refused to consider the offer, Ken's half brother Howie started his Tokitae Foundation.

Because of Lolita, and so many other whales now dead, Ken and Howie never forgave Griffin and Goldsberry for the havoc they wreaked upon the Southern community. J, K, and L pods were still trying to recover their numbers after so many family members had been taken from the Salish Sea.

Ken had been hired by NMFS back in 1976 to conduct the first-ever complete census of the Southern Resident community. Part of his assignment was to assess the impact that culling by SeaWorld and other facilities had on the stability and sustainability of that population.

But the following year, the government moved to quash his operations.

"It was insane," Ken told Jeff one day. "In 1976 the feds hired me to study whales, and in 1977 they tried to put me in the slammer for doing the same thing." Don Goldsberry himself had tipped Ken off that trouble was brewing. By then, Goldsberry was wealthy and he worked for SeaWorld. He also had a home on San Juan Island, and Ken would run into him in Friday Harbor.

"I was at the hamburger stand in town, and Don was there," Ken told Jeff. "And he said, 'Watch out, Ken, they're going to arrest you.' And I was like, what the hell? I mean, he's such a blowhard, this guy. He likes to tell tall tales. So I kind of wrote him off."

Ken went home and began planning his orca survey for that summer. His study permit was still awaiting final authorization at NMFS, but Ken was ready to start preliminary observations. The next morning, he got in his boat and went out to look at the whales.

"They were everywhere that day, hanging just offshore and going around in circles like pollywogs," Ken recalled. He was out on his Boston Whaler along with a forty-foot sailboat called *Bakers Dozen,* with eight volunteers on board, ready to do a follow-up study to the previous year's. Their motors were off and the sails were down. "We were off the front of the house and the whales were heading right for us, coming from the south," Ken said. "And then I see a helicopter coming round the point."

The chopper flew in low circles as officials with telephoto lenses photographed Ken and his little operation. It then came in even lower, driving the whales right next to Ken's boat. More photos were snapped.

"Then, after they got all the whales around us, they dropped a note on the sailboat with a piece of wood tied to a paper that said, 'People, you are in violation of the Marine Mammal Protection Act. Call this number in Seattle,' " Ken said. "And I'm just sitting there, blown away that all of this was happening. But the beauty of it was I had a tape recorder running on the hydrophone. Our engines weren't operating at all; you could hear the helicopter."

The feds threatened Ken with $20,000 in penalties and a year in prison. But he wasn't going down without a fight. "I shot a sixteen-millimeter movie of the whole thing," he told Jeff with a big laugh. "I got him herding whales around my boat. I got it all."

Ken had a friend in Seattle, Rick, who ran an adult-movie theater and knew where to process the film overnight. He gave Rick the negative. "And then I called up the enforcement guy. And he said, 'We'd like to set up a meeting at ten tomorrow morning to discuss this harassment incident.'" Ken said he would try to make it. He then told the officer he had filmed the whole raid. "I said, 'I'm not guilty of anything. I didn't have any engines running. We weren't doing anything wrong and there was *no* harassment. You'll see. Can I show you the movie?'"

The officer asked where the film stock was. Ken gave him the name of the company that was processing it that night. "And God, as soon as I said that, I thought, 'Oh, shit! What the *hell* did I do that for?'" he told Jeff. "So I got ahold of Rick and told him to have them make two prints and put the original in the vault and take the other one back to his theater. And sure enough, at one o'clock that morning, federal agents showed up with a subpoena and seized the theater print. They didn't know we had a copy."

Ken showed up at the meeting as planned. The agents calmly informed him he would be charged with harassing marine mammals, whether he had a permit or not. "And I said to these guys, 'Well, I filmed the whole thing. And I happen to know that you people have the film.'"

Jeff was loving this story. "So what happened then?"

"Well, they just said, 'We have the film, but it didn't turn out. There was nothing on it. And I told them, 'Well, gee, gentlemen, that's really too bad. But I happen to have a copy myself. Let's watch mine!'"

The charges were dropped.

Years later, Ken ran into Don Goldsberry in the Bahamas. Over a bottle of Crown Royal, Ken asked him, "What the hell happened back in 1977? How did you know, Don, about the raid before it even took place?" Don said he had attended a meeting with the head of the NMFS permit office in Washington, DC, where the entire operation had been planned, he told Ken. He claimed to be at a second meeting with the local enforcement officers who were going to do the actual bust. They had planned to intentionally drive the whales toward Ken's boat. It was all a setup, designed to stop his survey project.

Ken was appalled. "What was a SeaWorld representative doing at a planning meeting in DC to arrest me for studying whales? My deep suspicion was that SeaWorld had a hand in the whole raid."

"So SeaWorld didn't want the study to go through," Jeff said. "I can believe that."

"They absolutely did not want me out there counting whales. They did not want *any* data on longevity or anything from the wild, because obviously it would make them look bad. And they sure didn't want data on what their culling of the population had done to the Southern Residents. Besides, these are vengeful people. If you cross them, they'll come after you."

"What do you mean?" Jeff asked.

"Well, we had organized up here to get them kicked out of Washington State because they'd caught too many whales. And I had personally collected the evidence to prove it. But they didn't want my evidence out there, showing they had permanently affected this population. To this day they don't want to believe it. But they can't deny it anymore, even though it's certainly not part of their education program."

It hadn't been Ken's only interaction with SeaWorld. In 1993 several company executives—including Thad Lacinak and Orlando park president Bill Davis—went to San Juan Island to film a promotional video with British actress Jane Seymour. They came to gather footage of wild killer whales swimming in Haro Strait. Ken invited them over to the center to have a chat and take a little tour.

"We were all a little embarrassed for these people," he told Jeff. "I mean, they didn't know the *first thing* about wild whales." Astrid had tried in vain to explain the photo-ID science to them, "and she said they were totally dense about the whole thing. They just didn't believe you could track whales the way we do over such long periods of time."

One day, when a matriline from J pod was milling around in the cove, Thad Lacinak jumped in a kayak and paddled out to commune with the wild beasts. But he got too close: Much to Ken's amusement, the whales nearly whacked the SeaWorld executive into the icy water with their flukes.

Jeff laughed at the story. "I heard about that trip when I was still at SeaWorld!" he said. "I thought it was wrong that top officials would come here and see what wild orcas are really like while keeping the rest of us at SeaWorld in the dark. It seemed a bit like book-burning, you know? You will learn only what you need to learn."

Astrid gave Jeff a brief orientation on what the regular summer duties entailed. He was going to participate as a team member—with all the dishwashing and food-prep duties that came with membership. The survey volunteers would get up "super early," she warned, and hit the ocean on the *High Spirits* to seek, photograph, and identify Southern Residents. The *High Spirits* was an ideal stable platform for taking photographs, especially using long lenses. It being the beginning of the season, everyone was anxious to document newly arrived infants and recently departed adults.

The next morning, Astrid took Jeff and the other volunteers out on the *High Spirits*. It was a windy day—the air cool and clear, the water a bit choppy. The Olympic Mountains glowed white against a sapphire sky.

Suddenly there were whales *everywhere*. "J pod!" Astrid shouted, then started calling out the ID numbers one after the other. "There's J5. And J11." The animals traveled in twos and threes, the males easily distinguished by their tall, triangular, and, most of all, perfectly erect dorsal fins. They moved toward the north in an unhurried fashion, rolling slowly up and out of the water with their poofs of breath, and sinking back down again. Jeff was amazed by their graceful movements, their slow rising and falling beneath the surface. The *High Spirits* followed along for miles, passing a solitary minke whale on the starboard side. At Lime Kiln Point, a beautiful state reserve bordered by cliffs and kelp beds, the J's stopped to rest and mill about in the seaweed.

The oldest orca of all, J2, abruptly leapt from the water in magnificent breaches like nothing Jeff had ever seen before at SeaWorld. "That's Granny!" Astrid told him. "She's about eighty-five years old, the matriarch of the whole pod." Next to her was J1, better known as Ruffles because of the wavy pattern on the trailing edge of his dorsal fin. Ruffles was probably about forty-seven at the time.

Jeff could feel emotion well up inside him. It was like a melodramatic movie, when the John Williams–style Hollywood music rises to a heart-pounding crescendo. Teardrops began streaming from Jeff's eyes—some for joy; some for sorrow and shame. Jeff was now sobbing silently, embarrassed that others might notice. He was grateful for the dark, wraparound glasses he had on, hidding his face.

It was one of the most profoundly bittersweet moments of Jeff's life.

I am so lucky to be here at this spot, he thought, witnessing this amazing sight. So, what on *earth* was I doing at SeaWorld for eight years? Damn it, we were monsters for keeping those animals locked up in pools. Jeff's head was spinning with the conflicting feelings.

Howie was waiting for the boat when the crew returned to shore. "You know something?" Jeff said to his friend. "I just realized that, all of those years at SeaWorld, I was helping to perpetuate this insane captivity industry. What a tragedy. It's hard to believe that one can be so naïve. It's a tough pill to swallow."

26

The Strange Case of Daniel Dukes

Back in 1987, a string of trainer injuries rocked SeaWorld San Diego, culminating in the crushing of John Sillick by Orky II. In response, the company implemented a raft of safety measures designed to keep employees out of harm's way, especially during water work. By 1988, when trainers got back into the pools for the first time after the Sillick affair, SeaWorld had a rigid system of trainer seniority, one that required employees to attain specific levels of skill and experience before they could swim with the whales.

The new safety regime also required more spotters to be on duty during all interactions with the killer whales. And it required trainers and their supervisors to complete an official incident report each time a whale went "off behavior," displayed precursors to aggression, or engaged in aggressive behavior. Those reports would be circulated among the four SeaWorld parks for comment and review by supervisors and training curators and made available to all orca trainers as learning tools. Kept together in a master Incident Log, each report was supposed to describe the incorrect behaviors, determine their cause, and offer recommendations for preventing such problems from happening again.

For years, the new precautions were working. Although safety incidents did make it into the log, no serious injuries (at least among those reported) occurred at any SeaWorld park between 1988 and 1999. Indeed, the century would close without any further major problems involving captive killer whales and their trainers.

One near miss did occur before the new millennium dawned, however. It involved the park's twenty-two-year-old dominant female Kasatka (Russian for "killer whale"), an Icelandic orca taken in a 1978 roundup that had also netted Katina, the matriarch of Orlando.

Unlike Katina, Kasatka was not easy to work with. She was also fiercely protective of her calf, Takara, sired by Kotar and born in San Diego in July 1999. At night, when Kasatka was put in a separate pool from Takara, mother and child spent most of their time rostrum-to-rostrum on opposite sides of the metal gate. Sometimes when Takara called for her mother, Kasatka would split from the trainer's control to spend time with her calf at the gate or swim in angry circles around the pool until finally responding to a callback signal to the stage.

Kasatka grew more aggressive with trainers in the water after Takara was born. Her water-work relationships were deeply impacted by the whereabouts of her calf. Though she was impatient with rookie trainers in the water, she was protective of her control trainer. When others would engage in a little horse-play with a control trainer onstage, she might express her dismay at the perceived aggression with a high-pitched shriek or quick pop of her jaws.

However, Kasatka was known to display aggression without any noticeable precursors, apart from separation from Takara. She had engaged in mouthing behaviors when switching control trainers during water work and mouthed the feet of trainers during foot pushes. She pulled bootees off trainers' feet, grabbed diving gear in the water, and squirted water at maintenance workers when they made excessive noise around the pool. She also squirted trainers *and* guests when she became frustrated or agitated. At times, Kasatka jaw-popped at people or slid out of the water to challenge strangers.[1]

Kasatka was also aggressive toward subdominant whales, especially the males, whom she would repeatedly rake or even bite with her teeth. She also learned how to catch seagulls. Sometimes she brought them to the pool's edge. Other times, she ate them, ripping the birds apart in a spray of blood and feathers.

Her aggressive tendencies became more pronounced when social conflicts erupted among other whales, or when trainers had her switch behaviors with little or no positive reinforcement.

According to her official company Animal Profile, Kasatka had racked up a long rap sheet by the end of the century. In addition to minor foot-mouthing incidents, she had also jaw-popped (snapped down on) a trainer's foot and whacked a trainer in the back with her fluke.

In April of 1993, Kasatka had progressed to mouthing the legs of a trainer during a hydro. Three months later, she did the same thing. But this time, she grabbed a trainer by the knee and dunked him underwater, then grabbed a foot and dunked the trainer a second time. Neither incident led to a trainer injury.

Despite these reports, SeaWorld continued to allow water work with

Kasatka. "After establishing a relationship with Kasatka, most find her a great animal to interact with," her Animal Profile stated.

Then, on June 12, 1999, an experienced trainer named Ken "Petey" Peters was doing the 2:30 Shamu show with Kasatka and her calf, Takara, in the main pool. Takara unexpectedly split to a back tank during the performance. Kasatka then left Peters in the water and began circling the perimeter of the pool at high speeds—a known sign of frustration and a precursor to aggressive behavior. She then opened her jaws wide, moved in to grab Peters's legs, and tried to push him out of the pool. Fortunately he was pulled from the pool by the spotter trainer before she could reach him. The show was canceled, but later reopened with another whale.[2]

Peters was unhurt, allowing SeaWorld to enter the new century with its trainer-injury record unblemished since 1987. (Kasatka, however, was by no means finished with her unpredictability in the years to follow, especially when it came to water work with Ken Peters.) Though the company had dodged any serious trainer injuries for twelve years, its safety record was about to sustain an unlucky blow.

There is no way to know how many people have successfully snuck into a SeaWorld park without paying the admission fee. But in July 1999 a man named Daniel P. Dukes apparently did just that.

The twenty-seven-year-old grew up in Columbia, South Carolina, but never fit in with the staid mores of his conservative hometown. His parents sold real estate for a living. Dukes ambled from state to state, with stops as far afield as Washington and Texas, before he returned East and headed for Florida. He rarely stayed anywhere for more than a couple of months.

With his thin, five-foot-ten frame, scraggly beard, light brown hair pulled back into a long ponytail, and red-lettered *D* tattooed above his left nipple, Dukes was an odd-looking young man, but he considered himself a private soul who loved nature, animals, and music. Perhaps it was hard for Dukes to get or hold a job. His record of petty-crime convictions could not have helped matters. Dukes had done jail time on theft and drug charges and carried more than a dozen convictions in a number of states. On Christmas Day 1998, he was arrested in Marion County, Florida, on a misdemeanor marijuana charge, and again for pot possession in March 1999 in Volusia County, where he told the arresting officer he was heading for a "rainbow fair"—a type of New Age hippie gathering—in Ormond Beach.

Dukes then headed for Miami, where he ended up staying at a Hare Krishna temple in the tony Coconut Grove district. He lived there peaceably with the community of six worshippers, including a temple priest named Paul Seaur. Dukes told Seaur he had come from Jacksonville and was seeking "spiritual tranquillity." Seaur agreed to take the young man in as a religious volunteer.

Dukes seemed to adjust to temple life. He took part in its feed-the-homeless program and worked in the communal garden. He spent time writing in his journal and giving scraps of leftovers to wild birds. He never became a member, but while at the temple he did try to adhere to its rules, including a ban on alcohol, drugs, gambling, sexual activity, and eating meat or eggs.

Not everyone liked having the stranger around. One member, Lilakara Das, called him an "odd bird" who always seemed "in his own world." He also had a hard time getting up at 4:00 a.m. for prayers and began to skip out on assignments. Das would sometimes find him AWOL, meditating in the chapel while listening to heavy metal bands like Metallica. One day Dukes proclaimed a vow of silence, something that puzzled the others, given that the Hare Krishnas encourage chanting and singing.

A month later, Daniel Dukes left the community. "I want to be free," he told one member, "I want to travel around." Dukes made his way north to Indian River County, where he was arrested in Vero Beach after stealing a 3 Musketeers bar for breakfast. Dukes was still honoring his vow of silence and denied the charges by writing down on a piece of paper that he had not been inside the store. He was sentenced to three days in the county jail.

Dukes was released on Friday, July 2, and headed for Orlando.

On July 4, and again on July 5, amid the happy families strolling around SeaWorld, the disheveled wayfarer was spotted by security staff at the park. They noticed his fixation with both girls and killer whales. Dukes spent a lot of time at Shamu Stadium, staring at the animals. Still, his behavior did not warrant intervention from security.

On July 5, after dark when SeaWorld finally closed, Dukes hid somewhere behind Shamu Stadium. When no one else was around (it's not clear if night-watch staff were on duty at the time of the incident), he headed toward the "Dine with Shamu" area and its large, rectangular G Pool. Dukes climbed over a three-foot Plexiglas barrier and walked toward the water. He removed his cap, shirt, and trousers and laid them on the deck. Clad in black nylon trunks, he walked over to the low rock wall that enclosed the pool and approached the ledge.

Of all the places at SeaWorld, G Pool had to be the worst choice for a swim with the whales. On that night, it was Tilikum's pool.

At 7:05 the next morning a worker named Arturo Cordoba arrived at Shamu Stadium to begin cleaning the area around G Pool. As he was working, Tilikum swam by. Cordoba noticed something white on the whale's back. He asked a coworker, who said it was probably a toy. But then Tilikum surfaced from the water, and they could see it was no toy. As Cordoba ran to notify security, physical therapist Michael Dougherty was arriving for work. He glanced through the underwater glass and saw Tilikum floating calmly in the water, staring back at him. Then he noticed two legs dangling from the giant whale's flank, and they were not moving. A nude person was draped across Tilikum's back.

Tilikum allowed himself to be herded onto the medical lift, and SeaWorld staff retrieved the corpse. The young male had puncture wounds, multiple bruises, and abrasions across his face and body.

His swimsuit was still in the tank. The crotch and one leg were torn: Tilikum had nimbly opened Duke's scrotum and removed his left testicle. Divers had to retrieve it from the bottom of the pool.

Police could find no admission ticket among Duke's meager possessions (to this day SeaWorld refers to him as a "trespasser"), which included a white metal ring, a beaded bracelet, beads, a hair scrunchie, and $2 in cash. Orange County sheriff spokesman Jim Solomons said the cause of death was not clear. "The whale may have been surprised by the man in the tank," he said. "He may have dived suddenly, creating a vortex that pulled him under." Officers did find what seemed to be a Marlboro cigarette that had been restuffed with marijuana.

The media devoured the sensational story. Naomi's phone started ringing before she got to work. She spent the next several days reacting to the bizarre incident. First, HSUS issued a statement arguing that the death proved "why whales should not be held in captivity." It included a lengthy statement from Naomi, noting, "The fact that a SeaWorld patron was able to gain access to the whale pools after the park was closed demonstrates that SeaWorld does not provide enough security for whales and visitors alike."

But there was much more to it than simple security precautions. "SeaWorld's programs are not properly educating spectators about the wild nature of these five-ton animals," Naomi declared. Tilikum was smart enough to know this was not one of his trainers, she said. He was also desperately in need of something, anything, new.

"Putting orcas in captivity creates an environment in which boredom is often the norm, and a reaction to a novel 'object' in the tank is to be expected," Naomi said. But that critical safety message was not reaching SeaWorld audiences. Instead they were told that "the animals are friendly and enjoy being around humans. This man obviously believed what he heard at SeaWorld."

Later that day, Naomi told reporters that Tilikum was not at fault for Dukes's death. "He didn't kill him on purpose. This happened because SeaWorld put him in the way of people. This would not have happened in the wild." The destruction of the family structure so critical to orca mental health and well-being had turned some whales into sociopaths. "They're all socially warped, because they didn't swim with their mothers long enough to learn to be orcas."[3]

In yet another interview that hectic day, Naomi claimed that SeaWorld was portraying killer whales as benign cartoon characters: "They've turned them into Mickey Mouse." On her tour of SeaWorld, she added, she had been deeply disturbed by Tilikum's situation. "They are strictly controlling his behavior to suit their purposes. Everything he does is mapped out by humans. If that's not going to make him crazy, I don't know what will."

Ken Balcomb also did interviews from San Juan Island that day. He blamed the stress of captivity for Tilikum's aggressive behavior: "You pull them out of their element and you've already got a potential delinquent. You put them in a captive situation where they are locked in a small space with limited contacts. Basically, you're building a psycho."

SeaWorld officials criticized the critics. Dukes's death had been a "senseless, stupid situation," said Brad Andrews, director of zoological operations for the SeaWorld and Busch Gardens parks. "And then the Humane Society wants to make hay of it. It's sad when they want to make issue of this when they are not experts at all. We make sure people know these are wild creatures. We create opportunities for people to get close and appreciate them in a safe manner."[4]

SeaWorld suggested the cause of death was a combination of hypothermia (subnormal body temperature) and drowning, as though Tilikum had little to do with what befell Daniel Dukes. But the coroner's report contradicted that.[5] There was no mention of hypothermia; and Tilikum had caused injuries to the young man even while he was still alive.

Among his "pre-mortem" wounds were scrapes and bruises to the forehead, nose, eyelids, left lower eye socket, lower jaw, earlobes, left ear, chest, abdomen, shoulder, arms, right leg and foot, left hip, right knee, and left lower thigh. Mild hemorrhaging was also found beneath the skin of the skull and the chest and also in the brain and bronchial tubes.

After Dukes died, Tilikum continued to bat him about and nibble at parts of his body. He surgically removed the skin of Dukes's pubic area, including the scrotum, and left a three-centimeter bite wound behind the right knee, three bite wounds on the lower left leg, and two more tooth incisions above the left ankle. The deceased's blood showed no evidence of any drug use, including marijuana.

G Pool had been installed just six years before, and now someone had died in it. A decade later it would be the site of more destruction, again by Tilikum.

27

Transatlantic Ties

It was time for Keiko to head home.

In early 1998, the Free Willy-Keiko Foundation had big news. They unveiled designs for the one-of-a-kind floating pen that would house the famous whale in a North Atlantic bay or fjord during the next phase of his lengthy and costly rehabilitation. Made of heavy-duty, plastic pipe frame filled with foam, with mesh netting along the bottom and sides, the state-of-the-art enclosure would be 60 percent larger than Keiko's current tank at the Oregon Coast Aquarium. It featured a walkway around the pen, a medical pool, food-prep area, dive locker, and generator room.

Exactly where the floating pen would float was still unclear. Iceland continued to resist the idea of any captive killer whale returning to its waters, so other venues were also being considered, including sites in Norway and the United Kingdom—five thousand miles of North Atlantic coastline had already been scouted.

Plans to push forward continued to be attacked by industry and many members of the scientific community. For one, although Keiko could be commanded to catch a live fish, he wasn't that eager to forage for himself. Even foundation members were growing disheartened that he might ever make it on his own.

Brad Andrews from SeaWorld was disgusted by the entire spectacle. "I don't think it's fair and humane to the animal to try this operation just to make a few people happy," he complained to the AP. Keiko was not appropriate for release, Andrews said. He had been "imprinted by humans" for too long.

But then, in June 1998 the Free Keiko movement received an unanticipated shot in the arm. The government of Iceland did an about-face on Keiko. The country's chief veterinary officer traveled to Newport, Oregon, to examine the whale. He declared that medical tests showed no grounds for banning Keiko's return home. Next, Icelandic prime minister David Oddsson sanctioned the US government for holding a resident of another country against his will. He demanded that Keiko be returned to Icelandic waters by September 19 or Iceland would commence civil action against the government of the United States.[1]

That month, the foundation announced the location for its floating sea pen: a well-protected cove made from the remnants of a volcanic crater rim in the Westman (Vestmannaeyjar) Islands, about eighty miles southeast of the capital. The site, surrounded by six-hundred-foot volcanic walls on three sides, was called Klettsvik Bay, near the town of Heimaey.

Again SeaWorld blasted the project. Brad Andrews said that schoolchildren around the world had been misled, and "maybe now's the time to start telling them the truth—that they should find a companion for him and keep him where he's at." Keiko was headed for "an ocean pen where the weather conditions are ferocious. It's cold, it's miserable, it's dark. There's no contact with other whales. It doesn't make a lot of sense."[2]

Naomi snorted with derision. "What an utterly stupid statement!" she said to a colleague. "Andrews lives in *Florida*. Of course he's going to think that Iceland is miserable and dark. But to killer whales, and to a hell of a lot of Icelanders, it's home."

Meanwhile the Alliance of Marine Mammal Parks and Aquariums filed a complaint with NMFS alleging that the Keiko Foundation was about to "unfairly exploit" a loophole in the Marine Mammal Protection Act to send Keiko from the United States to Iceland.

SeaWorld and its allies were not the only ones opposed to the move. Some Icelanders were also trying to keep the whale out of their waters. One infuriated islander threatened to kill Keiko by feeding him poisoned fish. Who sent the threat to local media and why were unclear, though some people speculated the motive was envy. The Westman Islands were chosen over another site, the town of Eskifjördur (near Keiko's *and* Tilikum's capture site), whose residents were hoping for an influx of tourist dollars from the globe-trotting attraction.

In September of 1998, Keiko finally went home. He was flown aboard a US Air Force C-17 cargo plane, paid for by the foundation. It was the only aircraft in the world that could both carry the huge payload and land on a short runway. The eight-hour flight carried Keiko directly from Newport to the Westman Islands, with two in-flight refuelings. The plane suffered a major mishap upon landing and broke a wheel strut, sustaining a million dollars in damage. Neither the crew nor their precious cargo was injured.

As soon as he was lowered into his new pen, Keiko swam clear of the stretcher and instantly dove down into the chilly ocean—the first time he had been in the sea for two decades. He surfaced a full minute later, circled the perimeter, and vocalized with gusto. Though he returned to the edge to greet his human trainers who made the trip with him, Keiko seemed more intrigued by his new home than the humans. Within hours he began vocalizing with a pilot whale that swam into his cove.

The project veterinarian, Dr. Lanny Cornell, was astounded by the changes. He got choked up when talking about it. "As a veterinarian, Keiko's medical supervisor, and a human being," Cornell said, struggling to regain composure, "it can't get any better."[3]

That fall, when 135 mph, hurricane-force winds slammed the Westman Islands, Keiko proved naysayers such as Brad Andrews wrong in their predictions that the whale would founder in the dark, tempest-tossed Atlantic. Keiko not only rode out the storm with no problems, he began to spy-hop above the surface, letting his pectoral fins blow high in the air. He also breached from the water to get pelted by the salty spray.

The following spring, the Free Willy-Keiko Foundation partnered with the Jean-Michel Cousteau Institute to form a new group called the Ocean Futures Society. Jean-Michel, the son of Naomi's idol Jacques Cousteau, said at the time that Keiko was "a locomotive, an engine pulling the education about marine mammals with it."[4] Naomi loved that: It turned SeaWorld's entire raison d'être on its face. Clearly, keeping killer whales captive was not necessary to inspire and educate the public about the species.

Soon after arrival, Keiko's trainers began introducing live Atlantic salmon into his pen. He appeared to be eating at least some of the fish because their numbers, and his hunger, declined. He chased smaller fish that entered his pen through the nets. Keiko also began making deeper dives and ever stronger breaches. His posture and body movements began to resemble that of an oceanic whale rather than a trained performer. He spent less time at the surface than he had in Mexico City or Oregon, and he extended his attention to what was going on beyond the mesh of his confines, rather than merely on the people caring for him.

In late 1998, Ocean Futures hired two former SeaWorld trainers as lead consultants on the project: Robin Friday, the former curator and director of animal training in Orlando, and Mark Simmons, the close friend and former roommate of Jeff Ventre and John Jett. The two men quickly declared that efforts to retrain Keiko in ocean-survival skills were failing. Keiko was still far too attached to humans to fend for himself in the wild, they said. Meanwhile, Ocean Futures staff was venturing out into nearby waters to observe killer whales in the area in the initial stages of a planned photo-ID study of local populations.

When Jeff Ventre got word that Friday and Simmons had flown to Iceland to work on the Keiko project, he didn't know what to think. It sounded like an interesting place to be and, on the surface, a noble mission. He and Mark were still close. Was Mark really there to "free Willy"? Jeff e-mailed his friend Howard Garrett to alert him to the new development. Howie had flown to Iceland in October of 1998 to meet with the Ocean Futures staff and check on Keiko's progress on behalf of the Center for Whale Research. Now, he was spending two years in Miami, still trying to win the release of Lolita (Tokitae) from the Miami Seaquarium.

Howie, who had come to know Naomi through their mutual work on captivity issues, e-mailed her at HSUS. "I have some inside information about the new managers of the project," he wrote. "I can't give you my sources, but their names are Robin Friday and Mark Simmons and they are 100% veterans of SeaWorld. They consider the entire Keiko project a misbegotten movie-inspired fantasy by Craig McCaw and an amateur attempt to do the impossible."

Robin and Mark were "certain that they learned all that is worth knowing at SeaWorld," Howie said. They have had "zero experience in studying wild orcas; and they have no doubt accepted the disdain and denial of the results of long-term [longevity] studies. This is the belief system that is now in control, and JMC [Jean-Michel Cousteau] apparently has complete faith in these individuals.

"So in effect, SeaWorld has taken over the Keiko project. I don't think they take direct orders from Brad Andrews, but they are his apostles. Under his influence, Keiko will never see the open ocean unless he breaks out. The recent announcement by Ocean Futures Society that Keiko's chances of ever going free are 50/50 confirms my suspicions. They are trying to diminish hopes. To me this is a betrayal of the promise made to the children and the public, and to Keiko."

Howie complained that the Free Willy-Keiko Foundation had tried to exclude experts on wild orcas, including Naomi, Ken Balcomb, Erich Hoyt, Paul Spong, and John Hall. Instead, they had relied on the opinions of "marine park employees." But now Keiko was on the doorstep of freedom. It was imperative that experts on killer whales in the wild be allowed to participate in the next and final steps of his rehabilitation, Howie argued. "My hope is that enough of us can come together in the near future to dispel some of the myths promoted, however unwittingly, by Friday and Simmons and the entire industry, and convince JMC to announce that, barring some unforeseen setback, Keiko WILL be allowed to venture out in the open ocean, radio tagged and monitored of course and trained to come to a signal if he is hungry, etc., but that he will be given a chance to join up with wild conspecifics."

Naomi commiserated with Howard but wrote back to tell him that neither she nor HSUS had any control over the situation. David Phillips at Earth Island Institute, which helped form the Keiko charity, had never been very responsive to HSUS suggestions (despite their million-dollar donation), and

HSUS had no direct line of communication to Cousteau. Naomi passed the information up the chain of command, but she could do little else.

Besides, her plate was full. Over the past four or five years, Naomi had been busily engaged in an exhausting number of other marine mammal topics that had nothing to do with killer whales—and many of them had nothing to do with captivity.

For one, exposure to midfrequency sonar has been responsible for several stranding incidents since its introduction in the mid-twentieth century. Many beached animals have been found with signs of physical damage, including hemorrhage of the brain, ears, and other tissues, and air bubbles in their organs—not unlike the bends, an illness afflicting scuba divers who rise from deep water too swiftly.[5] Sonar and other human-made noise, such as shipping, air guns, pile-driving, and navy detonation of bombs and live fire during exercises, can disrupt foraging and communication among cetaceans, reducing their ability to navigate and locate prey through echolocation.[6] Naomi was working with environmental groups on improving the regulation of human-caused marine noise and minimizing its impacts on marine life.

Other issues and species that filled Naomi's animal-protection portfolio included the spread of swim-with-the-dolphin programs, especially in the Caribbean; consumption of endangered salmon by seals and sea lions; dolphin-safe tuna laws; sea otters and their recovery in California; manatees and their protection under the Endangered Species Act; polar bear sport hunting and conservation; commercial whale-watching regulation; and pushing for a ban on the gruesome and antiquated global whaling trade once and for all.

By the end of the decade, Naomi had become a recognized expert on many of these issues, earning her an invitation to participate on the Scientific Committee of the International Whaling Commission (IWC) in 2000. The year before, in May of 1999, the Scientific Committee had its annual working meeting—where much of the world's whale science gets done—in the Caribbean jewel of Grenada. Naomi flew down to the island as an observer, meaning she couldn't actively participate. She was expected to sit in the back and keep quiet, something that never came naturally to her. She was literally sitting on her hands during much of the twelve-day meeting.

Naomi enjoyed the experience immensely, and not only for its lively scientific discussions and gorgeous tropical venue. A friend at the meeting, Mark P. Simmonds of the Whale and Dolphin Conservation Society, introduced her to a young cetologist from England named Christien Parsons. Chris had floppy brown hair, a dark goatee, and wire-frame glasses, dressed in sandals, T-shirt, and khaki shorts—and sported a friendly smile.

Chris, almost eight years younger than Naomi, grew up in a working-class family in southwest England. He graduated from Oxford in 1991 with a degree in zoology, then earned his PhD at Hong Kong University, where he studied the biology of dolphins in the local waters. After that, Chris spent a year teaching at a community college in southwest England. He then went to Scotland to take a job as scientific director for an environmental group called the Hebridean Whale and Dolphin Trust. Chris was now a member of the UK delegation to the Scientific Committee, and this was his first meeting as well.

Naomi delighted in Chris's humor, accent, and dedication to whales. She also thought he was cute. But Chris had a girlfriend back home. So, despite the romantic tropical venue, he and Naomi became friends but nothing more.

The following year, in May of 2000, Naomi got approval from HSUS to travel to Iceland for a site visit with Keiko. She flew from Dulles Airport to Reykjavík, then caught a local bus to the ferry terminal for a nausea-inducing boat ride to the Westman Islands. She arrived at Heimaey on Monday afternoon, May 15, and met with most of the Ocean Futures crew, including project director Jeff Foster and head behaviorist Robin Friday. Naomi was able to observe Keiko from a site overlooking the sea pen. She went there at least twice a day and spent about ten hours total watching Keiko. She also went out to the pen itself twice during her visit.

Despite Howie's warning about Robin Friday and Mark Simmons, Naomi reported back to HSUS headquarters that she was "very impressed" with the operation. "It is very high end (that is, its infrastructure is excellent)," she wrote in a memo. "The crew seems to have good morale and seems positive about releasing Keiko (given the captivity industry background of most of them, this is a significant positive sign). If anyone can successfully release and track this whale, it is this operation. They are not doing anything by halves here."

Naomi was also pleased with Keiko's health. True, he still showed some negative behaviors, such as focusing on boats when he wasn't supposed to, approaching people on the Ocean Futures vessels or the sea pen with his head up and mouth open, demanding food, "but only infrequently," Naomi said.

Privately, however, Naomi did worry about the way that some Ocean Futures staff interacted with the whale, speaking to him out loud in baby talk, and making eye contact and physical contact with him when it wasn't necessary. To wean him from humans, they needed to show Keiko more tough love.

By this time, Keiko was being allowed to swim out of his pen and move about freely in the cove that housed it. A net stretched across the mouth of the inlet prevented him from leaving the area entirely.

"For the most part, he is behaving fairly independently, keeping himself entertained by teasing birds and swimming around the bay, being stimulated by his daily boat desensitization and boat follow sessions, eating live fish about 20% of the time, and showing good health," Naomi wrote back to Gaithersburg. "I believe he is as ready to be released as he ever will be. The Ocean Futures directors apparently concur."

The staff was preparing to lead Keiko for his first "controlled walk" (allowing him to swim just beyond the net barrier blocking off the bay to get him sensitized to going through the gate), and the Heimaey harbor authority had given permission for that next step. "They hope to take him for a longer 'walk' out of the harbor and into the channel between the island and the mainland by Friday, when there will be a dynamite blast set off by the harbor authority to prepare for a pier extension," Naomi wrote.

During that summer, the Ocean Futures team received a permit from the Icelandic government to begin taking Keiko out on "walks" farther into the open ocean, followed by a tracking boat equipped with an underwater recall system and equipment to pick up the radio signal from the whale's tag attached to his dorsal fin. Keiko made forty sorties that season outside his cove and into the North Atlantic, covering a total of six hundred miles during those trips. On fifteen of his excursions Keiko and his party encountered wild killer whales. Keiko showed interest in them. The wild whales tolerated his presence, but didn't express much desire to be with the lone stranger, save for a few curious juveniles who swam over to check him out, before being recalled by their mothers.

At the close of summer 2000, the native Icelandic orca pods moved out of the area to follow the herring. Keiko still exhibited no desire to join the wild population. Once the weather turned foul, he was confined to his sea pen until spring 2001, at the whopping cost of $300,000 a month for the food, support staff, and research vessels.[7]

The following summer, Keiko completed a large number of successful ocean walks once again, including several interactions with Icelandic orca pods, but nothing lasting more than a few moments. On some days, Keiko would swim off by himself, away from the other orcas and the herring they were chasing. Robin Friday was gloomy about Keiko's future. Keiko had failed to pass two key milestones needed for his release: He had not been seen interacting with any one pod consistently. Even worse, he had never been observed foraging on his own in the open water.

This was going to take more time.

The summer of 2001 sent Naomi back across the Atlantic to attend the IWC Scientific Committee meeting once again, this time in the somewhat shabby London suburb of Hammersmith, England. Chris Parsons was there and Naomi was surprised by how happy she was to see him. They had not communicated since meeting in Grenada two years earlier. Chris gave Naomi a warm hug.

"My," he said to her, "your hair smells nice."

Naomi felt a wave of attraction. For the first time, she wondered if something might come of this. From that day on, the two of them spent a good deal of time together, including every meal. They went off on lunches alone and sat together in all the meeting sessions, often laughing—too much, some of the others told them—at their own silly jokes. They discovered a mutual love for single-malt Scotch and geeky sci-fi movies and TV shows. Both were vegetarians. Naomi gave up all meat, poultry, and fish after a few years at HSUS; Chris had been a vegetarian since he was four.

Everyone at the meeting assumed they were together, but Naomi and Chris were only friends—nearly inseparable, but still only friends. A few days before Naomi's flight home, Chris left for Scotland. Naomi was crestfallen, but said nothing. She wasn't sure of his feelings.

But the ticket machine at the tube station swallowed Chris's credit card. With little cash in his checking account and just sixty pence in his pocket, he couldn't make it to the airport on time. Stranded and broke, he wandered back to the meeting venue, hat in hand. Chris asked if he could crash in someone's room until he could get some funds transferred into his bank account. Naomi told him he could stay in her room for the night.

The next day they hung out together and went for a long, enchanting walk, pub-crawling through Notting Hill. They were falling for each other. Chris postponed his departure by four days, when Naomi was scheduled to leave.

On their final day, as they were saying good-bye, Chris said, "Well, I guess we're going to get married then?"

"Yeah," Naomi replied, "I guess we are."

They made plans to see each other in September, stateside.

When Chris flew over to see Naomi, in mid-September 2001, the world was still reeling from the atrocities of 9/11. His flight was nearly empty and the terminal at Dulles was deserted when Naomi went to pick him up. Chris had brought something with him to present to Naomi: his great-grandmother's engagement ring.

The Al Qaeda attacks did little to halt the scientific research going on around the Westman Islands in Iceland. Ocean Futures had managed to photo-ID some 250 killer whales in the area, with many of the animals' vocalizations recorded and some biopsied for DNA samples, using a small dart gun that pulled a sample of skin and blubber from an animal with a retrievable plug that floated.

But Naomi, and many others who had worked on the Keiko project, had grown disillusioned with Ocean Futures and their SeaWorld-based staff. Keiko had been in Iceland for more than three years, and he was still far too attached to his human trainers. Naomi learned that Robin Friday, Mark Simmons, and others would grow nervous whenever Keiko demonstrated too much independence on his "walks." If he wandered too far from the tracking boat, they would call him back with underwater tones. Despite the tracking tag on his fin, they seemed afraid to "lose" him, even though, in many ways, that was the whole idea.

Naomi now suspected that Keiko's trainers did not want him going off on his own. Howie had been prescient in the warning: This was a subtle form of sabotage. Robin and Mark were training him the way the navy did its dolphins—to follow a boat and remain focused on it, so they did not go AWOL.

Then the funding from Craig McCaw dried up.

The billionaire had taken some hefty losses in the dot-com bubble bust and decided it was time to get out of the whale-rehab business. McCaw had already contributed $10 million of the estimated $15 million spent on Keiko.[8] In 2001 alone, the bills had reached about $1.8 million to underwrite twenty-five staff members, plus fuel, food, and equipment. The 2002 budget would be drawn down to $500,000 with a staff of fourteen. The helicopter and the large tracking vessel would have to go. Around that time, Robin Friday and Mark Simmons left the project and left Iceland.

With no one else stepping in to pay for Keiko's care and training, HSUS took over the project. Naomi and her colleague Richard Farinato—HSUS's captive-wildlife specialist—helped usher in new rules that barred team members from making eye contact with Keiko unless they were asking him to do something that contributed to his development. If Keiko approached their boat, they were instructed to go belowdecks before he saw them, so he would get no reinforcement for that behavior. Rubdowns purely for affection were likewise stopped, as was hand-feeding the whale.

Keiko began spending more time in the ocean, often in the company of other killer whales. He passed days out there at a time without returning to his pen. Naomi and Richard Farinato made another visit in June of 2002 to check on his remarkable progress. It was all good news. While under Ocean Futures' care, Keiko had never done anything nearly so adventurous as spending days on end away from his pen. Naomi did not think he would ever have left without the new tough love being shown him to nudge him from the nest.

Keiko spent much of his time near pods of whales that were foraging on fish, but there was no hard evidence that Keiko himself was eating. On some occasions when he was recalled to the tracking boat, the staff would perform a "stomach lavage" but never found any sign of fish.

Naomi and the HSUS team working on Keiko consulted the project veterinarian, Dr. Lanny Cornell, to get this assessment. "The only way he'll ever go off and eat in the wild is if he gets hungry enough to do it," he told them. "Leave him out there. Let a couple more weeks pass and see what will happen before getting worried. He needs to be pushed to the limit if he's going to be free."

Keiko continued to swim and travel with wild whales, usually keeping a hundred yards or so away, though it was still unclear whether he was eating. Killer whales can go for some weeks without food—though they start losing weight after two—and Keiko was still returning to the pen for meals. Even so, his team saw that as a positive development. Exercising his choice to come and go from his pen as he saw fit was a major advancement in Keiko's rehabilitation.

On August 4, 2002, Naomi arrived on the site for her second visit of the season, this time on her own. But Keiko was not in the bay—he was offshore swimming with wild whales. The next day, the pod he was following began heading east, out into the open North Atlantic.

Keiko went with them. He continued swimming eastward, occasionally making deep dives down to sixty meters or more, reminiscent of foraging behavior among wild killer whales and suggesting he might be pursuing schools of fish.

Naomi, meanwhile, spent much of her time in the company of the accomplished writer Susan Orlean, who had arrived in the Westmans to write a feature about Keiko for *The New Yorker* magazine. But now her superstar subject had taken off, leaving Orlean to write a disjointed profile published under the fitting title "Where's Willy?"

When the former Ocean Futures workers got word of what had happened, they complained that Keiko had been abandoned. They said Keiko had decided to take off because his caretakers were being "cruel" and ignoring him, denying him any social interactions with them. Keiko left, as they saw it, because he was lonely and "his humans" had betrayed his trust.

Naomi was bemused by such disingenuous platitudes. Keiko had not been abandoned. He was being tracked daily. "These people literally do not get that the whole point is to dehabituate him to humans," she told colleagues back in Gaithersburg. "The goal is to break those very bonds that the SeaWorld crowd goes on and on about. They see this kind of 'detraining' as cruel and cold-hearted, rather than a means to allow him to regain independence. They believed they had to take care of Keiko, when the point was getting him to where he could take care of himself. The whole thing seemed to go completely over their heads. That's why he never ventured far in three and a half years."

Naomi never saw Keiko again.

The whale was tracked by satellite all the way across the Atlantic. His path was fairly steady and straight, with only the occasional zigzag, which may have been Keiko following fish schools. A few days into his journey, the wandering cetacean turned up in a fishing harbor in the Faroe Islands, a remote Danish colony about midway between Iceland and Norway. Two of the caretaking team were sent to "get a visual" on him, but he didn't stick around long enough. Keiko continued his mysterious eastward trek. On September 2, 2002, after spending some five weeks swimming across the North Atlantic, Keiko followed a fishing boat into Skålevik Fjord on the coast of Norway.

His arrival attracted a lot of attention, to put it mildly. People boated and even swam out into the inlet to play with the Hollywood celebrity. Tour operators began charging people to go out on the water aboard "Keiko safaris." Keiko started to beg for handouts from tourists. Still, he seemed vibrant and healthy, despite the long journey. A stomach lavage showed no sign of fish, but his good health and steady weight and girth indicated he must have been eating *something* in those weeks at sea, according to vet Lanny Cornell.

But even as Keiko splashed and posed for photos with the good people of coastal Norway, a local scientist named Nils Øien of the Institute for Marine Research in Bergen warned that Keiko would probably perish during the coming winter if he didn't detach himself from people. Øien told Norwegian media that Keiko might pose a threat to Norway's booming salmon-farming industry. "Reports are coming in now that the killer whale is disturbing fish farms within the Skålevik Fjord," Øien said. "If there are more such episodes, he should be destroyed."[9]

The researcher insisted he was also thinking of what was best for the animal, unlike those sentimental Americans. "I think the entire Keiko story is more or less crazy," he told reporters. "Millions have been used to tame him, in order to turn him into a movie star. Then millions more are used to try and make him wild again. They should have let him live and die in captivity. Now that they have decided not to keep him in captivity, they should put him down. Those who think that they are helping Keiko by setting him free are actually doing just the opposite." Øien doubted Keiko could fend for himself in the wild and asserted, "It is better to destroy him now instead of letting him starve to death."

Naomi was horrified when she heard about all the human interaction Keiko was having. It could wipe out everything they had worked hard to achieve—complete or at least partial autonomy from people.

Though Naomi was transfixed with the Keiko saga at work, as was much of the world (he made headlines around the globe), she was also happily distracted by another matter.

On September 14, 2002, she married Dr. Edward Christien Michael Parsons in Scotland's iconic Eilean Donan Castle, a thirteenth-century fortress built on a small island on Loch Duich, in the northwest corner of the country. Naomi's mother, her father and his second wife, her brother Greg, an uncle, and some cousins all made the long journey there. Naomi's side of the wedding party also had some old friends, a few coworkers, and people from college and grad school—including one of her field assistants from Johnstone Strait.

Chris and most of the men in the wedding were dressed in kilts with traditional Scottish regalia. Naomi wore an elegantly simple ivory dress with her hair swept up and dramatically back. A simple silver circlet graced her head, but no veil. For a pair of geeky whale scientists, they cut a glamorous, romantic figure on the parapet of the medieval citadel.

Soon after that, Chris landed a job teaching marine environmental conservation and marine mammal biology in the environmental science and policy department at George Mason University, in Fairfax, Virginia. He moved to the United States to be with his new bride—and to continue advocating for cetaceans along with her from the North American side of the "dark and miserable" Atlantic.

28

Abnormal Activities

When killer whales were kept in captivity for extended periods, what effect did it have on them? Did confinement in a concrete tank morph them into some other category of animal? Some creature that is more neurotic, unpredictable, and much more dangerous, unlike anything found in nature? Naomi believed that to be the case.

Keiko, for one, was not a "natural" killer whale. Despite several years of rehabilitation, he was still a very different animal from the wild whales he was timidly trying to mingle with in Iceland. Keiko was abnormal because he could not find a family to bond with.

But at SeaWorld, the opposite dynamic was true: Animals that were completely alien to one another were compelled to form artificial "pods." Their bonds were involuntary; Keiko's bonds were nonexistent. In both cases, it was the pinnacle of abnormality for a killer whale in nature.

In late July 2002—just days before Keiko headed off eastwardly through the North Atlantic on a curious expedition we'll never understand—SeaWorld suffered its first serious trainer injury in fifteen years. This time, the incident went beyond the mere mouthing of a bootee.

A twenty-eight-year-old San Diego trainer, identified only as Tamaree, was doing dry work at the pool's edge with two killer whales, the female Orkid and the male Splash. Suddenly, one of the orcas lunged up and pulled Tamaree down into the water, where both continued to tussle with her. She would not escape unharmed.[1]

If there was one thing that experts on wild and captive orcas agreed upon, it was that killer whales' pulling humans into the water was far from a normal behavior—in *either* environment. Then again, neither Orkid nor Splash had experienced what could reasonably be considered "normal" upbringings.

To begin with, they were both born in captivity and had never experienced the natural world. Orkid, who was thirteen at the time of the Tamaree incident, was the offspring of Orky II, a Northern Resident male, and Kandu V, an Icelandic female. Orkid was extremely intelligent, perhaps the smartest whale in SeaWorld's collection, many people speculated. But she also had rather aggressive tendencies. Then again, Orkid had witnessed a lot of unnatural events in her young life, beginning the day she was born in San Diego, smack in the middle of the Shamu show, before thousands of awestruck strangers and their flashing cameras.

Three days after her arrival in the glass-and-concrete barrenness of her new world, Orkid's father, Orky, wasted away to death in an adjacent tank, a victim of pneumonia and kidney disease.[2]

Just eleven months later, Orkid was an orphan. The nursing infant witnessed her mother, Kandu, slam her head into the peduncle (the area between dorsal fin and fluke) of Corky II during a fierce altercation in San Diego, then bleed to death. The orphaned calf was essentially raised by humans, and two surrogate orca mothers, Corky II and the erratic Kasatka. Orkid also formed a tight bond with a male in San Diego named Splash.

Just about her age, Splash already had a lengthy history of abnormality. He was the first killer whale born at Marineland Ontario in Canada. His delivery, however, came prematurely, and he developed a form of epilepsy, requiring lifelong medication to control his seizures. At two years of age, Splash was taken from his mother and flown to San Diego, where he could receive superior medical attention at SeaWorld's facilities.

In 1995, Splash and Orkid were playing around when Splash suffered a serious seizure and slammed into a gate. Orkid and Kasatka kept him afloat until trainers could get the wounded youngster into the medical pool. His lower jaw had been severely injured, and despite a stiff regimen of antibiotics, an acute infection set in.[3]

SeaWorld vets decided to remove his lower teeth. The extractions left his mouth so deformed he could never fully close it again; the bottom part was now an unsightly mass of extruded flesh, with black spots where the roots of his teeth once were.

On the day that Orkid and Splash assaulted Tamaree, "she was playing with the whales, talking to them," according to SeaWorld spokeswoman Darla Davis. "The next thing we know, as it appears from the video, she was pulled into the water," Davis told the media, after she'd watched footage taken by tourists and security cameras at Shamu. Tamaree managed to climb out of the pool on her own, suffering scrapes and bruises. But soon her arm throbbed with pain and Tamaree learned it was badly fractured. A nearby hospital reset it with a pin.

SeaWorld insisted that neither of the whales had shown aggressive behavior in the past. But Orkid's Animal Profile (leaked many years later) showed another story entirely.

Between 1990 and 2002, her profile listed twelve aggressive incidents—one per year—which SeaWorld considered serious enough to include. She had head-butted a trainer's head, bumped a trainer's body, whacked several trainers with her fluke, and rammed a number of thighs, all during water work.

Trainers engaged in dry work with Orkid were not safe from her aggressive behaviors either. In several interactions with people standing on the relative safety of the ledge, she had lunged from the water to push a trainer in the stomach, mouth someone's thigh, bump a hip, and jaw-pop an arm.

In 1998, "Orkid was performing a hydro-hop behavior during a night show," her Animal Profile said. "The trainer accidentally hit her tail flukes with his hand upon his reentry and she responded by hitting him in the stomach with her head. She responded to stage call calmly." Orkid began to displace other whales by leaving her control trainer and "aggressed upon trainers after or during tactile reinforcement," according to her Profile. She also enjoyed "baiting seagulls and often catches them," it said, noting, "She is good at retrieving them alive."

Despite the incident with Tamaree and the record of Orkid's aggressions, SeaWorld decided that both Orkid and Splash would continue performing, but they would not be used for water work until officials completed a thorough safety investigation. The half measure did little to mollify SeaWorld's critics.

Meanwhile, the dispute raged over what constituted "normal" behavior for a killer whale in regards to Keiko and the strange saga his life had become. The whale's unexpected entrance on the scene in the coastal waters of Norway only made the arguing more frequent and bitter. Critics—and there were many—charged that Keiko's preference for the company of humans and lack of proof of any foraging in the open ocean were clears sign that his "freedom" had been a failure.

On the pro-release side, advocates argued that Keiko had spent nearly sixty days on his own at sea—a strong sign of self-sufficiency—and traveled more than a thousand miles without incident or weight loss. Even in Iceland, he was content to spend days on end in the ocean without returning to his pen. No matter what, his conditions were better now than they had been in Oregon, and far better than anything he endured in Mexico City or Ontario, Canada.

The fjord that Keiko had wandered into, however, was not suitable as an overwintering site because it would freeze over. Some people suggested taking Keiko back to Iceland. Others said he should be sent back to a display facility, where he could be cared for again by human handlers.

Leading the fight to repatriate the Icelandic whale from Norway to the United States was the Miami Seaquarium, who wanted a companion for their lonely and aging Southern Resident, Lolita (Tokitae). On September 13, 2002, Arthur H. Hertz, chairman and CEO of the Seaquarium, wrote to Florida senator Bob Graham expressing "concern" about Keiko because he was not in the company of wild whales and was "exhibiting solicitous behaviors toward people and vessels."

The harsh Nordic winter was quickly approaching, Hertz warned the senator, a Democrat. It was critical to "become pro-active regarding his recovery." The aquarium had already written to the Norwegian Directorate of Fisheries, "offering to rescue Keiko and bring him to the Miami Seaquarium® where he could live out his remaining years in the companionship of Lolita our killer whale," Hertz said.

It was not clear if Senator Graham's office intervened on behalf of Miami's popular aquarium, but the US government backed up the effort to return Keiko to the United States for a life in captivity. On September 19, 2002, NMFS sent a letter to the Norwegian fisheries director offering its support—political, diplomatic, scientific, and material—for the capture, transport, and relocation of Keiko to a marine park in the United States. NMFS said that Keiko was not acting the way a wild whale should act.

"We believe his behavior toward people and boats and the absence of interaction with free-ranging killer whales suggests that he is not exhibiting behaviors expected of a wild killer whale," NMFS said, adding that US public display facilities were standing by to take him. If Oslo agreed, NMFS would be available to consult with the Norwegians on the best way to rescue and transport the wayward whale back across the sea.

The Norwegians offered a speedy response. On September 24, the Norwegian ambassador to the United States, Knut Vollebaek, wrote to Senator Graham. The message was blunt and clear:

"Keiko is doing well and he is getting a lot of support. There is no immediate need for a rescue."

But the Norwegians added insult on top of rejection. "We are skeptical to keeping huge animals like whales in captivity," the ambassador said. "In Norway there is no tradition for that." It would be "problematic" from an animal welfare perspective to transport Keiko to Florida, he added. "We do not doubt that Keiko would get good support in Miami, but it would be a great step back to put him in an aquarium that will stress him."

The whale would stay in Norway, the ambassador insisted. Keiko was enjoying a "freedom that makes it possible for him to make choices." When Naomi read that, she felt hopeful and a bit vindicated.

Keiko's journey was not the only remarkable return-to-freedom story capturing the world's attention at the time. There was also the moving saga of Springer, a little female from the Northern Resident killer whale community who one day found herself horribly lost and all alone.

Springer (A73) was born to a female known as Sutlej (A45), a member of the A4 pod. In the summer of 2001, neither Springer nor her mother returned to Johnstone Strait with the rest of the A4s. Most people assumed the two of them had died, but in January of 2002, Springer was alone off Vashon Island, in south Puget Sound, far from her native waters. Sutlej was never seen again.

When Springer first appeared, her origin was a mystery. But researchers taped a few of her vocalizations. Helena Symonds, of OrcaLab, and John Ford, of the Canadian Department of Fisheries and Oceans (DFO), analyzed the recordings and determined that Springer belonged to the Northern Residents, some 250 miles north. At OrcaLab, a fourteen-year-old recording of calls made by A45, Sutlej, were compared with calls from Springer. She was confirmed to be A45's offspring.

During the spring of 2002, a heated debate ensued over what should become of Springer. Most scientists, environmentalists, and animal advocates were adamantly opposed to sending the whale to a marine mammal park. But life in Puget Sound was hardly ideal for a lone juvenile orca. Springer was underweight and in poor health. Ken Balcomb and John Ford discovered she had "ketone breath" (the smell of burning body fat for fuel), as well as intestinal parasites and a skin condition. Her prognosis was not good.

Later that year, NMFS, DFO, the Vancouver Aquarium, and groups such as OrçaLab, Orca Network, and the Center for Whale Research created a plan to transport Springer back to her native waters in Johnstone Strait. She was first transferred to a government research center at Manchester, Washington. There she was fed live salmon and underwent a battery of medical tests. Over the next four weeks, her health improved and she ate more fish. Tests found her clear of any genetic or communicable diseases, and she was deemed fit to return to Canada.

The timing could not have been better. On July 9, Paul Spong at OrcaLab announced the detection of underwater killer whale calls that closely resembled Springer's. The orcas were moving toward Johnstone Strait. Four days later, Springer was brought to a netted-off pen in Dong Chong Bay, on Hanson Island, near the OrcaLab location. As darkness fell late that evening, Springer settled in. She chased and fed on wild bioluminescent salmon, provided by local First Nations fishermen, as they zigzagged across the black water.

Suddenly, Springer heard the calls of other killer whales passing through the sound. She spy-hopped repeatedly, called out loudly, and pushed at the edges of her net. The OrcaLab staff soon determined that the whales out in the sea included her grandmother, aunts, uncles, and cousins. "She was vigorous and vocalizing and obviously interacting with the other whales," Paul Spong told reporters. "We were listening practically with our mouths hanging open all night."

In the early morning hours of July 14, 2002, part of Springer's pod cruised past Dong Chong Bay. Springer grew increasingly excited and exchanged calls with another whale. But after a while, the whales left. Springer calmed down and spent the next few hours alone. Then her family quietly entered Dong Chong Bay. Springer was now screeching and flailing about with excitement. The decision was made to drop the net. With a salmon grasped in her jaws, presumably to share with her relatives, she porpoised toward the other orcas waiting at the mouth of the bay. But then, for some reason, Springer hesitated and slowed down. The other whales moved on.

Now free, Springer spent the next few days wandering around the local waters, showing a disturbing attraction to boats and people. But gradually she moved closer to the other killer whales. She was spotted trailing her own pod about a half mile behind the others.

Three days later, Springer was accepted back into her community. Even as Keiko struggled to socialize and form bonds thousands of miles away, Springer was now home. A sixteen-year-old female relative, A51, or Nodales, began mentoring Springer in proper pod behavior. She even began to steer the young whale away from boats. "She is in excellent condition," Paul Spong said at the time. "There can now be no question about the success of the return project as it is clear that Springer has resumed living a normal social life among her kin and community." Springer remains with the pod to this day.

Back in Norway, things were not going quite as well for Keiko. He needed to get out of his fjord before it froze over. In October 2002, HSUS won permission from the Norwegian government to relocate Keiko to a site called Taknes Bay near the town of Halsa. Keiko arrived there on November 8. When winter set in, Keiko ended up under an ice floe anyway while swimming in a nearby fjord. When he came up to breathe, he was blocked from reaching air and had to struggle his way through the ice. Finally, he succeeded in breaking through the frozen cover, but not before scraping a good deal of skin off the top of his head. The wound eventually healed completely.

Throughout 2003, HSUS discontinued its attempts to sever Keiko's nearly lifelong umbilical cord with people, directing his caretakers to return to daily interactions. Though he still went out for "walks," Keiko mostly stayed in the bay, where he again assumed status as a local celebrity and tourist attraction. The long-term plan was to help him find a home of sorts with the Norwegian whales, but their travel patterns had shifted along with the herring, and they did not swim past Taknes that year, as they had in the past.

In early December 2003, Naomi was called into the Gaithersburg office of HSUS president Paul Irwin, along with a few other senior officials. They all sat down.

"I've just gotten a phone call," Irwin said. "I'm afraid Keiko is dead." He had learned the day before that Keiko was off his food and receiving antibiotics. But he decided not to tell anyone until it was clearer what was happening. But on December 12, Keiko had beached himself—attended by his loving caretakers—and died.

Naomi sat in stunned silence. She had sincerely hoped for more time to see what Keiko would do in Norway. It took her a moment to gather her bearings.

"Will there be a necropsy?" she finally asked, trying to remain as scientific and poised as she could. Irwin said no. Lanny Cornell had said Keiko's symptoms were consistent with pneumonia, and they'd have to leave it at that. Keiko would be buried beneath a pile of rocks on the shore at Taknes, so that people could quietly pay their respects. The spectacle of conducting a necropsy on the beach—slicing open the world's most beloved whale in front of the public—would be too much for everyone to bear, it was thought. Naomi vigorously opposed that decision. Critical data could be collected from Keiko, including diet, nutrition, muscle tone, infections, immune health, and, of course, cause of death. A curtain could be raised around Keiko to avoid offending anyone, she argued. Naomi lost that fight. Her insistence cost her some points among her superiors.

Immediately, the critics began circling in, accusing Naomi, HSUS, and the entire campaign of "murdering" Keiko. It was tough to take. HSUS staff worked overtime, struggling to make public the entire timeline. That way, Naomi thought, people could see for themselves that it had been seventeen months since Keiko took off on his own from Iceland, and that Keiko was among the oldest male orcas ever to have been in captivity.

But the criticism was deafening. The I-told-you-so retribution was barbed and heavy. Naomi felt compelled to respond, which she did in a post to the HSUS website, titled "Keiko's Sudden Exit from the Limelight Caps a Long, Strange Trip."

"From a lonely, broken whale in a semi-tropical climate wholly foreign to him, he was once again in the vast northern wild of his birth, robust, adventurous, alive," Naomi wrote.

> *On his "walks" in Iceland, he was free to leave or stay. And leave he did—taking off across the open waters separating Iceland from the rest of Scandinavia, swimming steadily for at least five weeks, almost certainly feeding himself, without human contact.*

What must he have thought, in those immensely wide spaces? Following the mackerel, hearing the whistles of dolphins and the booming of sperm whales for the first time in years? Was he frightened? Perhaps a little, as one is when one first leaves the comfort of the familiar for the unknown of "real life." But he kept his head, and he steered true and came at last to another shore, in Norway.

In his final months in Taknes, he was free to come and go as he pleased. He explored his surroundings, interacted with his human caretakers, mugged for the occasional camera, even chased birds and fish. He breathed clean sub-arctic air, watched the days grow shorter and then longer again, felt the fury of storms and the joy of a boisterous breach, with no walls confining him. He lived longer than almost any other male orca ever has in captivity. He lived at least as long as most male orcas do in the wild. He died as many might wish to die—suddenly, with little warning, happy and free and among friends. What a long, strange, amazing trip Keiko took. We will miss him.

Naomi's preemptive defensiveness was well-founded. The attacks kept growing. Clive D. L. Wynne, an associate professor of psychology at the University of Florida, unleashed an arch and accusatory op-ed in *The New York Times* on December 27, 2003. The story of "probably the most expensive animal in human history" had come to a close, Wynne said. By the time Keiko died, seven years of work and more than $20 million had been expended "vainly—and unwisely—trying to return the whale to the wild."

Though Paul Irwin and his HSUS minions had committed themselves to offering Keiko the "chance" of freedom, "there was never a shred of evidence to suggest that freedom was an aspiration that Keiko shared with the humans who cared for him," Wynne said. "Indeed, what we know about Keiko's response to his attempted liberation suggests quite the opposite."

Mark Simmons was even less charitable. He suggested that Keiko had died after he was prematurely jettisoned from the rehab project when the McCaw money, allegedly needed to keep him alive, dried up. He said Keiko was forced to rely on handouts from Norwegian fishermen up until he died. It incensed Naomi. Keiko had lived in Norway for fifteen months before he got sick. His caretaking team was right there with him the entire time. It was well established that, once captive cetaceans presented outward signs of illness, their clinical pathology had usually reached a life-threatening stage.

But the attacks continued. Opponents of the project said Keiko had turned up in Norway famished and sick. They alleged he had suffered needlessly and would still be alive if he had been returned to captivity.

Those challenges did not go uncontested. Dr. Lanny Cornell shot back at the critics. Keiko had shown no reduction in girth levels since leaving Iceland, he said, which would have been expected after such a prolonged period of active swimming if he had been fasting. Cornell had also reviewed extensive dive data from the period between Iceland and Norway and found that Keiko "exhibited deeper diving than he had ever done before."[4]

Such deep-water diving consumed a lot of energy and "would likely not be continued without the reward of successful foraging." Since there was no sign of weight loss, to sustain his weight Keiko would have needed to maintain an average daily intake of 125 to 150 pounds of fish, Cornell said. If he had not consumed that, "it would have manifest in emaciation and severe indentation behind the head and in the rib areas.

"There is overwhelming and conclusive evidence that Keiko foraged successfully and was able to sustain himself in the wild during the summer of 2002," Cornell insisted. "I am unaware of any scientist or veterinarian with even a rudimentary understanding of orca whales who could credibly claim that Keiko could undertake such activity and arrive in the condition that he did without being able to feed himself."

Bitter as it was, the fight over Keiko's death was soon overshadowed by another attack at SeaWorld, this time in San Antonio.

Trainer Steve Aibel was performing water work before a packed house with the male Kyuquot, or Ky as he was known. Ky was born to Tilikum and Haida II in 1991 while they were still at SeaLand of the Pacific. He was the reason that Tilikum had been isolated in the medical pool in Victoria and then sent to Orlando on an emergency import permit from NMFS.

Ky had been very attached to his mother.[5] When Haida died in 2001, it sent him into a deep bout of depression. He became moody and refused to perform certain behaviors. Ky assumed the subdominant position within the social structure in San Diego. In 2002, his closest ally, another subdominant whale named Winnie, died and left Ky more ostracized than ever from the rest of the animals.[6] The female Kayla and the male Keto, in particular, took out their frustrations on Ky, raking him repeatedly with their teeth.

In August of 2003, Ky had refused to let an unnamed trainer out of the water during a hotdogging session. He did not become aggressive, according to his Animal Profile, but he refused to obey callback tones and hand slaps on the water and persisted in blocking the trainer's exit. The trainer was able to move over to the Plexiglas wall and flee from the tank. When Ky noticed the escape, he began fast-swimming around the pool's perimeter, then slid out of the water and onto the main stage, ignoring all attempts to bring him under control. Water work with Ky was halted for a brief period after that.

"During summer 2003, this avoidance behavior became more prevalent, and also more unpredictable," his Animal Profile said. But, it added, "Overall, Ky is a very gentle natured animal."

Then, on July 27, 2004, Ky and Steve Aibel were thrilling the crowds with another hotdog session in the "Shamu Adventure" show when the whale suddenly refused to perform a rocket hop. Ky then refused to let Aibel exit the pool. Instead, he swam over Aibel the way a cargo ship might swamp a sailboat. He blocked Aibel's attempts to escape. Whenever the trainer came up for air, Ky would shove him back down under the water. Other trainers rushed to the pool's edge, but they could do nothing to stop the aggression.[7]

It took the audience a few moments to realize the attack was not part of the show (the entire incident is posted on YouTube).

Justin Lecourias, of Houston, was there with his family when it all began. "I said 'This ain't right, something's going on,'" he told reporters. "And then it got real bad." The whale spun out of control. "The whale would come up and the guy would go under before he hit him, so I guess it wouldn't hurt him that much. And then one time he didn't come up for quite a period of time, and that really freaked me out then. I told the kids to close their eyes, something was wrong . . . was like, you know, exhausted, coming around, the trainers and everyone around him was going nuts."

Ky continued to leap halfway out of the water and land on top of Steve Aibel. He refused repeated tries to call him back to stage, including tones, hand slaps, and other commands from trainers around the pool. When Aibel began to climb out of the water, Ky spotted him and pulled him back in. After several frightening minutes, Aibel was able to calm the rampaging adolescent and escape without injury. The crowd applauded.

Despite the tantrum, Ky never opened his mouth on the trainer. "It looked like Ky lost a little bit of focus," Aibel said later. "Seventeen years of training with animals and I've never had an experience like that." He said he wasn't frightened, though it sure looked that way on the video. Nonetheless, he did remain remarkably composed. Aibel said that years of training had taught him the patience needed to resolve the situation and get out of the pool. Ky was reaching sexual maturity, he explained, and a hormone surge might have triggered his abnormal behavior.

SeaWorld canceled the Shamu show for the rest of the day. Three days later, Aibel returned to water work with Ky.

The incident was followed by yet another attack, in Orlando in 2005, which sent an unidentified trainer to Sand Lake hospital for treatment of minor injuries. The cause was an "overly excited" killer whale, according to SeaWorld official Becca Bides. Later that same year in Orlando, an orca named Taku began bumping a trainer in the slide-out area. The other whales refused to perform after that incident, witnesses reported.[8]

April of 2005 brought even worse news to SeaWorld. Splash, the male orca that had fractured the arm of San Diego trainer Tamaree, was ailing once again. This time, he was not recovering quickly. Splash languished in the med pool, not responding to antibiotic treatment for his infection, which reportedly began in a tooth. Off-duty trainers came in to work just to be with the whale and his distraught caregivers.

"They were trying hard to be professionals at the same time they were losing a friend," recalled animal-training supervisor Al Garver. Watching his trainers trying to conceal their sadness "was the toughest thing for me to see."[9]

Splash was slowly disappearing. His two best friends, Orkid and a seven-year-old male named Sumar, tapped at the med pool's gate for him. Sumar's life had been no less abnormal than that of Splash and Orkid. He was born in 1998 at SeaWorld Orlando, the offspring of Taima and Tilikum. But Taima rejected Sumar and attacked him when he was just three months old.

Whatever Taima's motive, her son had been taken away from her and placed in another pool with the park's other females, Katina and her daughter Kalina. Nine months later, Sumar was sent to San Diego, where the matriarch Corky once again adopted an infant newcomer as her own.[10]

Now, Sumar watched in helplessness as his best friend expired quietly in the pool next door.

Splash's demise and its aftermath were covered in an in-depth feature in the *San Diego Union-Tribune*. In captivity, the death of a killer whale was as unnatural as its life. "Within the normally upbeat environment of SeaWorld, personnel who had worked closely with the 16-foot-long killer whale were devastated," the article said. SeaWorld veterinarian Tom Reidarson called the death "as hard a loss as when I lost my best friend," and Robbin Sheets, the supervising whale trainer, called Splash "a big sweetheart," who needed "a little extra attention."

Bereft trainers "put on brave faces and returned to work," the article said. "The killer whale shows at Shamu Stadium continued uninterrupted." Hours later, "trainers were so upset they could only perform a 'dry' show"—they could not focus on doing water work.

Few members of the public knew that Splash had died. Most of them didn't even know there *was* a Splash. The "Shamu" stage name had made it nearly impossible for people to tell one black-and-white whale from another. The *San Diego News Tribune* asked Naomi about that practice. "It's a great technique for keeping people unaware of just how many animals die," she said, adding that, during the previous nineteen years, on average one orca died each year at SeaWorld's parks combined.

But SeaWorld argued that Splash had lived far longer in captivity than he would have in the ocean. His life span "was a testament to everyone's hard work on his behalf," chief vet Tom Reidarson remarked. Naomi was not given a chance to respond to that claim. But she considered what she might have told the press. It was possible, she thought, that captivity started Splash's problems in the first place, causing his premature delivery, which led to the calf's developing epilepsy and other health problems.

Even if captivity wasn't the cause, one could argue that prolonging the animal's life was no mercy. In the heartless calculus of evolution, a diseased or deformed infant would be detrimental to the gene pool. In nature, the best thing for such an animal was for it to die of its maladies quickly.

Splash's necropsy report indicated that he died of stomach perforations.

The year 2005 was not over. More aberrant behavior among captive killer whales lay ahead.

On October 9, SeaWorld San Antonio's female Kayla gave birth to a new calf. Kayla was the offspring of the Icelandic female Kenau and the Northern Resident male Orky, both now deceased. The calf's father was Keet, offspring of Kalina (Baby Shamu) and Kotar (who died in 1995 when a gate crushed his skull).

This was Kayla's first pregnancy, and she had never spent time around nursing mothers before. She'd never learned any proper maternal skills because no female orca had ever taught her. Even though her calf was normal and healthy, Kayla rejected it, refusing to offer it milk.[11] For the first time, SeaWorld employees had to take on the nursing responsibilities of a captive killer whale who had failed her own calf.

More than fifty staffers were assigned to provide around-the-clock care for the calf, a female, including bottle feedings every two hours with infant formula, mixed with milk expressed from her mother. SeaWorld called the infant "K-calf" after her mother, but later renamed her Halyn. After separating Halyn from Kayla, veterinarians at San Antonio put the infant into a holding pool along with a bottlenose dolphin, but that animal shunned her as much as her mother had. She was moved back to her original tank.

SeaWorld took the maternal rejection in stride. It was all in a day's business, seemed to be the message. "We don't know all of the specifics of why an animal rejects a baby," Dudley Wigdahl, vice president of zoology, told reporters in Texas. "A first-time mother may have been confused with the birthing process or mechanics of birth."

Some staff at SeaWorld, meanwhile, relished the chance to play surrogate mother to a newborn orca. They viewed the episode as a *plus*. "This is a career goal to work with a calf from day one," said Julie Sigman, who had worked at the park for a decade. Usually, she added, it took several months.

Naomi shook her head when she read that. These people are delusional, she thought. They should be asking themselves why Kayla rejected her calf in captivity when something like that was highly unlikely to occur in the wild. Instead, they were doing a cheery public-relations dance about how *lucky* they were to hand-raise a newborn killer whale—an animal they were now anthropomorphically referring to as SeaWorld's newest "baby."

29

Ken and Kasatka

If, one day, SeaWorld is forced to remove trainers from its killer whale pools once and for all, historians may look back on 2006 as the beginning of the end of water work. Two attacks at the end of the year, both in San Diego, were frightening and fateful.

First came an incident with the aggression-prone Orkid. On November 15, she was doing water work in the newest Shamu show, called "Believe." The thirty-minute multimedia spectacular celebrated children's fascination with the sea and its creatures. "Believe" was divided into six acts: Discovery, Share the Joy, Ballet, Immersion, Pass the Torch, and Celebration. SeaWorld boasted that its choreography showcased one hundred separate orca behaviors, fifty-two of them newly developed. On average, water-work performers were now doing thirteen behaviors with the whales, as opposed to previous shows that averaged about eight. The extra workload heaped additional stress onto humans and orcas alike.

On this day, Orkid was doing a water-work segment along with Sumar—the juvenile son of Tilikum and Taima—and orca trainer Brian Rokeach. Rokeach had worked with killer whales for less than three years. This was his first time doing water work with the unpredictable Orkid.

Three other trainers were onstage, including Ken "Petey" Peters, who had already experienced aggressive incidents with Kasatka in 1993 and 1999. Rokeach had just completed a successful "fluke splash ride" and dove off Orkid, who was now at one end of the pool. Then she swam up beside him, which was not in the playbook. The other trainers called Orkid and Sumar back to stage. The whales dove under the surface and disappeared, presumably to head back to the control trainer onstage. Rokeach began swimming the short distance to the stage when Orkid sneaked up behind him and took his left ankle into her mouth.[1]

Orkid executed a barrel roll and, with Rokeach in front of her, pushed the trainer all the way to the bottom of the thirty-six-foot-deep main tank, A Pool. Sumar dove to the bottom with them, but did not take part in the attack.

Back onstage, Ken Peters scrambled to slap the water's surface in an effort to recall the off-behavior whales to stage. When that failed, he ran in the back and retrieved the remote audio recall box and signaled to them three or four times to return to stage. Neither orca obeyed the tones. Peters slapped the surface one more time, and it worked. Orkid and Sumar swam up to the stage and rested their chins contritely on the curved concrete like errant children.

Finally, and much to the relief of staff and audience members alike (a video of the incident is also playing on YouTube), Rokeach broke the surface and took a deep breath. He had been underwater for about twenty-six seconds.

With the two rambunctious whales under control, Rokeach paddled over to the stage behind them and, with help from other trainers, escaped from the water. The trainer waved gallantly to the anxious crowd and slowly limped backstage with a torn ligament in his ankle. It took a lot of physical therapy for him to recover.

Rokeach and Peters later said that Orkid did not appear to be acting aggressively. She was simply presented with the chance to grab Rokeach's foot, so she did. Both trainers said Orkid tended toward opportunism and had grabbed feet before. They could recall no precursor to explain why Orkid went off behavior.

Brian Rokeach took the fall for the incident. It was his fault, he said, not Orkid's. It had been a mistake for him to swim toward stage before Orkid and Sumar were under control. But he explained that he knew that other trainers were about to bring another whale into the pool and wanted to exit quickly so as not to delay the next segment of the show. He could have left the pool from the side, but decided to go for the stage instead because it was closer.

After the attack, SeaWorld added new rules to its safety protocol, including a requirement to have one trainer stationed with the callback device at all times during water work, and increasing the number of trainers in the Shamu show from four to five. Orkid, meanwhile, was permanently taken out of water work.

These measures would do nothing to prevent another terrifying whale attack, in two weeks, in the pools of San Diego.[2]

For the third time, Kasatka turned on Ken Peters, this time with far more serious implications. On November 29, 2006, a small midweek audience of about five hundred people were scattered about the fifty-five-hundred-seat stadium to see the "Believe" show. Corky had been put on light duty that day because she had recently been raked by Kasatka and the wounds on her flukes had not healed. Orkid was not doing water work because of her incident with Brian Rokeach, so Sumar and Kasatka were called upon to perform much of the water-work segments in the show.

Backstage, however, Kasatka's newest calf, Kalia (sired by Keet and just under two years old), was getting rowdy in one of the pools. One senior trainer, John Stewart, noted that Kalia was being extra playful and "acting a little goofy" during the performance, yet nothing out of the ordinary for a young calf. But a supervisor/trainer at the stadium, Tucker Petrzelka, disagreed, saying that Kalia was out of control during the show.

Two other trainers, Lindy Fordem and Matt Fripp, also noticed some type of commotion backstage between Kalia and her mother. Fordem saw Kasatka "head bobbing" the calf. Kasatka often behaved that way, but Fordem thought the older whale was being extra-stern that afternoon, describing Kasatka as acting like an "angry mom." Fripp witnessed the same incident, but thought nothing of it.[3]

When Kasatka's turn came to perform in the show, Lindy Fordem "walked" the whale through the connecting backstage pools and handed off Kasatka in the main pool to Ken Peters. "Mom was being very vocal with the calf," she said of Kasatka, not as any kind of warning, just a point of information. Regardless, Peters did not hear her.

Kasatka began the segment by performing perfectly. Peters asked her to do a surf ride, followed by a foot push ending in a slide across the stage. He did some dry behaviors with her from stage, then dove into the water for the big finale, the rocket hop.

As Peters dove under the water to meet up with the whale, the four other trainers were onstage dancing and clapping to the loud music. Before long they could tell something was not right. Petey had been under the surface for far too long.

Peters had been waiting for Kasatka to touch his foot, the beginning of that particular behavior. He was about ten or fifteen feet down. Suddenly, he heard a killer whale vocalizing loudly. Peters described it as a distress vocalization or cry.

He later learned the wailing was Kalia's screeching for her mother from the other pool.

Kasatka instantly pulled her rostrum away from Peters's feet. Then she grabbed his ankles, pulling him underwater for several seconds. When he surfaced, she grabbed him again, this time "rag-dolling" her trainer violently by shaking him back and forth with her powerful neck muscles. Kasatka took him under again, for a minute or more.

Then slowly and deliberately, as if performing a bizarre underwater pas de deux, the whale began to spiral upward with Peters's foot in her mouth. She exhaled a cloud of white bubbles from her blowhole.

When they finally resurfaced, Tucker Petrzelka heard a shout for help. He slapped the water, trying to bring Kasatka back to stage. Matt Fripp grabbed the callback device and deployed it while John Stewart slammed a metal bucket against the pool's side. Kasatka was having none of it.

She decided to take Peters to the bottom once again. They could be seen beneath the surface. She still had a foot in her mouth, and she dragged the trainer around, dunking him periodically and ignoring all signals to return to stage. Peters remained unbelievably calm, as he was trained to do (much like Steve Aibel when Ky had begun porpoising on him in San Antonio two years earlier).

Finally Peters told his colleagues to abandon the recall effort since it only seemed to be making Kasatka bite down harder. Each time he tried to extricate his foot from her huge jaws, she bit down again.

By now the audience was terrified.

Kasatka was careful to keep Peters in the middle of the pool and away from the other trainers, who were trying to rescue him at the edges. Peters managed to hold his head above water and gently stroked the whale in an effort to calm her down. For a while, she did just that.

Then a trainer threw a "scubacuzzi" (a life preserver with scuba gear and oxygen supply) onto the surface. Kasatka slowly swam over to inspect the object, keeping Peters in her mouth and away from the oxygen.

As park employees ushered the shell-shocked audience from the stadium, several staff members threw a net into the water and began pulling it across one end of the pool. Kasatka let go of Peters's foot to go have a look at the net. He was treading water in the middle of the pool. The whale swam underneath him. Peters tried to keep her from grabbing his feet again by kicking her.

Peters realized the tactic was futile. Resigned that Kasatka was going to take him under yet again, he drew in air and waited. It didn't take long. She grabbed his foot, thrashed him around a little, then dove to the bottom anew.

This time, she laid her entire five-thousand-pound body on top of the trainer, pinning him to the concrete for a minute or more.

Peters went limp. He felt his breath being forced out. He wondered when, or if, Kasatka was going to let him up.

Mercifully, she grabbed him and brought him to the surface to breathe. Peters began rubbing her sides again. Finally, she let him go and began drifting toward the stage.

By now, the staff had got netting across part of the pool, from stage to slide-out area. Kasatka and Peters were about three feet from the net, close to the slide-out. He backed away slowly from the brooding Kasatka, gingerly patting her flank the entire length of her body, then turning away quickly. He went over the net and, with a few powerful strokes, beached himself onto the shallow area. He sat in the few inches of water, now protected by the net, catching his breath.

Kasatka noticed he had escaped. She turned away from the stage and charged toward the net, clearing the barrier without any problem.

"Look out!" someone yelled. "She's coming over the net!"

Peters saw the whale coming back for him. He scrambled backward in the shallow water and tried to stand up to run. But his feet—wounded, numb, and bleeding—would not carry him. Peters was freezing and still out of breath. He felt ready to pass out. But his colleagues grabbed him just as Kasatka moved in only feet away. They pulled him to safety and a waiting crew of rescue personnel.

Kasatka turned and cruised away slowly. For several minutes after the attack, she swam around the perimeter with one of Peters's socks in her mouth, making pathetic-sounding vocalizations.

"She didn't show me any precursors. She didn't tell me, she didn't show me," Peters later told his colleagues. The aggression had come as a total surprise, he said, without any signals that Kasatka was about to go off. (Later he would recall young Kalia's pathetic cries from backstage, just before Kasatka went berserk.)

Ken Peters suffered puncture wounds to both feet and a broken metatarsal ligament in his left foot. He was transported to UC San Diego Medical Center for surgery and three days of IV antibiotics to prevent infection of his many bite wounds. He said that, although this was not normal behavior for Kasatka, he would not swim with her again.

The attack was big news. Networks descended on San Diego to cover the aftermath. CNN interviewed several guest eyewitnesses who talked about how shocked they were to see something so brutal at an entertainment park.

But SeaWorld's chief trainer, Mike Scarpuzzi, told CNN that Kasatka had never before displayed "this particular unwanted behavior to this extent." Yes, he admitted, "There are times like this. They are killer whales. She did choose to demonstrate her feelings in a way that was unfortunate." But, thanks to Peters's ample training and strong swimming abilities, he succeeded in calming the whale long enough to end the incident. "It turned out exactly the way we would want it if something like this were to ever happen."

Back on the East Coast, Naomi watched the whole drama unfold with a boatload of aggravation. First of all, what did Scarpuzzi mean by "if this were to ever happen?" It *had* happened in the past, she thought, shaking her head in disbelief. And it was clearly going to happen again.

Only next time, someone might lose his or her life.

Naomi's phone was now ringing constantly. She was the media's go-to source for all things captive killer whale. She told reporters that the frightening attack was yet more evidence that orcas were unsuited for public display and performing tricks for audiences. "The risk of a tragic outcome is too great," she told the *Los Angeles Times*, "for the trainers and the whales."

But Mike Scarpuzzi "rejected Rose's assertion and said that the public's concern for the welfare of large marine mammals comes in large part from shows and research at SeaWorld parks," the *Times* said. "He noted that although the whales are trained, they remain wild animals, with a degree of unpredictability. 'They *are* killer whales. Even though we've had thousands and thousands of good interactions, we are going to have some that don't go well.' "

Well, now *there's* an understatement, Naomi mused.

When Jeff Ventre heard about the accident, he was a third-year medical student doing an internal-medicine rotation at St. Luke's Hospital in Kansas City, Missouri. Ken Balcomb had referred a reporter from ABC News' *Nightline* to Jeff. They wanted a former trainer to come on and talk about the dangers of working with killer whales. But Jeff declined the interview. Not only was he busy in school, he didn't know Ken Peters, and he had never worked with Kasatka.

Meanwhile, the State of California had become increasingly interested in events at SeaWorld that did not "go well." They opened an occupational safety investigation of the latest Peters/Kasatka encounter. The state agency in charge was the Department of Industrial Relations, Division of Occupational Safety and Health, otherwise known at Cal/OSHA. In 1970, when Congress created the Occupational Safety and Health Administration (OSHA) within the Department of Labor, lawmakers included a provision allowing states to opt out of the federal program, provided that their own workplace safety requirements and enforcement equaled or surpassed those of the feds.

The Cal/OSHA investigation was thorough, and Naomi and other SeaWorld opponents were pleased to discover its official findings hard-hitting. Cal/OSHA released its verdict on February 28, 2007, three months after the Peters/Kasatka incident. Naomi was sent a copy. The report was not exactly a full condemnation of SeaWorld, but its implications were still something the multibillion-dollar industry would find disturbing, if not outright threatening.

The agency issued two documents: a two-page Information Memorandum, and a more detailed eighteen-page Narrative Summary of the attack and its historical context.

The best news for SeaWorld was that Cal/OSHA did not issue citations for any serious safety violations. "The employer has taken reasonable and responsible steps to correct the hazard in accordance with industry practices," the agency said. "Suggested modification of existing emergency procedures and training responses were also identified."

But the Information Memorandum drew SeaWorld's ire. Cal/OSHA duly noted that SeaWorld had relied on training to reduce aggression and developed emergency procedures for employees who became endangered. "The employer has been following industry standards and is recognized as a leader in training killer whales for performance. The employees are well trained and followed existing emergency procedures in this instance."

But then the agency stated the indisputable: "These procedures were not entirely effective at stopping the unwanted behaviors of the killer whale during this attack."

If the procedures weren't effective, Cal/OSHA concluded, then being in a pool with killer whales could not be classified as a safe work environment.

"Short of eliminating all of the water work in interactions with the killer whales, there is no guarantee that employees can be kept safe from an attack by a killer whale once they get in the water with the animal," Naomi read. She took that sentence to mean that the only true remedy to the hazard was an end to water work, period—something she felt would carve deeply into SeaWorld's thrill-driven bottom line.

Instead of a fine and a citation, however, Cal/OSHA's Information Memorandum offered eight "suggestions" for improving safety. Some of them SeaWorld was already implementing, or at least considering. Others, to the industry, were outrageous and out of the question:

1. Improve control of the behavior of the killer whales by reducing their environmental stressors and stresses from performances, interactions, and activities that take away from the time that the trainers and the whales need to reinforce social bonds, and allow the trainers sufficient time to focus on the whale and training of the killer whales.
2. Having more orcas within the population to choose from for various interactions would reduce the stress on individual animals and help to keep the trainers from having to rely on one or two animals for the majority of their performance tasks.
3. The presence of calves within the population can be a stressor both to the mother and to the other whales. A separate, non-performing population used for reproduction only could eliminate this stress within the performing population.
4. Animals not used for reproduction may benefit from surgical or chemical sterilization, demonstrated as effective in controlling unwanted behavior, particularly aggression, in a variety of domesticated mammals.
5. Additional shepherd's hooks and/or rescue and resuscitation devices should be located at the slide-out with the emergency net.
6. The employer should continue to investigate portable breathing devices such as but not limited to the "Spare-Air" model that could be worn by any trainer before they enter the pool with a killer whale.
7. Improve communication methods in the stadium so that trainers

located at the slide-out and trainers on the stage can effectively communicate with each other during a show.

But this suggestion was what most alarmed SeaWorld officials:

8. The employer should also consider that lethal force against a killer whale may have to be used to save the life of a trainer where an aggressive whale is out of control and not responding to other available control measures. This could be either above the water or in the water by means such as but not limited to a bang stick deployed by one or more divers in scuba gear.

Cal/OSHA's Narrative Summary provided fascinating insight into what went on backstage at SeaWorld, including the factors that drove captive killer whales to aggression—and how SeaWorld planned to combat the problem. It suggested that the SeaWorld parks had been pushing their killer whales and trainers too hard to continually outdo themselves in their daring water work feats—all for the purpose of drawing crowds back to the park each year. The report also included a blow-by-blow description of the Peters attack. It read like the shooting script of a Hollywood horror movie.

A leading cause of aggression in captive orcas, of course, was stress. "Common environmental stressors are food, territory or space, reproduction, and social interactions," the summary said. "Of great concern in any captive population is the lack of activity and a sterile, unchanging environment. These are stressors that can result in boredom and subsequent abnormal behaviors."

If killer whales do not have enough to do, the tedium will cause stress and aggression. But the opposite extreme might be true as well. All of the trainers interviewed by Cal/OSHA officials complained that the extra workload created by special VIP events and other interactions outside the regular shows were reducing the time they could spend on enrichment activities.

The relatively small tank size might also be stressing the whales. "These animals have responded and adapted fairly well to their spatial limitations," the summary said. "But the extent to which it is a stressor cannot be entirely known. An animal capable of traveling hundreds of miles is going to have some physical and psychological issues with being confined to a swimming pool. The drooping dorsal fin of a captive male orca (as seen in Ulises, San Diego's only adult male) has been attributed to a lack of long distance swimming and reduced muscle strength."

The presence of orca calves within earshot of performing whales was also deepening the dilemma. Some trainers told investigators that 90 percent of the problems they had with orcas were caused by calves. "The dam is always going to be protective of her calf," the report said. "And that kind of behavior is instinctual, and no amount of training is likely to supersede those instincts."

The criticism of SeaWorld's killer whale program only got worse from there. "Orcas are still wild, whether they were captured in the wild or bred in captivity, and as wild animals they can be completely unpredictable and unpredictable is dangerous. It is true that dangerous wild animals are often kept in zoos all over the world, but their keepers aren't at risk of being attacked because they don't perform, interact or get in a cage with the animals in a zoo." The show-business glitz of SeaWorld made it "closer to a circus than a zoo," Cal/OSHA said. "And yet even a six ton tiger might not be as fearsome if you didn't have to swim with it. The humans that swim with and perform with orcas in this setting are putting their lives in danger every time they jump into the pool."

Though only a tiny fraction of interactions had resulted in orcas going "off behavior" and threatening or attacking their trainers, "these two incidents in one month alone are somewhat alarming," the report warned. "In combination with the attack in Texas in 2002 [it] might be indicative of an alarming trend, possibly related to the demands of the new show and increased water interactions."

All of those incidents offered another frightening lesson: "Once the animal goes 'off behavior,' they are essentially out of control of the trainers." As one employee put it, trainers lose control of the whales until "they are done being off behavior." In other words, the whales, not the trainers, decided when the danger would pass.

Fortunately for Ken Peters, his "training and sense of calm" allowed him to wait out the rampage and finally escape. But "not everyone would be able to have such an extraordinary presence of mind while in the jaws of an animal that is capable of killing him," Cal/OSHA said. It was foolhardy and dangerous to "disallow our own primal fears [that is, the fear of being killed] to entirely rely on the actions of the victim to free himself."

The report minced no words in questioning the financial motives for SeaWorld's tolerance of water work hazards to its employees:

> *This is a marine park with a goal of conservation and education, but it is also in the business of entertaining. Shamu the killer whale has always been at the forefront of their marketing and advertising and is probably the main reason why people visit the park. . . . In order to maintain interest in the Shamu show, the employer has to make sure that they are doing something that the audience hasn't seen before. So each year is bigger, brighter, more spectacular, etc. Part of that is marketing but they have also increased their demands on both the orcas*

> *and the trainers to bring in something new and exciting each year in order to keep audiences coming back year after year.*

Cal/OSHA knew that the use of lethal force to stop an attack would not be adopted at SeaWorld. "Animal trainers believe that they know the animal, and they also believe that they can control the animal. This is an emotional response to their close association, bonding and training with the animal. It is clear that they have not considered use of lethal force as an option. However, at some point they may have to make a choice between a dead human or a live animal."

But even if trainers did make that choice, how would they kill an incensed orca with a trainer in its mouth? (It was hard to shoot an animal underwater, and the bullets might strike the victim.) Loud explosives (which can kill cetaceans but spare humans) were one option. "Bang sticks of sufficient caliber could be used by a scuba diver if he could get close enough," Cal/OSHA said, adding that such an extreme step would "naturally" come as a last resort.

Of course, that was assuming that SeaWorld allowed its trainers to resume water work, the report stated, "because the best solution is to not get in the water."

The report's final conclusions were damning:

> *The contributing factor to the accident, in the simplest of terms, is that swimming with captive orcas is inherently dangerous. The trainers recognize this risk and train not for if an attack will happen but when. The orca is capable of tearing off an arm, a leg, or a head, and if that is against its nature it could easily drown a human or trap it in the cold waters of the tank until the human expires from hypothermia. Even if the animal does not have the intent to kill, the bulk and weight of its body is enough to smash a person against the sides of the pool, knocking them unconscious or crushing them to death. Unfortunately, even with the best procedures and the best trained staff, there was little that the employees in this case could do to rescue their coworker when the tools they had at their disposal failed to sufficiently distract the orca or cause it to return to normal behavior.*

In both the Orkid and the Kasatka incidents, the animals "were probably gentler in these cases than they would be in the wild," the report surmised. "But even their 'gentle' bites still managed to do considerable damage to frail human flesh. Any less gentle and these injuries would have been much worse." For example, if the orcas shook trainers underwater and caused their heads to be "smacked against the sides of the pool," Cal/OSHA said, "they might not have even survived."

The safety agency then issued a highly prophetic warning: "If someone hasn't been killed already, it is only a matter of time before it does happen."

Extraordinary as the document was, the political maneuvering and corporate cloak-and-dagger intrigue that ensued was even more dramatic. Naomi learned from a friend and colleague, Bill Rossiter, of Cetacean Society International (CSI), that SeaWorld was trying to get as much of the Cal/OSHA citation overturned as possible. CSI had been following the entire Cal/OSHA investigation since November, and Bill was incensed at the undue influence SeaWorld was exerting to dilute the report.

SeaWorld was strongly objecting to the entire "Information Memorandum" and demanded its withdrawal, arguing that such a document should only be authorized when an actual violation of safety standards was identified. As for the "Narrative Summary," company officials complained that most of its contents went far beyond the expertise of the investigators and should likewise be deleted (even though it was based on interviews with SeaWorld's own trainers).

Bill and Naomi were contacted by a Cal/OSHA supervisor, who told them how adamantly the company was fighting to have as much of the report retracted as it could. "This whole request seems so stupid to me," the supervisor said. Even if the document was redacted and vastly changed, the original document would remain online for all to see, he said.

SeaWorld met in San Diego with Cal/OSHA officials to launch objections to the report. During a follow-up conference call the next day with senior Cal/OSHA leadership in Sacramento, the state agreed to revise the document. "Much of the information in the report reflects a complete lack of understanding of the complexities of marine mammal biology, behavior, and husbandry," Mike Scarpuzzi told reporters.[4]

Even so, he said, SeaWorld was reviewing safety procedures and had already adopted three of the eight recommendations: installing shepherd's hooks, an emergency intercom system, and the exploration of emergency "spare-air" supplies that could be carried by trainers while in the water with killer whales.

On March 2, 2007, Cal/OSHA sent out a press release announcing it had "revisited" the Peters attack reports. The agency had issued the Information Memorandum "in violation of its own policies and procedures," the statement said. "This error is being addressed and Cal/OSHA regrets the difficulties it may have caused SeaWorld, its staff, and its patrons."

The turnabout was not only breathtaking, it was unprecedented.

"Cal/OSHA realized its error when it met with SeaWorld officials yesterday to discuss these reports," the state's mea culpa continued. "As a result, Cal/OSHA will take corrective action, including withdrawing the 'Information Memorandum.'" The agency added that it would "thoroughly review" the Narrative Summary and make any revisions to that document as needed. "Cal/OSHA acknowledges that many of the statements made in the 'Narrative

Summary' require expertise in animal behavior, which Cal/OSHA does not have. Also, some of the expressions of opinion and other statements contained in the report stray from describing the evidence and are clearly inappropriate."

The official about-face in just two days was a staggering triumph for SeaWorld and Anheuser-Busch. Naomi would later be told that SeaWorld had put extraordinary pressure on the state to rein in what the company felt was the excessive meddling of Cal/OSHA. Governor Arnold Schwarzenegger, a friend and political ally of August Busch III's, reportedly intervened personally to ensure that SeaWorld got what it wanted.

Naomi learned that the withdrawal of an Information Memorandum was not unprecedented at Cal/OSHA, but the rewriting of a Narrative Summary was positively unheard of. It was time to fight back. Bill Rossiter told Naomi he was determined to keep the Narrative Summary of the Peters case intact, and he wanted her help.

On March 12, Bill wrote to Cal/OSHA's San Diego district supervisor, Luis Mireles, in a letter of protest signed by thirty-nine organizations in the United States, Canada, Japan, France, Mexico, and elsewhere, urging the agency not to retract any of the Narrative Summary.

"This event was neither unique nor unusual," Bill wrote. "We urge Cal/OSHA to maintain a strong position on the issue of trainer safety. If Cal/OSHA fails to act for human safety, the agency becomes complicit in the next tragedy, which is, as initially stated, 'only a matter of time.' "

But the redrafting of the original Narrative Summary proceeded. Naomi learned that SeaWorld attorneys were working alongside Cal/OSHA officials to make the changes. She requested copies of the original and redrafted documents and compared them side by side. The edits were mostly deletions, with few additions or revisions. More than half of the original document was simply gone.

The redacted text excluded any references to killer whales being inherently dangerous and unpredictable; information on orcas displaying individual personality differences that made careful evaluation of their "mood" daily or even hourly essential for trainer safety; admissions by trainers that stressors in captivity did exist and contributed to unavoidable risks of animals going "off behavior"; and that trainers had "no tools at their disposal to punish an orca that is misbehaving. There is little that they can do to punish an animal of this size anyway."

All descriptions of previous "off behavior" incidents at SeaWorld and other facilities (both injurious and non-injurious), save for two previous incidents with Kasatka and one incident two weeks earlier involving another whale at SeaWorld San Diego that resulted in a minor injury, were likewise deleted.

And of course any words about it being only a matter of time before someone was killed were likewise excised from the text.

"The original narrative summary made it clear that . . . in-water interactions were inherently risky and incidents such as the one between Kasatka and Peters could and should be anticipated," Naomi would later write. "The final version implied the opposite, leaving the reader with the impression that in-water interactions were inherently safe, that 'off behavior accidents' and attacks were completely aberrant, and that the routine safety precautions taken by trainers were good practice but almost never needed."

After the report was withdrawn, SeaWorld went on the PR counterattack. The curator for SeaWorld Entertainment, Julie Scardina, appeared on CNN's *Larry King Live* to defend orca captivity and—true to the park's party line—suggested that whales were better off in protective tanks than the big, bad open ocean. She repeated the corporate mantra that the lucky whales were receiving "pampered care" while also educating the public about orcas in the wild.

"You know, a city person living in the city is very different from a person living in the country," Scardina said. "They are so well taken care of, and what's happening out in the wild right now certainly isn't the best situation for the animals. Animals are barely surviving in many places out in the wild."

"Are you saying they're better off at SeaWorld?" King asked incredulously.

"I'm not saying they're necessarily better off one place or another. Populations are declining in the wild. We're taking great care of them in captivity. They're educating people. They know us. They're learning from us. We're learning from them. And people didn't even know about killer whales not that long ago. They were shooting them out in the wild." SeaWorld had an opportunity to educate people with animals that were "getting the most pampered care that you have ever, you know, experienced. If you've ever had a dog that has a veterinarian on call twenty-four hours a day that has ten trainers to take care of it every single day."

When Naomi watched the broadcast, she was struck by Scardina's use of the word "job" to describe what killer whales "do" at SeaWorld. They were being worked, quite literally, to death, Naomi thought.

SeaWorld's victory against Cal/OSHA was, however, dampened. The original Information Memorandum and Narrative Summary both appeared on various websites for all the public to see.

SeaWorld was not required to do anything to make its workplace safer for orca trainers. Cal/OSHA had only made suggestions, and those had been withdrawn.

Four years later, critics would charge that if Cal/OSHA hadn't bowed to political pressure, Dawn Brancheau might still be alive.

30

Tenerife

Not long into the new century, SeaWorld officials must have realized they had a reproduction problem on their hands: If anything, the corporation's orca breeding program was *too* successful. With only three parks to house its two dozen killer whales (SeaWorld Ohio closed in 2001), or eight per park, SeaWorld needed some spare real estate.

That's where "breeding loans" came in. The arrangements—killer whale rentals, more than loans—provided yet another revenue stream for SeaWorld while alleviating its overcrowding problems. Breeding loans also built stronger relations with other display enterprises while increasing genetic diversity in the collections, where some inbreeding was becoming almost inevitable.

In February of 2006, SeaWorld flew four of its youngest captive-bred orcas to a marine park in Spain on a breeding loan. The Spanish facility, known as Loro Parque, or "parrot park," was on Tenerife in the Canary Islands, the volcanic archipelago off the coast of western Africa.

Loro Parque was founded in Puerto de la Cruz, Tenerife, in 1972 by German entrepreneur Wolfgang Kiessling as a tourist attraction featuring tropical parrots, but had since grown to include dolphins, sea lions, sharks, penguins, alligators, and land mammals such as gorillas, chimps, tigers, and panthers. It had just completed construction on a new stadium, Orca Ocean, featuring a large postmodern amphitheater covered in a graceful series of white panels resembling the sails of an Armada galleon.

Financial details on the transaction were not made public, but SeaWorld earned a certain percentage of every ticket sold as long as its whales remained at Loro Parque. The loan was for twenty-five years, with a ten-year renewal option. Loro Parque agreed to pay for transport costs of the whales and provide business-class tickets and accommodations for all SeaWorld staff who needed to visit Tenerife during the construction phase and the first year of operations. SeaWorld would provide a supervising trainer to Loro Parque for the first year or so, and the Spanish park agreed to cover the trainer's five-star-hotel costs. Loro Parque also agreed to serve Anheuser-Busch beer products exclusively for its premium and imported beer.

The four young whales in the loan—two males and two females—had led lives that could best be described as "interrupted." There was Kohana, three and a half years old, who was born at SeaWorld San Diego to Takara and (via artificial insemination) Tilikum in May 2002.[1] When she was just shy of two years old, Kohana and her mother were transferred to Orlando. Eighteen months after that, in 2006, Kohana was separated from her mother and sent on her way to the Canary Islands. The other young female, Skyla, was born in Orlando in February of 2004 to Kalina and Tilikum but, at just two years of age, was dispatched to Spain.[2]

Skyla was related to both of the young males sent to Loro Parque. First there was three-year-old Tekoa, who shared a father with Skyla, Tilikum. Tekoa was born in Orlando to the hybrid Taima in 2000.[3] Unpredictable Taima, who had attacked her firstborn calf, Sumar, showed the same aggressive tendencies toward Tekoa. He also had to be permanently separated from his mother after multiple attempts by Taima to hurt or kill him. So far, Taima was proving to be a poor mother.[4] In April 2004, SeaWorld sent Tekoa to live in San Antonio, before he was flown to Tenerife in 2006.

Skyla was also the half sister of Keto, the son of Kalina and Kotar. He was by far the oldest and perhaps most dysfunctional killer whale of the Loro Parque quartet. Keto, ten, had been born in Orlando in 1995 but proved to be rowdy and somewhat unpredictable.[5] Before he was four, Keto was sent to San Diego, where he spent just ten months before being transferred to Ohio for ten months, and then to San Antonio.[6] Five years later, he was on the plane to Spain.

When Naomi learned about the "breeding loan" between SeaWorld and Loro Parque, and the young ages and disrupted lives of the four whales in question, taken from their mothers and other adults and sent by themselves somewhere else, she said to herself, "Nothing good can come of this."

Naomi discovered that SeaWorld had been a critical component of Loro Parque's new venture into killer whale training and display. Loro Parque "provided us greater flexibility in managing our collection of killer whales," SeaWorld spokesman Fred Jacobs explained.[7] Clearly, SeaWorld would be intimately involved in the care, husbandry, and training of its four whales. Meanwhile, several Loro Parque trainers had spent a year at SeaWorld San Antonio to learn all the highly complex aspects of animal training, husbandry, medical care, nutrition, enrichment, safety, stage production, and more that make up the components of a killer whale show.

The imminent transfer of the four whales—whose care and welfare (as Naomi read the law) were still covered by the US Marine Mammal Protection Act while the property of an American facility—sparked an outcry among anti-captivity advocates. Naomi and other anti-cap groups voiced their concerns about the loan, and Naomi worked behind the scenes with her contacts at NMFS to glean more information about the deal. She wanted to see what legal mechanisms, if any, might be in place to block the transfer. But the options were flimsy. Unlike the import of a marine mammal to the United States, the *export* of animals did not require a permit or a period of public comment—thanks to the 1994 MMPA amendments. HSUS legal counsel conceded there was not much they could do.

But a colleague of Naomi's at the UK-based Whale and Dolphin Conservation Society, Cathy Williamson, continued to pursue the matter with NMFS. She kept Naomi in the loop on her work, including a letter she sent to NMFS outlining why the deal should be struck down.

"We are extremely concerned at reports that the export will involve two very young female orcas, including one calf less than 22 months old that may still be suckling," Cathy wrote. "Separating calves from their mothers at an early age can lead to infant mortality and directly contradicts the advice given by marine mammal veterinarians, including Jay Sweeney, who states, in the CRC Handbook of Marine Mammal Medicine that 'attempts at removing a juvenile cetacean under two years of age from its mother frequently results in significant stress . . . [since it] remains emotionally dependent on its mother.' "

NMFS refused to take any action on the matter, and on February 14, 2006, the four young killer whales were flown to Tenerife.[8]

Opening ceremonies at Orca Ocean stadium in Loro Parque were a splashy affair, with three thousand guests, scores of VIPs, and more than two hundred reporters from the islands and the Continent, all gathered to see the first killer whales ever brought to the Canaries. The park's founder and president, Wolfgang Kiessling, played proud host to the head of the Canary Islands autonomous government, and several American dignitaries, not least among them August Busch III.

Many senior staff from SeaWorld had also been flying in and out of the Canaries. Thad Lacinak, now vice president and corporate curator for animal training, was present when the whales arrived and oversaw their introduction into the new 7-million-gallon complex. Chief veterinarian James McBain traveled to Tenerife at regular intervals to check on the animals, and Brad Andrews committed himself to visit twice a year. Senior orca trainer Mark Galan was sent from Orlando to Loro Parque for the first eighteen months of operations to oversee training and animal-care programs at Orca Ocean, whose video surveillance system was transmitted back to the United States for monitoring by SeaWorld staff.

On March 17, 2006, the first show open to the public was performed by Loro Parque's inexperienced staff and their young brood of whales. But problems with the new pools—the disturbing details of which would only come to light after the death of Dawn Brancheau—shut down the operation immediately. It took more than ten weeks to fix the pool issues: Loro Parque officially reopened on June 3. By that fall, things were running fairly smoothly with the orca shows.

Whenever SeaWorld's on-site supervising trainer, Mark Galan, left the islands, he was temporarily replaced by other supervisors from SeaWorld. In September 2006, Dawn Brancheau got the chance to go. Dawn had ascended through the ranks and was now an accomplished trainer. Congenial, athletic, and extremely attuned to the animals, she was one of the most respected figures at Shamu Stadium in Florida. Dawn also knew Keto, Tekoa, Skyla, and Kohana, as all four whales had lived in Orlando at some point in their unsettled lives.

Dawn was as popular at Loro Parque as she was back home. The staff was dazzled by her craftsmanship in the tank. She worked closely with the other trainers and became good friends with a young man named Alexis Martinez—a Canary Islands native with short-cropped, black hair and a charismatic smile. Dawn would often hang out with Alexis and his fiancée, Estefanía Rodriguez, after hours in a local tapas place.

Back in North America, meanwhile, SeaWorld kept moving its whales around like multi-ton chess pieces. In November of 2006, SeaWorld dispatched a young male killer whale named Ikaika on another international breeding loan, this time from Orlando to Marineland Ontario. Meanwhile, SeaWorld officials also switched Kayla, from San Antonio, and Taku, from Orlando.

The following October, Loro Parque had its first serious incident between a killer whale and a trainer. Claudia Vollhardt, a twenty-nine-year-old from Germany, was working a preshow warm-up session with the three-thousand-pound juvenile Tekoa.[9] The SeaWorld supervisor on duty at the time was Steve Aibel, the Texas trainer who had successfully calmed an amok Kyuquot and escaped from the pool unharmed in 2004.

Claudia Vollhardt had got her start at Loro Parque's dolphinarium and then transferred to Orca Ocean. This particular day, she was having difficulty executing a foot push with Tekoa. After a few aborted tries, Tekoa became frustrated and took Vollhardt's arm into his mouth.

Then he dove to the bottom.

After holding the trainer underwater for a moment, Tekoa dragged her back to the surface. He raced toward a metal gate separating the show pool from the back pools, and as others looked on in distress, Tekoa rammed Vollhardt against the gate several times.

Steve Aibel shouted to the other trainers to unfurl a net into the pool. A distracted Tekoa let go of the young woman, now unconscious, and Steve Aibel got her out of the water and administered CPR. Teeth marks clearly perforated her wet suit. Even as Vollhardt lay injured and bleeding by the pool's side, Tekoa tried to lunge out of the water at her. An ambulance raced the wounded woman to La Laguna Hospital, where she was admitted to the ICU. Her right lung had been punctured and her forearm was fractured into three pieces.

Loro Parque said the attack was an "accident" brought about by bad luck. Water work was suspended for more than six months at Loro Parque and extra restrictions placed on working with Tekoa.

After the attack, Alexis Martinez confided in Vollhardt about his growing worries over Loro Parque's whales and their increasingly unstable behavior. He was alarmed at how often the orcas were banging their heads on the metal gates. He had also witnessed many close calls, any number of which could quickly have turned hazardous.

More dreadful news slammed the industry in the coming months. On October 17, 2007, less than a year after his relocation from Orlando to San Antonio, the male Taku, born in 1993, suddenly died.[10] He was fourteen. Taku was the son of Tilikum and Katina.[11] Taku later impregnated his mother to produce Nalani in 2006,[12] the world's only fully inbred orca.

Eight months later, death claimed Taku's niece, two-and-a-half-year-old Halyn, the female calf whose mother, Kayla, had refused to nurse her after her birth in San Antonio, a source of rejoicing among some staff at the prospect of hand-raising an orca calf from day one. The cause was acute encephalitis.[13] "While we recognize that death is part of the life cycle," SeaWorld said in a news release, "we are saddened over the unexpected loss of this animal."

Then in the spring of 2009 Loro Parque's youngest killer whale, five-year-old Skyla, joined her half brother Tekoa (both sired by Tilikum) in banishment from water work during shows at Tenerife. Skyla was in the pool with trainer Rafa Sanchez when she began shoving him around the pool with her rostrum during a show. She then rammed him into the pool wall.[14]

Now, only Keto and Kohana were allowed to perform water work at Loro Parque.

That summer, Dawn Brancheau's friend Alexis Martinez turned to his personal diary to describe his growing worries about the erratic behavior among the American whales. Their seemingly bottomless sex drives were on the verge of upending the fragile social order imposed upon the hormonally charged adolescents.

"Keto is obsessed with controlling Kohana, he won't separate from her, including shows," he wrote. "Tekoa is very sexual when he is alone with Kohana (penis out)." Meanwhile Keto was also engaging in homosexual behavior with Tekoa.[15]

In September of 2009, the rotating SeaWorld supervisor at Loro Parque was Brian Rokeach, the San Diego trainer who had been attacked by Orkid in 2006, resulting in a torn ankle ligament. Alexis wrote, "Brian had a small incident with Keto the first hour of the morning," and said it had been "a very bad day for Keto." A few days later he added this entry: "All the animals are bad. Dry day for Kohana."

Back in the States, rumors were flying about a possible sale of the SeaWorld franchise to another corporation. Anheuser-Busch had been acquired by the huge Belgian conglomerate InBev. That company was reportedly only interested in the core drinks operations at Anheuser-Busch. On October 10, 2009, InBev announced it had sold SeaWorld Parks and Entertainment to a huge private equity firm, the Blackstone Group, for $2.7 billion.[16] Blackstone's other assets included part of SeaWorld Orlando's competitor, Universal Studios, and a European company called Merlin Entertainments, which, ironically, had openly questioned the suitability of holding cetaceans in captivity.[17] Blackstone also owned Legoland and Madame Tussauds. The SeaWorld deal included ten parks in total, with 25 million visitors a year, including Busch Gardens, Aquatica, and Discovery Cove.

SeaWorld officials said no immediate changes would be made to their operations, and the vast majority of visitors had no idea that the sale had even taken place. "It don't matter," park guest Lisa Dunagan told the *Orlando Sentinel*. "Everyone is here to see Shamu."

Four months later, on February 24, 2010, Blackstone investors would be forgiven for pangs of buyer's remorse. Perhaps they did not realize that their killer whales might actually live up to their names.

31

Death at SeaWorld

The early sky loomed gray and foreboding. A strong gale blew out of the north and rippled across Live Oak Lake with gusts up to thirty miles per hour, slamming the windchill down to the low forties. For Dawn Brancheau and her husband, Scott, February 24, 2010, started as just another cool winter morning in their modest one-story home on lakefront property outside St. Cloud, Florida, a quiet Orlando suburb near the honky-tonk tourist mecca of Kissimmee.

Scott and Dawn lived in the house with their two chocolate labs, Ruger and Maggi. It was a good life, full of the comings and goings of family and friends from town, church, and of course SeaWorld. Dawn had 250 people on her Christmas card list. She was also growing out her long, straw-colored ponytail, which she was about to snip off and donate to Locks of Love, a charity that provides human-hair wigs to cancer patients.

The former high school class president and homecoming queen treasured her job at SeaWorld—she had never tired of her lifelong dream. Having worked her way up the ladder at Shamu from trainer to assistant trainer in her first two years, Dawn continued to rise with a promotion to assistant supervisor in 2004 and supervisor in 2008. Now in a salaried position, she was one of the most senior people in the entire animal training department.

Dawn continued to garner favorable annual performance appraisals. Her most recent review, in 2009, listed among Dawn's competencies that she "passionately supports department, park and corporate initiatives."

Dawn had worked with nearly all the killer whales that came through SeaWorld Orlando during those years and had been assigned to various animal teams in the past, including for Takara, Taku, Skyla, Tekoa, Unna, Sumar, Ikaika, and Taima. She was now on the teams of Katina, Kayla, Trua, Nalani, and Malia for water work whales, and the teams for both dry work whales, Tilikum and Taima (Taima had been removed from water work in 2007 because of her aggressive tendencies).

Now approved for "Level 3" water work, Dawn was one of the few people in the park who were given the okay to perform every behavior in a whale's repertoire, including the hotdogging stunts such as hydros and rocket hops. Dawn was a star in Orlando. A picture of her, smiling next to a killer whale, was part of SeaWorld Orlando's advertising campaign. It was included in a giant composite mural that welcomed arriving passengers to Orlando International Airport. Dawn was often featured in TV news stories about goings-on at SeaWorld and was profiled in Orlando newspapers and magazines as well.

Back in 2006, she had been written up in the *Orlando Sentinel* when SeaWorld was getting ready to launch its new "Believe" show, which the newspaper said was designed to be "inspirational" and to leave audiences "with the notion that if people can swim with killer whales, they can achieve anything."

Of course, if people can swim with killer whales, they might also die with killer whales. Even Dawn sensed that. "You can't put yourself in the water unless you trust them," she said, "and they trust you."

After Dawn died, many people would wonder if she had grown to trust Tilikum more than she should have.

But in February of 2010, Dawn Brancheau must have felt extremely safe at work. After all, SeaWorld had come a long way since the bad old bone-crushing days of 1987, when injuries in the killer whale pool were rampant. Much to its credit, SeaWorld had continued to implement and reform training, safety, and emergency procedures designed to provide several layers of protection to reduce accidents, attacks, and the impact of other aggressive acts by the killer whales in its collection.

In addition to compiling Animal Profiles and incident reports, SeaWorld had also created something called the Behavior Review Committee (BRC), whose purpose was for senior staff at all SeaWorld parks to constantly review animal and trainer safety records before approving whales and trainers for advancement to more complicated behaviors, including water work.

Trainers learned how to go limp when grabbed by a whale, and whales learned how to keep their mouths closed and to keep swimming if someone fell in the water, and to swim toward one of the gates whenever a net was dropped into a pool.

Meanwhile, new trainers were walked through a lengthy checklist of steps they needed to complete, along with a supervisor. The complicated and arduous undertaking started with the basics—such as how to walk safely around a pool and over gates with and without buckets—and went through every emergency procedure, animal incident report, and safety check. Working at Shamu Stadium required a dive physical, a scuba certificate requirement, CPR classes, spotting directions, "water extrication" classes, instructions in needle disposal, and monthly net-deployment drills.

All staff were fully schooled in the locations and proper use of scubacuzzis, oxygen tanks, shepherd's crooks, scuba gear, backboards (stretchers), pony-bottle air equipment, air horns, emergency alarm switches, and emergency nets.

Then there were the safety procedures specific to Tilikum.

On day one at Shamu Stadium, all new trainers were required to learn the Tilikum Safety Protocol. Part of that procedure included giving newcomers what was casually referred to as the "Tilly talk." The talk was short and simple: If you get in the water with Tilikum, you will likely not survive.

The rules on conduct around Tilikum were pages long. Stadium employees were required to stay behind metal barriers when working with the whale, while guests and non-Shamu personnel had to remain at least five feet behind the pool walls. In the slide-out areas, trainers had to remain positioned in front of his pectoral flippers. When Tilikum was in the back pools, swimming in the front pool was permitted only when all gates were locked and doubly secured (and vice versa).

Some of the rules were quite extravagant, especially when it came to walking over bridges that crossed the gates between pools. In the "Tilikum General Guideline," for example, rule number 14 stipulated, "An approved non-tactile team member or above, in a wetsuit, may walk over the front side of a gate surrounding a pool to which Tilikum has access only if Tilikum is under stimulus control and gate has a safety rail."

In case of emergency "in which trainers, guest, or non-department personnel fall or otherwise enter the water with Tilikum without authorization," an extensive and separate protocol was also devised. The first step was to initiate the "A.A.R.M" sequence: Hit the *alarm* to activate the siren and call 911; try to get *air* to the victim via a pony bottle; unfurl the appropriate *reel* of netting; and *move* all animals to other pools.

Net pulls were used to control, separate, or remove the animal, preferably toward D Pool, where the false bottom could raise the whale quickly out of the water if needed.

If Tilikum responded to the net properly by heading toward a gate, "the victim should be instructed to exit the water if capable," the special protocol stated. If unable to exit, the victim should be assisted with emergency equipment. If Tilikum did *not* respond to the "recall stimulus" efforts, then staff was instructed to repeat all the same methods: spare air; hooks and poles "to assist victim when in reach"; opening pool gates for animal egress; bringing "a large food bucket poolside"; or using the net to "aid with animal control, separation, animal removal, or rescue of victim."

But it's unlikely that on the morning of February 24, Dawn Brancheau's mind was overly concerned with rescuing someone from Tilikum's pool. The park was quiet that morning; a Wednesday in February was never peak time, and the cool, gray weather wasn't helping. A heavy rain was forecast for the afternoon. It was going to be a fairly easy day. During February only two "Believe" shows took place per day, though Dawn was also on deck as control trainer during that day's "Dine with Shamu" lunch in G Pool.

A little gloomy weather was not going to keep everyone out of the park that day. Many of them were at SeaWorld for the first time, including John Kielty, a Massachusetts native now living in Myrtle Beach, South Carolina, who had driven down to Orlando to spend some time with his vacationing sister, her husband, and her son. John, in his early forties, with a soft, round face, light-brown hair, and South Boston accent, had headed south for the winter after his construction inspection business began to fall off with the economic downturn. John was excited by the diversion. He felt good about SeaWorld and its efforts to save the world's oceans, and he was anticipating seeing the "Shamu whales," as visitors to the park that day were calling them.

John, his sister, and her family walked across the long wooden causeway that bisects the water-ski lagoon and leads visitors to Shamu Stadium. There was no line to get in. They ignored the young people hawking official ponchos and towels in the Splash Zone and headed for drier seats farther up in the stadium.

John wasn't impressed with the show. It seemed kind of lame to him, and the whales didn't appear to be enthusiastic about performing that day. In fact, they didn't seem to be behaving well at all. After some foot pushes and perimeter swims and a couple of stand-ons, that was about it. Before he knew it, the show was wrapping up. It was time for the big splash finale.

The finale was a flop. One whale came out into the main pool, gave a modest push of its flukes, and soaked a small handful of people on the far left-hand side. That's when John noticed another whale shoot out of the gate from the back pools and swim directly for the splash whale, ramming that animal and then chasing it around the tank.

Clearly, this was not part of the show. The audience waited patiently, not sure what was supposed to happen now.

Dawn Brancheau and her colleagues had been well trained in handling these kinds of situations. Once in a while, the whales would show more interest in playing or sparring for dominance among themselves than doing backflips for tourists—no matter how attractive the reinforcement proffered for conditioned behavior.

First of all, the trainers knew *never* to say that a show was being "canceled." The proper buzzword, approved at corporate headquarters, was "postponed." The list of prohibited buzzwords had grown over the years since Jeff Ventre and Carol Ray had started at SeaWorld, as documented by memos sent to staff in a series called "Let's Talk . . . adlibbing." The key to using the approved vocabulary was to "stay positive and keep it on a 5th grade level," the memo said. The official taboo terms, it added, "are the stuff of the Dark Side and could have a negative connotation."

The company handbooks even had prepared scripts for trainers to follow when the animals acted up and caused an interruption. The approved chatter was designed to play down an incident, reassure the audience, and ad-lib as if nothing out of the ordinary were occurring, no matter how serious the situation with the whales. Trainers were told to "keep things lighthearted, always smile" and "keep an upbeat demeanor and voice." There was even a suggested joke: "Whales are around 5000 lbs, we are ____, and believe me, there is no way I'm going to get a 5000 lb whale to do anything he/she doesn't want to do!"

That corny yarn, just thirty minutes later, would prove to be accurate, but also entirely inappropriate for the situation.

The script continued, "We understand that things don't go right all the time, but we do our best to get things back on track and have them not occur again from the way we respond—they are animals, not robots."

During the ensuing downtime, as staff tried to sort out the animal problem, the narrator was instructed to banter about the whales. Quite tellingly, only one topic pertained to the natural history of orcas: "Killer whale society is a matriarchal society," the ad-lib script said. Everything else was about captivity: "Discuss the fundamentals of animal training. Start with positive reinforcement and the variety of ways to reinforce correct behavior. . . . Discuss HELPRS and explain why it is important to keep variety for killer whales. . . . Discuss the requirements to become an animal trainer."

During these unscheduled intermissions, SeaWorld was given a golden opportunity to educate their captive audience on the lives of killer whales in the ocean. But it was wasted: Trainers were not told to ad-lib about orca family relations, Transients and Residents, geographic ranges, foraging behaviors, vocalizations, echolocation, or natural life spans.

On February 24, when the "Believe" show broke down, the trainers began their "lighthearted" explanations. "Just like feuding siblings, our whales are misbehaving, and we're just gonna ignore their behavior until they simmer down a little bit," John Kielty heard one of the female trainers say with a laugh in her voice. "They're more interested in splashing each other right now than splashing you guys. We'll give them ten minutes and see what happens."

After the unplanned intermission, in which the feuding whales were separated, the show started again. It was supposed to be the big splash finale, but the whales were still not interested in performing. To John, they seemed agitated, uncooperative, and focused on other things.

"Okay, folks! Thank you for coming to the show," the trainer abruptly said. "We hope you enjoy the rest of your day at the park." With that, everyone got up to leave.

"Well, *that* was pretty bizarre, wasn't it?" John's sister asked.

"It sure was," John said. "Those whales wanted to fight, not perform. That show sucked."

They left the stadium and walked around the nearby area for about forty-five minutes. Then they heard an alarm going off, coming from the "Dine with Shamu" area, behind the stadium.

Suzanne Connell had a terrific surprise in store for her son, Bobby, for his tenth birthday. Months before she and her husband, Todd, had flown down to Orlando with their son, Suzanne had quietly booked a table for lunch at the "Dine with Shamu" experience for their magic day at SeaWorld. Bobby loved SeaWorld and he loved Shamu. This was the family's third trip to the park in almost as many years, and they never missed the killer whale show. Bobby had pictures of himself with several of the trainers he idolized, including Dawn Brancheau. As they strolled across the wooden footbridge toward Shamu Stadium, Suzanne pulled out the passes and sent the boy into fits of delight.

The Connells, visiting from Somersworth, New Hampshire, entered the restaurant at about noon and were seated at table 13, one of the best in the house—right next to the spot where the lead trainer would interact with "Shamu." The dining area was a covered outdoor café called The Cove, spread out along the north end of G Pool and the adjacent Green Room, where an all-American, all-you-can-eat buffet was laid out for guests. They had all paid $35 per adult and $18 per child to experience "Dine with Shamu."

As they ate their burgers and chicken fingers, a large bull with a flopped-over dorsal fin—Tilikum—swam around G Pool in rolling circles. The water was gray, murky, and opaque, a product of the overcast sky. Behind him, in an adjacent pool gated off from Tilikum, the Connells could see some trainers in their orca-like black-and-white wet suits preparing other whales for the 12:30 "Believe" show.

The whales were restless, churning up the water and making wild vocalizations. The Connells had never heard SeaWorld animals making such a racket. Then they noticed that Tilikum was equally agitated and making eerie, loud cries of his own. He was clearly reacting to what was going on in the other pool. The show got under way, but then was abruptly halted, something the Connells found extremely odd.

A little after 1:00 p.m., the narrative portion of the "Dine with Shamu" experience began. A young female stage manager came out and greeted the diners, following yet another carefully crafted script. "Dine with Shamu" was produced according to its own particular set of "Presentation Guidelines" for staff to follow. Given the more intimate setting, the low number of guests, and the opportunity to question trainers, uncomfortable topics stood a far greater chance of coming up here than during the main Shamu show.

"Please refrain from addressing the dorsal fin topic," the guidelines said. "We have many other fun, pertinent, and interesting topics to discuss." SeaWorld then instructed its trainers how to explain why some whales were fed large cubes of gelatin. Tilikum alone consumed up to ten gallons of the stuff (eighty-three pounds) every day, partly to prevent the dehydration caused by eating thawed frozen fish. "When discussing jello, please do not mention that it is offered to the whales to provide hydration. Refer to it as another type of reinforcement or enrichment—a tasteless, sugar-free, colorless treat! They seem to enjoy the texture!"

Then, without irony, the guidelines added, "Please make sure your information is accurate."

The banter should also be "entertaining and easy to understand. This should be fun and the REAL inside scoop for DWS."

Trainers were encouraged to focus their references on "our Shamu Family. We can introduce the whales by name but we want the guests to feel like they are watching Shamu/Baby Shamu during the interaction." What any of this had to do with educating the public about killer-whale natural history, the guidelines failed to explain.

And then there was the Twitter plug. "Just out of curiosity, how many of you are on Twitter?" the script said. "Awesome! Well, Twitter is so popular, that even Shamu is doing it! That's right, check out REALSHAMU at Twitter to find out what's on the big guy's mind. He'll answer questions, let you know what he's having for breakfast, his favorite English soccer team . . . seriously, it's very cool."

Such was the ease with which SeaWorld got around its own aversion to anthropomorphizing the animals.

Then it was showtime. The beaming stage manager returned to the approved patter: "Is everyone enjoying their meals? . . . Excellent! I'll tell chef. Now, I bet everyone here is ready to meet the stars of our show . . . our trainers and our whales!"

The diners were introduced to the narrating trainer, Jan "Jay" Topoleski; the spotting trainer, Lynne Schaber; and the control trainer for the show, Dawn Brancheau. For the first few minutes, triumphalist ballad music and a wailing female vocalist blasted through the speakers as Tilikum swam slowly around the pool.

Bobby and Todd Connell took turns videotaping the event. The images they gathered on that overcast afternoon would soon be seen all over the world.

"Welcome to SeaWorld!" the stage manager said. "Now, we'd like to teach you a little bit about the whales. Hi, Jay!"

Jay, who was standing on the south side of the pool, opposite the dining tables, began the narration. "Are you excited?" he cried. "Are you ready to meet Shamu?" Everyone cheered. "We're gonna show you guys how we actually interact and how we train our Shamu whale every day!"

As Jay spoke, Tilikum popped his big head out of the water right in front of the Connells' table. Dawn, standing just feet away, threw a few smelt into his gaping pink mouth.

A bit later, Todd Connell videotaped Dawn, now across the pool and standing on a smooth stainless-steel plate used to weigh the orcas. She was on her knees, inches away from Tilikum, shoveling fish into his mouth from a bucket. Then she kissed him on the rostrum.

"Guys, this is the largest male killer whale in any marine park facility in the world," Jay said as Dawn dropped more fish in Tilly's mouth. "I kid you not. He's about twelve thousand pounds. That's six tons of fun and love!" The audience chuckled. Dawn gave Tilikum another hand signal, sending the whale into a barrel roll and then straight toward the tables across the pool.

"There he goes," Jay said. "A big hello for Shamu!" Tilikum rose up in a spy-hop right in front of the delighted guests. "Whoa! Shamu!" Jay said. Tilly began screeching, like an EMS siren: *Uuuuuh-eeeeeee-uuuuuuh*.

"Of course, one of our main responsibilities is helping them to have fun," Jay continued. "Whales love to have fun!"

Dawn held up a silvery fish. Tilly nodded, as if to say, "Yes, please." Dawn signaled him to raise his left pec, and he complied. "Oh, he's gonna show off that fin a little bit! Look at that!" Jay continued. Tilly slammed his big black paddle onto the water's surface with a resounding *thwack*. He then swam along the tables, gingerly making little splashes with his fluke, but not enough to soak the guests the way he might do during the "Believe" show in A Pool.

A little later, Dawn was back across the pool, once again on the metal scale. "Now, to put this in perspective, our next biggest whale is only half his size, that's about six thousand pounds," Jay went on. "That's incredible! These guys are born at three hundred and fifty pounds and seven feet long. Ladies, that's like giving birth to Shaquille O'Neal!"

Dawn began pouring buckets of water into Tilikum's mouth, another secondary reinforcer he liked. Then she sent him on a "perimeter pec wave," his second of the show, where he cruised slowly along the tables on his left side, gently rocking his right pectoral fin in a back-and-forth motion, resembling a wave. "He likes waving so much it's beyond belief," Jay said.

Dawn and her spotter, Lynne, returned to the area next to the Connells' table. "The whole pod is ruled by the females," Jay continued, "if you can believe *that*!" Dawn turned around to the audience and cracked a knowing smile. "*I* can believe it!" she said, and laughed. She then rocked to the left and right, engaging Tilikum in a little dance. The diners laughed and clapped. "That's very nice!" Jay said. "There's a lot of dancing going on over there."

Dawn gave Tilikum the signal for a "raspberry" (the farting sound orcas can make with their blowholes). But he offered her only a halfhearted, long exhale instead, which earned him an LRS (least reinforcing scenario) or three-second "neutral response."

The behaviors and the banter continued. Dawn stimulated Tilikum by splashing water on his abdomen and left pectoral fin. He rolled over in apparent pleasure. Tilly then moved about fifteen feet away, intently watching his trainer. Dawn tossed him a good-size fish, and he caught it. Then another. "Oh, good shot, Dawn!" Jay said as she kept throwing fish to Tilikum. "Three for three! Wow, four—a high one!"

Then Dawn twirled around like a ballerina, smiling with one arm folded upward, and Tilly spun in the water in time with her. She bridged him with a whistle, then fed him some more fish. "Woo-hoo!" Jay said. "He's quite a dancer there."

Next, Dawn sent Tilly into the slide-out area at the west end of the café for a final pose, fed him a few more fish, then headed over the steep footbridge that spanned the gate between G Pool and the smaller holding pen known as F Pool. She was followed by Lynne Schaber, who walked downstairs to the underwater viewing area to get ready for the big photo opportunity with Tilikum, the denouement of the "Dine with Shamu" experience.

Dawn walked behind Jay as he continued the narration. "During the course of our interaction today you've seen us use the tools of the trade. The whistle, the target, and the bucket of fish. But in reality, the real secret to this relationship—the most important tool we use—is our heart. If we don't put our heart and soul into building our relationships with the whales, none of this would be possible."

Those were among the last words Dawn Brancheau would ever hear.

A few minutes later, the Connell video shows her lying on her stomach in six to eight inches of water, facing Tilikum; an unnatural smile is plastered across her face. What happened next remains the subject of intense argumentation and speculation.

Todd Connell switched off his video camera at that instant to join the last stragglers leaving the Cove restaurant, along with his wife and son, Bobby.

Then Suzanne Connell heard a splash.

"Hey!" she shouted to Jay. "He took her down! He took her down!" According to Suzanne's account of the event, Jay had not been looking at Dawn. When he turned and saw that she was gone, he ran to the water and started slapping the surface, desperately trying to call Tilly back.

Down at the underwater viewing area, Lynne was speaking with about thirty guests who had gathered for the "Photo with Shamu" opportunity, including some guests who had been at the "Dine" event upstairs. Lynne shouted for Dawn through an opening to the pool deck that it was time to send Tilikum down, but he had not yet arrived for the photo.

Then they saw a splash. Lynne and the guests watched in shock as Tilikum did a great barrel roll to his left and dove deeper into the pool. He had Dawn Brancheau in his mouth.

Jan Topoleski sounded the emergency alarm as "Dine with Shamu" staff hurried to usher the remaining guests away from G Pool. The Connells stood in shock as the terrible tableau unfolded before them.

Tilikum pushed Dawn around G Pool, rammed her twice head-on, then dragged her to the bottom and held her there for several seconds. Somehow, she broke free. She made a desperate swim for the surface. The Connells saw her head pop out of the water. She stared into their eyes with a look of panic and a plea for help.

Tilikum grabbed her once again and pulled her under. The Connells finally left the "Dine" area.

When the alarm went off at Shamu Stadium, it sent an emergency broadcast, Signal 500, over the park's internal radio network. SeaWorld staff began running toward G Pool, leaving behind their own duties with animal training and office work. More than thirty employees descended on the scene. Many began to unfurl and deploy the emergency netting, while others continued to try to recall Tilikum with water slaps and the tone box—all to no avail.

Tilikum would not release his trophy. He swam around the pool rapidly with Dawn in his mouth. When he dove again to the bottom, her motionless body drifted up to the surface. Tilikum swam to the opposite end of the pool, turned around, and moved toward Dawn once again. He gained speed as he approached, then rammed her body for the third time.

It took more than thirty minutes to corral Tilikum—with Dawn's body in his mouth—out of G Pool and eventually into the medical pool.

Once in the med pool, the false bottom was raised and staff managed to cover Tilikum with a net. The whale grasped Dawn by her arm and shoulder. He thrashed around violently, unwilling to relinquish his prize, causing Dawn's body to flail about like a doll. Several staff members walked on the raised floor to retrieve Dawn, but he continued to thrash about.

Rescuers spent ten minutes prying Dawn from his jaws. When her body was finally extracted, Tilikum was lowered into the water. But then someone noticed that part of Dawn's left arm was still in the whale's mouth. He had to be raised up out of the water once again so that the arm could be retrieved.

Someone else found Dawn's ponytail and scalp floating in another pool at the stadium complex.

Dawn's body was laid out on the deck next to the med pool. EMS crews on the scene cut away her wet suit and attempted to resuscitate her with a defibrillator. It was no use. Dawn was covered with a sheet to prevent hovering television news helicopters from photographing her body. This heartbreaking image was blasted around the world via satellite. CNN, Fox News, and MSNBC went to split-screen mode and, in a corner of the screen, showed a large killer whale floating alone in a small pool next to a body sheltered by a canopy.

Every so often, Tilikum would poke his head over the edge and gaze forlornly at Dawn, almost as if to see if she had recovered.

PART THREE

AFTER DAWN

32

Wake

The debate over keeping killer whales in captivity will grind on as long as new generations of *Orcinus orca* are bred and trained to do backflips in artificial oceans for the delight of the public. The battle has only intensified as industry defenders have circled the wagons against assaults from nearly all sides: the media, the courts, the federal government, and, worst of all for SeaWorld and its allies, members of the ticket-buying public.

It wasn't always like this.

Over the decades, the SeaWorld community grew accustomed to public adoration. They had come to expect admiration and respect for their state-of-the-art animal care and training, their wide-ranging projects to rescue and rehabilitate distressed animals, their backing of marine science research, support of habitat conservation, and efforts to educate the public about the richness of life beneath the sea. Despite a few "extremist" detractors such as PETA, HSUS, and other groups, SeaWorld had generally managed to steer clear of major controversy.

On the morning of February 24, 2010, the company's image had been as bright and shiny as the great wet Shamu himself. By day's end, however, the debate over captive killer whales would never be the same.

Dawn Brancheau's death that gray afternoon changed everything. The horrible event—tragic and perhaps preventable—forced SeaWorld to revamp its safety procedures for working with killer whales and invest millions in novel technologies such as fast-lifting, false bottoms in the pools. It opened the door to crushing media examination, the likes of which SeaWorld, or any other entertainment company, had never witnessed, and it brought the full force of the Obama administration to bear on what was happening in the pools of Orlando.

Dawn's death also upended the dynamics of the captivity debate. The worldwide notice it precipitated prompted people around the world to take a second look at SeaWorld and consider, perhaps for the first time, what life might be like for killer whales outside the ocean. They wanted to know why such an intelligent creature would commit such a brutal act.

For Naomi Rose, Jeff Ventre, and other captivity opponents, SeaWorld's protective veneer had been violently stripped away, allowing the sunlight of public scrutiny to penetrate the darker recesses of backstage life at Shamu Stadium. There was no going back now.

When the alarm went off at Shamu Stadium at 1:33 p.m. on February 24, 2010, it not only drew SeaWorld staff from all over the park, it mobilized a small army of emergency responders from outside the facility, including the patrol division, forensics unit, and homicide, robbery, and family-crimes squads of the Orange County Sheriff's Office (OCSO). Also on-site were authorities from the medical examiner's office, the Florida Fish and Wildlife Conservation Commission, and the Tampa office of the Occupational Safety and Health Administration (OSHA), the powerful inspection and enforcement agency of the US Department of Labor.

OCSO detectives took witness statements from more than forty people, though only a small number were on-site when Tilikum pulled Dawn from the ledge and into deeper water. The vast majority of "witnesses" were SeaWorld staff that had converged on Shamu Stadium after the emergency was already under way. Just a few of the interviewees were actually at G Pool when Dawn went in, and only two of them—trainer Jay Topoleski and security guard Fredy Herrera—said they saw the actual "grab." Their eyewitness accounts would soon contradict each other.

Inexplicably, OCSO detectives spoke with just two guests that day, including a tourist from the Netherlands named Susanne DeWit, who was downstairs at the underwater viewing area waiting for Dawn to send Tilikum below the surface for the after-lunch photo op. She did not see Tilikum grab Dawn, but told detectives that when the two of them appeared in the window, the whale had the trainer by her shoulder and neck.

Other guests who were at the underwater window, or still up on the terrace, when the attack began were never interviewed by detectives. Later that day they came forward to speak with the media, who descended on the park en masse. Local TV stations began wall-to-wall broadcasts on the killing, including the local NBC affiliate, which interrupted programming from the Winter Olympics in Vancouver.

Early that afternoon, SeaWorld officials braced themselves for their first of many difficult encounters with reporters. The grim-faced gaggle of executives tasked with the bleak job included Orlando park president Dan Brown, along with head trainers Chuck Tompkins and Kelly Flaherty Clark, her fists clenched into tight balls of grief. Chuck and Kelly were especially close friends with Dawn.

OCSO spokesman Jim Solomons opened the impromptu briefing, held outside the low-rise buildings that house SeaWorld's business offices.

"What happened was we had a female trainer back in the whale holding area," Solomons said incorrectly—and uncorrected by the SeaWorld executives standing by his side (there was no whale holding area in G Pool). "She apparently slipped or fell into the tank and was fatally injured by one of the whales." Again, this incorrect information was left unchallenged by the executives.

There was no sign of foul play and the death appeared to be an accident, Solomons said. He identified the victim only as a forty-year-old female trainer. The whale was likewise unnamed.

Dan Brown, shaken but composed, read a prepared statement from the company: "It is with great sadness that I announce that one of our most experienced animal trainers drowned in an incident with one of our killer whales this afternoon. We've initiated an investigation to determine to the extent possible what occurred."

Brown said such an incident was unprecedented in the history of SeaWorld. "Nothing is more important than the safety of our employees, guests, and the animals entrusted to our care." He added that all standard operating procedures would come under review during the ensuing investigation.

Reporters pressed the executive for more details on the trainer. His face wracked with pain, Brown pleaded, "It's still early. Please bear with us. We've just lost a member of our family. We'll get back to you as soon as we have some more information."

The press conference was beamed live around the world—and down the street. At their nearby hotel, Todd and Suzanne Connell watched the proceedings in disbelief. After witnessing the carnage in G Pool that so traumatized their son, Bobby, the Connells had been shepherded to a small holding room, where, they said, a pair of SeaWorld managers expressed no sympathy for their ordeal and showed no interest in their statements or the video they had shot in the moments leading up to the attack.

The Connells were already disgusted with SeaWorld. Now, they realized from watching the news, park officials were doing nothing to dispel the misstatement that Dawn had slipped or fallen into the water. Todd and Suzanne were outraged. By their own recollections, Tilikum had grabbed Dawn by her long ponytail and flipped her over his body. Todd picked up the phone and called WESH-TV, the NBC affiliate in Orlando. Producers accepted his offer to review the pre-attack video. Todd also called CNN and others. Before long, video crews where showing up at their hotel.

"We left in tears. Nobody approached us to ask, 'Are you okay? Is your son okay?' And that is what I had the hardest thing swallowing," Suzanne told the WESH reporter. "I got more infuriated when I saw the spokesperson from SeaWorld say she slipped and fell into the tank, which was not what happened at all."

Todd Connell recounted what he had witnessed. "I saw her hair in his mouth," he said of Tilikum. "And he rolled and she went over the top of him and went right up underneath and then he just took her down."

The killing had been particularly traumatic for the couple's son. "I saw her like went up and get some air and somehow back down," Bobby recounted. "And then twenty minutes after that, I didn't see her still." Suzanne said Bobby had been "hysterical." Todd Connell added, "Thank God I turned my camera off. I would not want to have shot any of what we witnessed."

The Connells weren't the only witnesses to dispute SeaWorld's official account. Victoria Biniak, visiting with her husband from nearby Clermont, Florida, told Orlando station WKMG-TV that Tilikum "took off really fast in the tank and he came back, shot up in the air, grabbed the trainer by the waist, and started thrashing [her] around." She said the assault was so violent that Dawn's bootee "flew off her foot."

A SeaWorld employee who requested anonymity confirmed Biniak's depiction of events and told WKMG that the whale's name was Tilikum.

Other witnesses told the *Orlando Sentinel* that Tilikum had grabbed Dawn by the upper arm before taking her underwater, racing to the other end of the tank, and turning her over like a helpless seal. Brazilian tourist Joao Lucio De-Costa Sobrinho said Dawn was bleeding from her mouth or face. "It was terrible," he said. "Very difficult to watch that image."

Later in the day, in an e-mail announcement, the sheriff's office identified the victim as Dawn Brancheau, who, it noted, "was reported to have 16 years' experience working with killer whales."

By late afternoon, SeaWorld officials realized that the "Dawn fell in the pool" story was not accurate. Chuck Tompkins corrected the record in a live interview with WFTV news in Orlando. "She was pulled in and she drowned," he said with a dour face.

"Did she fall in?" the reporter asked.

"No, she was pulled in. At the end of it, she was apparently rubbing the animal down and apparently the whale pulled her in." Asked if Tilikum would be put down, Tompkins looked horrified. "No, absolutely not, absolutely not. He's a large animal, he's a killer whale, and we understand the risks of working with these large animals."

Tompkins said later that everyone at the park was aware of the danger with Tilikum, but added that safety protocols for the whale would be evaluated and changed. "We interact with him a lot, but only under the safest conditions. Or so we thought."[1]

Naomi was in her cubicle in the Gaithersburg office catching up on e-mail when a message appeared in her in-box from Richard Patch. Richard, a colleague at the Humane Society Legislative Fund (the HSUS lobbying and political action arm), handled marine mammal issues. He was a reliable source for breaking news, and today's bulletin was no exception.

"Trainer killed at SeaWorld," the subject line announced.

The message was brief, short on details. Richard had just heard on the news that a trainer in Orlando had drowned in the killer whale pool. More details would be available soon. Naomi immediately began e-mailing around to colleagues and checking news websites.

Within the hour, media were coming to *her,* in search of comment and perspective.

In the past decade Naomi had focused most of her captivity work on dolphins, especially the swim-with attractions popping up around the world, large captures in the Solomon Islands, and substandard display facilities in the Caribbean. One recent victory was stopping live captures and the development of a dolphin park in Panama by Ocean Embassy,[2] the Florida company run by onetime Keiko team members Robin Friday and Mark Simmons, the former roommate and SeaWorld colleague of Jeff Ventre and John Jett.

Over the years Naomi had carved out a niche as one of the world's leading authorities on captive marine mammals. She was now a go-to person among major media whenever captive cetaceans made the news. However inadvertently, Naomi had become the Jane Goodall of marine life held in captivity.

She was stunned by the death in Orlando, but had no time to emotionally process the news. It arrived out of the blue. With mere minutes to brace herself for the media onslaught, Naomi shifted into a mode that she described as "sound-bite autopilot."

The next several days melded into a blur of telephone interviews and live appearances at radio and TV studios around the capital. She featured prominently in many of the thousands of news stories reported around the world.

In one interview with the *New York Daily News,* Naomi compared Dawn's case with the other two deaths on Tilikum's record: "What he did in the previous two incidents were not attacks. To him, the people were toys, and to a whale, being underwater for twenty minutes is nothing." But in Dawn's case, she added, "This sounds a little more like an attack."

The *News* asked Naomi what should become of the whale. "I dread to think of what they think their options are," she said of SeaWorld.

Some people were already calling for the perpetrator's extermination, even certain anti-captivity activists. "Tilikum is a casualty of captivity; it has destroyed his mind and turned him demented," claimed Russ Rector, a former dolphin trainer who ran the Dolphin Freedom Foundation in Fort Lauderdale, Florida. "If he was a horse, dog, bear, cat, or elephant, he would already have been put down after the first kill, and this is his third."

Naomi wasn't surprised that some people were calling for this outcome, but she knew that death was not the answer for Tilikum. "It's not his fault what happened, just as it wasn't Dawn Brancheau's," she told *The Guardian*. "The fault lies with using these wild animals as entertainment."

In an interview with the *St. Petersburg Times,* Naomi reacted to early reports that Tilikum would remain at SeaWorld—and stay in the "Believe" show: "A return to business as usual is terribly disappointing. It *will* happen again. It's not a question of if. It's a question of when. This is his normal behavior. It's what he does when someone falls in the water." To keep Tilikum in proximity to people "is like leaving drugs on the table for an addict."

Naomi told AOL news, "SeaWorld should have changed their policy in terms of Tilikum years ago. This was an accident waiting to happen. One of the biggest problems is that SeaWorld miseducates the public. What the public is seeing isn't a contented animal behaving naturally, but a caricature dragged into a box and not given any choice."

SeaWorld was "not without its defenders," AOL noted. Journalist Amy Sutherland, who studied exotic-animal training for two books, *Kicked, Bitten, and Scratched* and *What Shamu Taught Me About Life, Love, and Marriage,* said the company was "revered in the animal training world." SeaWorld had "pioneered some of the most progressive training techniques and is well-known for using positive reinforcement. They do a great job, both in caring for their animals' needs and for their trainers' safety." Sutherland had confidence that SeaWorld would conduct a thorough investigation and adjust its safety procedures to prevent such a tragedy from happening again.

One of the hardest-hitting articles in the media frenzy following the tragedy appeared in SeaWorld's hometown paper, the *Orlando Sentinel,* under the headline "Killer-Whale Experts Say: Reintroduce Tilikum to the Wild."

According to the paper, "a conversation has started within the whale research and advocacy communities about what ought to be done with an orca linked to three human deaths." Every whale expert the reporter had contacted "would rather see killer whales in their natural habitat—the earth's oceans—than in tanks at an Orlando theme park. One suggested that building a tank the size of Rhode Island wouldn't be large enough for a six-ton male such as Tilikum, an animal capable of swimming 100 miles in a day."

But reintroducing Tilikum to the wild "would be costly, would include serious risks for the animal and would not guarantee his survival," many experts cautioned. "Nonetheless, in the aftermath of veteran SeaWorld trainer Dawn Brancheau's death this week, authorities such as Naomi Rose argue there is a moral obligation to release Tilikum."

"There is absolutely a risk in keeping him where he is," Naomi told the paper, predicting Tilikum would kill again if kept in captivity. "I will take bets on that and win. Boredom, depression—these cause physical problems in human beings, chimpanzees and, believe me, killer whales."

Naomi said she hoped the tragedy would become "a tipping point" in the debate, but added, "SeaWorld is not going to change, at least not without some serious pressure from its customer base."

Howard Garrett, of the Orca Network in Washington State, told the *Sentinel* that Tilikum could be cared for much the same as Keiko had been in Iceland. He envisioned handlers taking Tilikum on "walks." If the whale failed to locate his family, he could be trained to return to places where humans could feed and help him. "They need companionship above all, more than food," Howie said of killer whales. "In the absence of family, they'll follow human friends."

A colleague and neighbor of Howie's, Jenny Atkinson, executive director of the Whale Museum in Friday Harbor, on San Juan Island, said the ocean might not be the best place for an animal such as Tilikum. "He has been in captivity since he was two. It's what he knows," she told the paper. "I can't say what I want for that whale is what the whale wants because I can't ask him. . . . You don't know how he would respond in the wild or how the wild would respond to him."

That night, Naomi appeared on *Larry King Live*. "I feel very strongly that certain animals don't belong in captivity," she said. "My personal opinion and my organization's opinion is that orcas are a species that doesn't belong in captivity. I think that, put into that kind of confinement—they're very large animals. They're very social. They're very intelligent. And those are very small enclosures for an animal of that size."

By the end of the long day on February 24, Naomi realized just how profoundly everything had changed. Suddenly, killer whales were part of the national conversation, and much of that discussion was firmly centered on captivity. No attack by a captive orca had ever prompted such serious questions about captivity. Naomi had seen nothing like it.

"It's extraordinary," she remarked to her husband, Chris. "The captivity issue in the US just went from zero to sixty in one second." Tough questions were now being asked by a national media that had, until now, uncritically celebrated SeaWorld for more than a generation. Dawn's death may have been tragic and pointless, but it certainly wasn't meaningless.

• • •

Dr. Jeffrey Ventre was working with patients at a clinic when he got the news. Now a full-fledged MD, he had moved to New Orleans in 2008 and entered a residency program at Louisiana State University's Health Sciences Center in New Orleans. Jeff was a physiatrist, with specialized training in physical medicine and rehabilitation. PM&R doctors work with patients with traumatic brain and spinal-cord injury, as well as those recovering after amputations, strokes, heart attacks, low-back pains, car accidents, and gunshot wounds.

John Jett called at 3:30 p.m. and broke the news. John and Jeff had kept in contact since their SeaWorld days. After quitting the park, John sold aerospace components before earning a master's degree in environmental science from Oklahoma State University and a PhD in the same field from the University of Florida. Now he was working at Stetson University, a top-rated private school in DeLand, Florida, north of Orlando, as a visiting research professor specializing in waterway management.

They both agreed that the killer whale in question was Tilikum. The dead trainer must have been someone senior, someone who had been there for a long time—someone they knew. They both found it difficult to believe claims that the victim had fallen into the pool and drowned. SeaWorld trainers were all excellent swimmers. There was more to this story, they thought, especially if Tilikum was involved. Both men felt awful for the trainer, whoever it was, and that person's family.

But they also knew that this calamity would be transformative: It would bring an immediate halt to all water work at Shamu Stadium, at least for now, and a high-stakes, high-profile investigation.

Even as Jeff and John were comparing notes, the Orca Network's Howard Garrett, up in Whidbey Island, Washington, was fielding calls from worldwide media. One was from CNN reporter Gabe Ramirez, seeking someone to comment for that evening's broadcast of *Anderson Cooper 360°*. Howie suggested that Gabe contact Jeff, who had far more experience with captive killer whales. Jeff was booked for that evening's show.

Jeff initially declined the request to go on television. He was hesitant to get involved. But after another phone call from John Jett, and one from Jeff's sister, Kim, now living in San Francisco, he somewhat reluctantly agreed to do the show, but only if he could call in via telephone. John tried to convince his old friend to go into the studio and conduct an on-camera interview, but time was running short. In the end, Jeff spoke with CNN by phone. The *AC-360* staff badly wanted a former trainer on the air that night, especially someone who had worked both with Tilikum and Dawn. The connections CNN gained from the interview would prove to be tremendously valuable later on. For the next two years, no network would cover the Tilikum saga more diligently than CNN.

Anderson Cooper was off that night. *AC-360* was being hosted by CNN correspondent Jessica Yellin, who opened the segment by saying, "People go to the aquarium to be amazed, entertained. But today, visitors at SeaWorld Orlando were horrified by what they saw. They watched a woman die." Yellin asked Jeff if he was surprised by the news.

"Sure. You don't expect something like this to happen. Tilikum is a great animal. And Dawn is a great trainer. I'm sad it did happen. I'm shocked." Jeff added that he was aware of Tilikum's history. In the back of his mind, Jeff was worried about a potential backlash against Tilikum, and a public cry for him to be put down. Jeff wanted to assure the audience that Tilikum was a good animal and should not be blamed for Dawn's death.

"And, do you think it's surprising, then, that Tilikum would attack again?" Yellin asked. "There are two other instances where people have died. Do you think this animal should still have access to humans?"

Jeff said it was hard to tell whether this had been an act of play or aggression. "All I can say is that Tilikum is a good animal, and Dawn was a great trainer." But, Jeff was asked, why did SeaWorld still want to keep Tilikum around? "He provides semen to impregnate females," Jeff replied. "He's a beautiful animal. He is huge. He is impressive. And people just see him and go, 'Wow.' And, you know, he is—he is a money stream, as well as a great animal."

Yellin noted that what happened at SeaWorld "raises some pretty basic questions, including whether killer whales should be kept in captivity at all, let alone with trainers anywhere near them."

Every good story has two sides, all journalists know, and this was a good story. Speaking for the other side that evening was Jack Hanna, the cowboy-hatted director emeritus of the Columbus Zoo and a passionate defender of animal display.

Yellin challenged him right out of the gate. This was the orca's third killing, yet most people "just assume these animals are domesticated; they're our entertainment," she charged. "But, really, this is an eleven-thousand-pound, twenty-two-foot whale. Should we be surprised that it would kill?"

Hanna acknowledged that Tilikum "of course" was a dangerous animal, but the previous deaths were of no consequence. Daniel Dukes had hopped a fence without authorization. "That's like going over the fence at the NASCAR race. You can't blame SeaWorld for that." As for the Keltie Byrne tragedy, Hanna praised SeaWorld for taking the rogue animal after she died. "SeaWorld, I take my hat off to" because the company "brought that animal to their parks to give it a life."

Yellin wasn't buying it. "Three deaths on this one whale's head. It seems to me—shouldn't one death be enough? Shouldn't SeaWorld have *retired* this animal?"

No one ever swam with Tilikum, was Hanna's answer. This was the fault of Dawn, the top trainer "maybe in the history of SeaWorld," he said. "Did human error happen? Yes, it did. So, I don't think that the whale, myself, should be punished or something."

The next afternoon, Jeff was invited back on CNN, this time on *The Situation Room with Wolf Blitzer*. In his opening teaser, Blitzer posed a question not typically heard in mainstream media before Dawn's death: "Are these giant predators or are they performers?"

Blitzer checked in with CNN's resident curmudgeon, Jack Cafferty, who cited an interview with Naomi Rose: "She says killer whales normally live in groups with their families. The males stay with their mothers their entire lives. They rely on their family for social structure and play and cover hundreds of miles of ocean. She says situations like the one at SeaWorld cause stress. *Duh*."

Cafferty ripped into SeaWorld, something that must have dialed up the anxiety level among embattled officials down in Orlando. The whales were merely performing "stupid tricks over and over," he alleged, with his trademark scowl. "Inevitably, when some human being is playing around with these wild animals and gets attacked, it's the animal that gets put down. It ought to be the other way around. Here's the question: Should wild animals be used as entertainment? . . . Post a comment on my blog."

Blitzer took over from there. "The killer whale attack raises new questions about whether these powerful predators should be kept as performers. I spoke with Jeffrey Ventre. He is conflicted on the captivity issue." As Blitzer spoke, an old video of Jeff hotdogging in Orlando played on the screen. "He says Tilikum gets excellent care at SeaWorld, and he believes that's the best place for him to be. But he makes some counterpoints." Blitzer discussed fin collapse in male orcas as CNN showed images of a wild and a captive whale. It was striking. Blitzer then aired a taped interview with Jeff, to make the point that "there is also financial motivation here," as he put it.

"I am sure that he cost millions of dollars, and I'm sure that he is worth many millions more," Jeff said in the interview. "He perpetuates the SeaWorld gene pool. He continues to perform and wow people. To lose that genetic diversity by eliminating one of a few successful male breeders would be a big, probably a detriment to the breeding effort in general, and would impact SeaWorld's long-term goals, which is to perpetuate itself."

CNN reporter Brian Todd, who had interviewed Jeff, said to Blitzer, "This is a debate that's going to be going on for a long time, this whole captivity issue."

"And it is going to be really heated," Blitzer added. "What does he [Jeff] think may have triggered this whale to go ahead and do this?"

"It is fascinating to hear an expert talk about this. He says that Tilikum has some potential stressors," including life in a small pool as a subdominant male in a female-run society. "He says those females, Wolf, are very, very dominant in these pods."

Near the end of the show, Cafferty read some of the responses to his query about using wild animals for entertainment. Judging by the replies, few viewers endorsed the idea.

"Killer whales should not be kept in captivity and used for entertainment in any way!" wrote someone named Chris, who saw a SeaWorld official "trying to convince everyone that the sole purpose for keeping these animals was for research. That is BS. . . . I mean, I just saw a picture of a woman surfing on top of one of these animal's heads. Is that research?"

Another viewer named Dan said he visited SeaWorld as a child and recalled the "magical atmosphere of the place." Then he returned as an adult. "I was sickened and sad. Anybody who cannot see the cruelty inflicted on these creatures hardly deserves to be called human. As badly as I feel for the trainer, I feel even more sorry for the whale."

The following morning Jeff appeared on the CBS *Early Show*, repeating his opposition to freeing Tilikum: "There's no better place for Tilikum than SeaWorld. He gets the best care possible for his particular situation." Why oppose his release? Because Tilikum spent most of his time resting at the surface and could not handle the rigors of being a wild orca, one that "swims pretty much its entire life," as Jeff put it. Tilikum's lack of viable teeth was another reason he was unsuited for the ocean.

A few days later, John Jett submitted an essay about Dawn and Tilikum to the *AC-360* blog, which published it. John wrote that the incident "should cause us as a society to ponder the wisdom of keeping large carnivores such as killer whales in captive environments." While working at SeaWorld, he said, his job had made him "emotionally impossible to participate." Now he counted himself among the "growing contingent who considers killer whale captivity grossly unjust."

John agreed with his friend Jeff on the question of freedom for Tilikum, for much the same reasons. But that didn't mean that captivity was justified, even though parks such as SeaWorld relied on their "loudly proclaimed statement" that seeing captive killer whales "somehow makes us better global citizens: greener, wiser, and more likely to support conservation efforts."

As for the industry's scientific contributions, John noted a paucity of "peer-reviewed articles published as a result of captive killer whale observation, especially within the past 20 years." And there was zero science to show that captivity was beneficial to killer whales or to human society. "The burden of proof is, or should be, on zoological parks where animals like killer whales are held. Show us the proof that making killer whales do tricks somehow leads to pro-environmental attitudes and behaviors among visitors."

Jeff and John knew that their public pronouncements would strike a nerve among pro-captivity types, but never imagined the most strident reactions would come from their old friend Mark Simmons.

"Jeff—Call me," Mark wrote in an e-mail. "I don't know why or what in the world is going on—I saw your interview and John's article on CNN—I'd like to know what the hell? If I've ever been a friend to either one of you—and if you know how to reach John—you'll call me or tell him to call me."

Jeff wrote back right away: "Mark—I didn't realize our friendship had ended. I miss you brother. My position on Tilikum has been stated. He can't be released secondary to dental issues. In regard to JJ [John Jett], I read his article and thought it was well written; have forwarded to family and friends. I encourage you to do the same. He writes beautifully."

Jeff told Mark to contact John at Stetson, and Jeff mentioned that he was coming "home" to central Florida that May. "If you'd like to speak, let's get together then, have a beer, after this has blown over. It would be great to see you and John again."

Mark quickly wrote back: "I'm not saying our friendship has ended Jeff—I'm not that shallow. I don't agree with John or an anti-captivity stance . . . but he or you are entitled to your opinion just as I am." Mark conceded that SeaWorld was not a "flawless institution." On the other hand, the anti-captivity side was "downright criminal. I can tell you things that would shock you about Keiko, about the economics of anti-captivity," which he witnessed while working for HSUS, Earth Island, and Ocean Futures.

Medical records and film footage from the Keiko project had been "doctored" to benefit HSUS and the others, Mark alleged. "I have in my possession every e-mail, every record on Keiko and photos, etc. that can back up everything I would tell you." Mark also blasted John Jett for having "jumped on this media bandwagon at Dawn's expense." If John had really felt so strongly about animal welfare and trainer safety, "he wouldn't have waited for such a sensational time to promote it."

Mark then wrote to John directly: "I read your article. I can't disagree more." Mark had sent a response to the *AC-360* blog, but it had gone unpublished. "It is abundantly clear that the editorial staff are openly anti-captivity, anti-zoological."

Mark assailed John for his timing. "Dawn believed passionately in the mission of parks like SeaWorld and was one of the most outspoken proponents of animal conservation. . . . Where have you been on this issue for the past 15 years? Speaking out now, claiming embellished killer whale credentials in the process and complimenting Dawn in the process is poor form." Jeff had been "no less guilty of grandstanding" in his own statements.

Mark finished by assuring John that he was, and always would be, a friend. "I wish we had talked, if for no other reason to avoid the same old tired arguments propelled by the fanatical groups for the last 30 years."

He attached his response to John's CNN blog and asked for John's help in convincing *AC-360* to post it.

In the essay, Mark attacked John over the idea of freeing Tilikum, which was bizarre because John had written that the whale was a poor candidate for release. "You don't have the first clue of how this would be done," Mark wrote. "Let me answer that from firsthand experience: it cannot be done with a long term captive whale." The Keiko project, he said, was "the most famous case of animal exploitation in history." Those who led the campaign (including Naomi Rose, presumably) "should have been charged with animal cruelty." It was an inhumane "experiment" that ended in Keiko's "cruel and drawn-out death. The same fate should never befall Tilikum."

Mark then embarked on the "ocean as scary place" argument: "I've seen wild dolphins with cancer and synthetic hormone toxicity from our waste water disposal." He had witnessed dolphins wrapped in fishing filament that caused body parts to fall off from stunted blood flow. He'd seen "emaciated groups of animals" and wild orcas "with amputated or bent dorsal fins." In contrast, John's depiction of captivity, was "flawed by a limited view of this world." SeaWorld was not a "prison" and trainers were not "wardens." The animals, he added, "were given the best care in mental, physical, social and environmental stimulation."

Not only that, but SeaWorld also managed to take care of "at least 100 endangered and threatened species the world over." It had spent more than $25 million in "active hands-on conservation," rescuing some seventeen thousand animals in its history.

On the flip side were "radical extremist groups" trying to capitalize on manufactured sympathy for whales such as Tilikum and Keiko. It was a coldly calculated "business strategy." Activist groups raised more cash on the backs of captive dolphins and whales "than any other single animal issue," he said, again without offering evidence. (HSUS, for one, did not typically mention marine mammals in its own fund-raising appeals, which were largely based on pets and farm animals.) "I am angry that you take this cheap and ignorant stance against SeaWorld and in effect, ALL quality U.S. permitted and fully accredited zoological institutions. You jump at the chance to be a critic, but fall short of offering anything material in Dawn's honor."

John read Mark's e-mail and essay and shook his head. He was not in any way going to help Mark get this published, even if John could pull strings at CNN. John confided in Jeff that their old friend was starting to concern him.

"I knew Mark was prone to wrong conclusions, but his diatribe was just over-the-top," John said, adding that Mark was angry at him because "I shared my informed opinion with the world; my words hurt because they were so hard to answer."

Was SeaWorld really an "important social institution," as Mark Simmons insisted? Or was it little more than a company whose main function was profiting from "parks and entertainment," as implied by its new corporate title under the Blackstone Group? That question was up for debate, and the argument continues to this day. But one thing was certain: Without profits, SeaWorld would not exist; and without killer whale shows, those profits would be greatly diminished.

Clearly, the show had to go on. How and when it resumed was a delicate issue requiring a fleet of public relations and crisis-response specialists. Dawn's death threw SeaWorld officials into a defensive crouch—unfamiliar territory for them. The company had already got off to a terrible start by permitting the falsehood that Dawn had fallen in the pool, until it was corrected not by staff, but customers. It was a colossal PR blunder: SeaWorld had lost some of its swagger. Naomi, as she watched events unfold in wake of the tragedy, had never seen a paradigm shift so rapidly in her life. "They're scared," she said to Chris. "Dawn really did change everything."

A business analysis from the Associated Press framed the problem well: "SeaWorld's online pitch for its Dine With Shamu show: 'Be part of an up-close and unforgettable adventure!' has taken on unintended and ominous meaning after the death of veteran trainer Dawn Brancheau. Now the company must reassure visitors the park and its sister locations are still places where they can enjoy a family friendly day."

As part of its "reassurance" campaign, SeaWorld Parks & Entertainment president and chief operating officer Jim Atchison held a somber news conference in Orlando one day after the killing.The venue was probably ill-advised. Atchison spoke in front of the viewing window of the killer whale tank, with several "Shamus" mugging for the camera behind him.

It was the normal venue for many company press conferences, but some critics took Atchison to task for speaking there. The whales were too cute and distracting to be used as a visual backdrop for such a grave topic, some of them charged. Others said it was insensitive to include the sight of *any* killer whales during a press conference about a whale that had killed someone, especially a beloved employee. It didn't help matters that Atchison's podium sported a large sign saying SEAWORLD PARKS & ENTERTAINMENT. This was a moment for contrition, critics howled, not branding.

Atchison announced that all three SeaWorld parks had suspended their

Shamu shows, but would resume them in two days. Trainers would stay out of the water indefinitely until an investigation was completed. In Orlando, Tilikum would return to the show for the splash segment soon, just not right away.

"He's been a part of our team and he will remain a part of our team," Atchison said of the giant bull. Pulling Tilikum from performances "would be a shame. This is really a wonderful animal, and his participation in our shows, his engagement in our interactions and so forth, is very important to his overall health and husbandry."

Atchison bristled when reporters pressed him on the wisdom of keeping a killer in such proximity to human trainers, and whether SeaWorld was holding on to Tilikum merely for his value as a sperm producer. Tilikum was the only breeding male at SeaWorld; the other males were too young or, in the case of Ulises in San Diego, apparently unwilling to impregnate females.

"Tilikum is a valuable orca, having fathered 13 calves during an extended period when parks have abandoned efforts to obtain more of the animals from the wild," the *Orlando Sentinel* reported. Nonetheless, Tilikum's reproduction value had been "vastly overplayed within the media," Atchison insisted. "If Tilikum never sired another calf, it wouldn't make a difference to us." The article was written by Jason Garcia, a beat reporter covering the region's entertainment parks. He would follow the Tilikum drama for years to come.

When asked about the ethics and merits of keeping orcas in captivity, Atchison became "defiant," according to the *Sentinel*. "We have created an extraordinary opportunity for people to get an up-close, personal experience; to be inspired and connect with marine life in a way that they cannot do anywhere else in the world," Atchison said. "And for that, we make no apologies."

SeaWorld officials had furnished investigators with a video of the attack captured by security cameras, though Atchison refused to provide details on what the images depicted. He also declined to discuss SeaWorld safety procedures for working with killer whales.

Meanwhile, the Alliance of Marine Mammal Parks and Aquariums, which counted SeaWorld as a founding member, had assembled a panel of industry leaders to help the company assess what had gone wrong, and how to prevent such an occurrence in the future. The experts hailed from several facilities, including the Georgia Aquarium, Miami Seaquarium, Marineland Ontario, and the US Navy Marine Mammal Program.[3]

But industry self-policing alone was not going to satisfy government officials, Naomi was relieved to learn. Several outside agencies were also investigating the death, including OCSO, OSHA, and the USDA's Animal and Plant Health Inspection Service (APHIS), which administered the federal Animal Welfare Act.

As part of its public relations defense, SeaWorld enlisted the help of Thad Lacinak, who left the corporation in 2008 to form his own company, Precision

Behavior, an animal-training consulting firm. Thad went before the cameras to defend his former employer, even if that meant blaming Dawn for her own demise. His words were strikingly similar to what Jack Hanna said on *AC-360*. "She was an excellent trainer, one of the best I've seen in my life," Thad stated. "But allowing her ponytail to drift into the water like that in front of the animal turned out to be a fatal mistake." Dawn had "laid completely down," a vulnerable position to assume with Tilikum. Once Dawn's ponytail had "apparently" drifted, "Tilikum just opened his mouth, sucked it in, and pulled her in the water." Either the rules for handling Tilikum had changed or Dawn had violated them, Thad said. When he was at SeaWorld, all trainers were prohibited from lying down so close to the whale.

When Tilikum dragged Dawn into the water, the situation turned into a big game, Thad speculated. "It was more novelty." Once other staff arrived at G Pool and unspooled the nets, Tilikum likely became even more aggressive in trying to maintain control over "his toy."

As for calls to release Tilikum, that wasn't going to happen. "Releasing any of these killer whales to the wild would be the wrong decision. We already tried it once, Free Willy. He died a terrible death." Thad insisted that captivity was not only the best way to study killer whales, it was the *only* way. "These animals are invaluable in terms of what we can learn from them. And you cannot learn about killer whales through a pair of binoculars."

Naomi scoffed at that statement. She had spent five summers observing wild killer whales, often with binoculars. Her work, and that of so many other field researchers around the world, had helped build the body of knowledge about the species, including its complex social behavior.

But according to Thad, the average person would hardly be interested in boring old cetacean science on its own: "We know for a fact that people do not learn in static conditions. They learn from these animals when they are entertained by them."

What a fool, Naomi thought. Most people who love to learn about animals do so because they love the *animals,* not because they want entertainment. Such a self-serving statement, she thought, cheapened both the animals and the people who were interested in them.

In the aftermath of Dawn's death, Thad and Chuck Tompkins emerged as two of the key public faces of SeaWorld. It was important, and only natural, to show sympathy to the victim and her grieving family and friends, but also critical to protect the family-friendly brand and reassure patrons that this incident was a unique outlier, a freak accident.

"What you need to remember is, we've done thousands of interactions with this animal with no incidents whatsoever," Chuck told the *Orlando Sentinel.* Tilikum and Dawn had a positive relationship, he said. Tilly was especially enthusiastic and responsive to Dawn. He knew her well and "liked working with her."

In yet another public statement, Chuck embarked on a lengthy treatise on "building a relationship with these animals and developing trust. Before we put anybody in the water, it's two or three years." But he acknowledged, "We're working with wild animals, they're not tame. If there's sexual activity in the pool, or maybe they're socially interacting with each other, we're very sensitive to those things. You don't pet your dog in the middle of a dogfight. We train our staff to observe these things."[4]

If something didn't "feel right," trainers simply got out of the water. They knew when to step back and "let them be animals," Chuck said. All trainers knew of the risk, he said. But working with killer whales was worth it. It had been "one of the most awesome experiences I've ever had in my life."

The great bull Tilikum, Thad added, was a "valuable asset." As the only breeding male, he was often kept separated from the other killer whales. The same separation happened in the wild, he said. That statement, Naomi thought, was simply untrue.

33
Battle Stations

On Saturday, February 27, 2010, SeaWorld tried to gain the upper hand in its PR recovery war. That morning at 11:00 a.m., the Orlando park resumed the "Believe" show, only now with all trainers remaining safely on dry land. SeaWorld had also scheduled a tribute to Dawn Brancheau. Loyal fans began lining up as early as 9:00 a.m. for the event.

Images of Dawn, working with killer whales, were projected on the JumboTrons. "It brought tears to my eyes. She died doing what she loved," park guest Billy Grady from Georgia told the *Sentinel*. "If it stopped, I don't think she'd be happy. She's looking down on us, and she's happy it went on," Grady said.[1]

Melodramatic, perhaps, but Grady was probably right. His sentiments were echoed by friends and family of the deceased at a private burial in her native Indiana on March 1. It attracted worldwide media attention, including from CNN's *Larry King Live*. Reporter Jason Knowles of WLS-TV Chicago covered the proceedings.

"The family really wanted this to be a very private moment" and was avoiding the media glare, said Knowles, who did manage to speak privately with some of the mourners. The subject of captivity came up frequently, and Knowles was surprised to find that many of the mourners wanted the Shamu show to continue because "this is what Dawn Brancheau would have wanted."

Naomi recorded the *Larry King Live* show and watched it the next morning. She had never seen anything quite like it. The segment quickly tumbled into a free-for-all quarrel. "Should whales be captured and forced to perform?" King asked in a loaded question. "We're going to debate it."

Among the debaters that evening: Thad Lacinak; Jack Hanna of the Columbus Zoo; Jane Velez-Mitchell of CNN Headline News and an animal welfare advocate; and Ric O'Barry, the former trainer who worked on TV's *Flipper* before becoming a leading advocate against captivity. O'Barry's transformation came about after he "got tired of telling that lie about education and research," he explained to King. "The fact is the show I was doing was nothing more than a spectacle of dominance and, in fact, a form of bad education."

Larry King turned to Jack Hanna: "Ric says the education part is—is a farce, for want of a better word."

"Well, it's not a farce, Larry," Hanna replied. Places like SeaWorld were educational, they helped people "grow and grow each year." People such as himself and Ric O'Barry were able to see killer whales in the wild, but "99.9 percent of the people, Larry, in the world" would never be able to.

When Naomi watched the show, she knew from the start that many of the industry talking points would be trotted out. Jack Hanna could recite them all, she thought. This shopworn canard about whale watching as an ultra-elitist pastime, accessible only to a minuscule and presumably wealthy sliver of the population, was bullshit. Hanna was claiming that it was easier and cheaper to go visit SeaWorld than to experience killer whales in the wild. Not necessarily.

Adults were paying $81.99 for a ticket to SeaWorld Orlando, plus $73.99 for each child. In contrast, in the Pacific Northwest, the Victoria Clipper company ran all-day cruises from Seattle to the San Juan Islands, including a full two and a half hours dedicated to killer whale watching (as opposed to the half-hour Shamu show), narrated by a certified naturalist, for $70 per adult. Children under twelve sailed for free. A family of four could spend $312 for a day at SeaWorld, or less than half that amount, $140, on a Clipper boat. And viewing spots from the shores of the Salish Sea cost nothing.

CNN's Velez-Mitchell picked up on the revenue theme: "Whenever you see animal exploitation, follow the money. Larry, this is big business. The Blackstone Group, which is traded on the New York Stock Exchange, recently bought SeaWorld along with a cluster of other amusement parks for—I know you're sitting down—2.7 billion dollars," she said, before abruptly switching gears to animal welfare. "If you were in a bathtub for twenty-five years, don't you think you'd get a little irritated, aggravated, maybe a little *psychotic*? This animal was speaking with its fins, saying, 'Get me out of this damn tank.' "

King asked Thad Lacinak for his response.

"Jane, you are—you are a *newscaster*. You have no knowledge of what you are talking about with dolphins, whales, or anything," he answered gruffly, without addressing the allegation that whales were getting "a little psychotic." To Naomi, Thad was being typically dismissive of anyone not inside the clubby world of marine mammal trainers. This common, lowbrow debate tactic, she thought, sought not to dignify any specific comment with a direct response, but simply to devalue one's opponent with an ad hominem attack.

Then Jack Hanna joined the fray: "How are you going to love something, Larry, unless you see something? You can't love something and save something unless you see it."

Naomi had heard this argument before. It was ridiculous on its face, she thought. What about dinosaurs? People, and especially kids, were crazy about dinosaurs. They *loved* them, without ever having laid eyes on a single one.

Velez-Mitchell was waiting for someone to make the "Shamu is here to educate us" argument. She pounced and read a statement from the SeaWorld website: "'This seasonal show is a rock 'n' roll concert of unprecedented proportions combining improvisational movements of killer whales with music remixed from some of the hottest rock stars in the industry.'

"That," she said, "is not education."

King drew a deep breath. "We're not going to resolve this tonight, but it sure is interesting," he sighed. "Thad Lacinak, how do you respond to Jane Velez-Mitchell?"

"Well, Jane was talking about the killer whales traveling a hundred miles a day," Thad said. "The only reason killer whales in the wild travel that far at any given time is because they're hunting for food." Some killer whales in Puget Sound "hang around in areas in small coves where the salmon population is high during certain times of the year." The whales "don't move from that area because that's where their food source is."

Naomi again bristled at the standard industry misinformation. Ecologically speaking, any species' home range was as large as needed to support its food requirements. Animals with all their energy needs met in the immediate area evolved to have a small home range. But animals with energy requirements that were met only by widely dispersed resources evolved to cover a larger home range. Restricting an orca's range was *precisely* the reason why killer whales did not thrive in captivity. You cannot switch off millions of years of evolution just because an animal is captive.

All carnivores that are wide-ranging do poorly in captivity. Naomi was thinking of one study in particular, a 2003 paper by Ros Clubb and Georgia Mason, who analyzed data from carnivores with small home ranges and densely distributed prey, and those with large ranges and widely dispersed prey.[2] The former did well enough in captivity—their health was good, they didn't develop behavioral stereotypies (pacing, etc.), and they had lower infant mortality rates. The latter fared much worse in captivity—their health was fair to poor, they often developed stereotypies, and their infant mortality rates were higher.

Killer whales—by Thad's own admission—were wide-ranging. According to the pattern that Clubb and Mason described, they would do poorly in confinement. Thad could not have it both ways: either a species was adapted to be wide-ranging and therefore confinement caused it to suffer; or it didn't need a large home range in the wild, making it more adaptable to captivity. What's more, exercise was good for a reason: If a species evolved to be active (even because of its food requirements), then taking that away could kill.

"Thad is only presenting half the story," Naomi complained to Chris, who watched the show later. "If it's because he doesn't know the other half, then he's being deceptively selective in what he learns. This is part of the miseducation I'm always going on about. SW chronically presents only half, or less, of the science. And they do it in an almost convincing way, just enough to mislead people."

Thad's contention that some killer whales "hang around" in small coves in Puget Sound all summer and "don't move from that area" ignored what was known about the Southern Resident community. If one was going to put oneself out there as an expert, Naomi thought, one should have much broader expertise in the subject.

Then, about midway into the broadcast, Ric O'Barry dropped a bomb that is still reverberating today in the captivity debate.

"Larry," he announced gravely, "on the twenty-fourth of December, a SeaWorld killer whale killed its twenty-nine-year-old trainer, Alexis Martinez. Did you hear about that? That was sixty days ago."

King was staggered. So was Naomi. "Why didn't we hear about that?" King asked.

"I'm talking to trainers now at Loro Parque, who tell me that SeaWorld showed up and had the body cremated on December twenty-fifth, the very next day," O'Barry said. "We're going to hear about that, and we're going to start hearing about many trainers at SeaWorld that have been very seriously injured." (SeaWorld denied Martinez was cremated.)

Naomi's jaw dropped. How could something involving yet another trainer's death, and orcas owned by SeaWorld, happen without the marine mammal community knowing about it?

It would take more than a year and a half before all the details on the killing of Alexis Martinez (a close friend of Dawn Brancheau's) by a SeaWorld orca would come to light. His death would figure prominently in the federal legal proceedings that were about to challenge the very idea of putting trainers in the water with the ocean's top predator.

This had been quite a show, and it wasn't over yet. King turned his line of questioning toward Jack Hanna. "Wouldn't it be obvious that whales would have a better time being free than being captured?"

"Not necessarily, Larry. No," Hanna protested. "Thad said it very well: A hundred years ago, man would go a hundred miles looking for food." Then, rather jarringly, he switched rhetorical gears. "When men go up to space in the shuttle and they come back and our astronauts are killed, that's because they wanted to discover space, find out more about what it can provide to the human world. We're here because we *love* it, and we know what the results will be, Larry. It will be millions of people educated about our animal world."

Was Hanna *really* comparing Shamu shows to the space program, in terms of their scientific and educational contributions? Naomi thought he was implying that the risks involved were equally justifiable and that the benefits were equally great.

O'Barry did not want to discuss the space program. He still wanted to talk about the Canary Islands: "A man lost his life here, and these two guys representing SeaWorld are being very quiet about that. You may not have been able to get your staff to confirm what I said, but that's a serious thing." He asked the two men if they knew about the death.

Thad said he did know, but not the details.

Jack Hanna also knew, but he still wanted to talk space. "What about our astronauts? What are you saying, Ric? It's called a *killer* whale. This is a dangerous animal. This is a job we have. Killer whale, Ric. You should know that."

Thad said even his five-year-old daughter "gets it: Habitat dictates behavior."

Now, Naomi thought, he has really stepped in it. If habitat dictated behavior, then a suboptimal habitat such as SeaWorld would dictate suboptimal behavior, such as Tilikum's.

After watching the show, Naomi sent an e-mail to an HSUS colleague who held particular sympathy for captive cetacean issues: "Frankly, I found the entire King show highly entertaining." They both agreed it was definitely worth watching. "We needed only some popcorn and a giant soda to complete the experience," Naomi joked. "What a donnybrook! Ric and Thad had some priceless moments."

All joking aside, Naomi had serious business to tend to. She wanted to find out more about the death at Loro Parque, and how such news could have been kept from so many people in the marine mammal community for a full two months. She reached out to colleagues such as Courtney Vail of the Whale and Dolphin Conservation Society, Susan Millward of the Animal Welfare Institute, Bill Rossiter of Cetacean Society International, and Mark Berman of Earth Island Institute. Mark had heard about the incident at the same time as Ric, who worked with Earth Island. The paucity of information in the wider animal community astounded even Mark, who always seemed to hear about everything before anyone else did.

The same day as the Larry King broadcast, March 1, 2010, CNN published Naomi's take on the killing in a blog post titled "Tragedy for Two at SeaWorld." The death had been a misfortune "for Ms. Brancheau and her devastated family, to whom I and my organization offer our condolences," Naomi wrote. "But it is also a tragedy for Tilikum."

That intelligent, social killer whales were suitable for confinement in "featureless enclosures" should, "on its face, be ludicrous." But people enjoyed their beauty and athleticism, leading society to "recast the image of these animals from 'killer whales' into 'sea pandas.'" We admire their power and grace, but ignore the "irony of forcing them into straitjackets of concrete," Naomi wrote. "We believe they are happy because they seem to be smiling."

It was not only tragic but "almost irrational" to assume that captivity would be in the best interest of this species. Some whales might adjust more readily than others, and some might find humans to be a decent substitute for their family, "but no orca captured from the early 1960s . . . through the late 1980s (when most orca captures ended due to public outcry) was ever best served by being taken from his or her mother," she said. But now that civilization had entered the twenty-first century, perhaps things might change. Tilikum might be allowed a chance at a better life, retired to a sea pen, and given "more space and choices and stimulation." Meanwhile, as the remaining captive killer whales aged and died, society should "let orca exhibits become a thing of the past."

The media furor continued as people from all sides weighed in on the tragedy. TV host and animal activist Bob Barker urged SeaWorld to release all of its animals,[3] while actress Tippi Hedren (of *The Birds* fame) said SeaWorld should halt its animal "circus,"[4] and actor Matt Damon said the place should be closed down altogether.[5]

People had lots of ideas about what to do with Tilikum, but the most macabre was offered by the American Family Association. The conservative group issued a decree that Tilikum should be stoned to death in accordance with biblical law.[6] "If the counsel of the Judeo-Christian tradition had been followed, Tilikum would have been put out of everyone's misery back in 1991 and would not have had the opportunity to claim two more human lives," proclaimed AFA leader Bryan Fischer. "Says the ancient civil code of Israel, 'When an ox gores a man or woman to death, the ox shall be stoned, and its flesh shall not be eaten.'"

Fischer didn't stop there. The Bible called for the owner of an animal responsible for killing someone to die as well. "The Scripture soberly warns," he said, "if one of your animals kills a second time because you didn't kill it after it claimed its first human victim, this time you die right along with your animal."

The New York–based Blackstone Group, owner of SeaWorld Parks & Entertainment, is one the world's largest private equity and financial advisory firms. Its chairman and cofounder Stephen Schwarzman has amassed an estimated worth approaching $6 billion. He is hardly shy of the limelight—or controversy.

Schwarzman was perhaps best known for a notoriously extravagant sixtieth birthday party he threw for himself in 2007 at Manhattan's cavernous

Park Avenue Armory. The multimillion-dollar fête, lavished with coverage by the New York press, included performances by comedian Martin Short and musicians Rod Stewart, Marvin Hamlisch, and Patti LaBelle—who was backed up by the Abyssinian Baptist Church choir as she belted out a gushing ballad to the tycoon. A large portrait of Schwarzman was reportedly couriered to the armory from the living room of his gargantuan, $37 million Park Avenue apartment (once owned by John D. Rockefeller) just for the occasion. Guests included Colin Powell, New York mayor Michael Bloomberg, and the city's powerful Catholic cardinal Edward Egan.

Schwarzman, a major backer of the 2008 McCain-Palin ticket, took some hits from some corners of the American media when he attacked President Barack Obama in 2010 for trying to raise taxes on private equity firms, such as Blackstone. "It's a war," Schwarzman said of the fight with the White House. "It's like when Hitler invaded Poland in 1939."[7]

With that caliber of opposition, Naomi and her colleagues at HSUS harbored few illusions about their chances of convincing Blackstone to let Tilikum retire to a sea pen far away from SeaWorld. But that didn't stop them from trying.

In early March, Wayne Pacelle, the charismatic and sometimes controversial president of HSUS since 2004, sent a letter drafted by Naomi to Stephen Schwarzman, entreating him for Tilikum's release. Pacelle began by expressing condolences on Dawn's death on behalf of the HSUS's 11 million members and constituents.

"I wanted you to have the benefit of our firsthand knowledge and experience concerning Keiko," he wrote. HSUS suggested devising a similar plan for Tilikum. "We are not proposing a release into the open ocean, but a transfer of the whale to a sea pen that will allow him to live in a more suitable environment."

This third fatal incident involving Tilikum "appears to have been an escalation of his previous behavior—he actively pulled Ms. Brancheau into the tank with him," the letter continued. "Clearly Tilikum is a hazard to his trainers." Putting more distance between the whale and his trainers might protect them, Pacelle wrote, but such measures would also "increase his isolation and decrease his well-being." Given the safety issues, and the prospect of his remaining in a small enclosure for the rest of his life, HSUS felt that a sea pen would be best, "in Iceland or a similarly suitable location, with all deliberate speed." Pacelle then requested a meeting with Schwarzman.

The response was predictably harsh and dismissive. SeaWorld Parks & Entertainment president and CEO Jim Atchison wasted no time in attacking HSUS for its role "in the tragic release experiment involving Keiko." That episode demonstrated the "cruelty" of trying to release a long-captive marine mammal and showed a "lack of expertise in matters of marine mammal care and the vast complexities of reintroduction involving even those animals well suited for return to the wild. Under no reasonable interpretation could Keiko be considered a suitable release candidate."

The Keiko "experiment" had cost tens of millions of dollars, money that could have saved thousands of other animals, but instead "cost an innocent animal its life," Atchison wrote. He said that virtually all experts on the release of marine mammals had agreed that the Keiko project was "a disgraceful act," and its execution, by any measure, "irresponsible and reckless." Keiko had suffered serious injuries in the wild and then died "prematurely and unnecessarily." He should have lived out the rest of his life with other killer whales "in an accredited and professionally operated zoological institution." Atchison did not mention that SeaWorld itself had rejected taking Keiko back in 1996.

Meanwhile, the SeaWorld parks would continue the vital work of rescuing wild animals in distress. More than fifty rescued animals were being treated by SeaWorld vets at that moment, Atchison said. In the past nine weeks alone, SeaWorld had rescued, treated, and released nearly a thousand animals.

Naomi noticed that Atchison did not specify what, if any, of that good work pertained to wild killer whales. She assumed that none of it did, or the CEO would have mentioned it. And, she asked herself, how many of those releases were successful? Did anyone monitor those animals to know if they survived?

"SeaWorld always uses its rescue operations as PR opportunities, rather than doing the right thing by animals and improving scientific knowledge about those species," Naomi explained to people at the office who read the letter. "The work that they actually do isn't as great as they claim. It *looks* good, but in the end, they can't offer evidence of success because they don't monitor most of the animals they release."

SeaWorld and Stephen Schwarzman were clearly going to win this round. HSUS did not stand a chance against the Blackstone behemoth: Tilikum would remain in Florida.

John Kielty, the construction inspector from Massachusetts who was at the "Believe" show before Tilikum dragged Dawn into G Pool on February 24, decided it was time to speak up. He was disgusted with SeaWorld. First there was the misstatement about Dawn's falling into the water. Now some of those same officials were trying to convince the public that there were no signs of anything amiss at Shamu Stadium prior to Tilikum's rampage.

John figured the right thing to do was to contact the company directly. He wrote to SeaWorld via their website's feedback forum and described the commotion he had seen during the "Believe" show that day, before it was abruptly ended. He also asked SeaWorld to share that information with the relevant investigative agencies working on the case.

The reply, a form letter, thanked John for his message and assured him that everything was being done to investigate the incident and prevent one like it from happening again. Nothing like it had ever occurred in SeaWorld's history, the letter said. Nothing was more important to SeaWorld than the safety of employees and guests, and the well-being of the animals in the company's care.

The nonresponse infuriated John. He would not be dismissed so lightly. Even though he was still in the final days of his vacation, back in Myrtle Beach, South Carolina, he reached out to media outlets to give them his account of the unrest during the "Believe" show. Several other witnesses, none of whom had been interviewed by the sheriff's office, were also speaking out.

"SeaWorld, I know, has repeatedly said there were no problems with the animals prior to this event, and everything had gone according to schedule throughout the day leading up to Dawn's death, and I disagree," John told a local TV news station. "I know there's hundreds of other witnesses who would agree with me that there was something wrong prior to her death."

When not speaking with reporters, John spent hours on the Internet, researching which government agencies were investigating the Brancheau death, and came across the Orange County Sheriff's Office, the Florida wildlife commission, the Commerce Department's NMFS, the Department of Agriculture's APHIS, and the Department of Labor's OSHA. John wrote to them all.

"I am dismayed over reports that continue in the news media by SeaWorld staff (primarily . . . Chuck Tompkins) that there was no indication that the killer whales were behaving abnormally prior to Ms. Brancheau's tragic death," he said in his letter. "While this may be true of the few minutes immediately preceding the attack during the 'Dining with Shamu' segment, problems were evident during the 'Believe' killer whale show." He then described what he had seen during the aborted splash section of the show and asked if SeaWorld had provided authorities with its own video record of what happened in the main pool just thirty or forty minutes before Tilikum went berserk.

Dawn might still be alive, John said, "if tighter restrictions were in place, and followed, when faced with this abnormal behavior."

The only reply John got was from OSHA. The lead investigator on the case got back to him right away. Her name was Lara Padgett, a no-nonsense and highly trained safety and health compliance officer who had spent years in the air force protecting service members against the hazards of working in stressful conditions around weapons and live ammunition, among other duties. Lara knew little about killer whales—other than what she and her husband

and kids had seen and enjoyed at SeaWorld—but she knew a great deal about workplace safety and employer negligence.

She thought she had a good case in Orlando.

Lara, a tall, forty-year-old strawberry-blond woman who favored sensible pantsuits when not on inspections, had arrived on the scene at SeaWorld at midafternoon on February 24. Lara was also OSHA's lead inspector in the 2009 death of a monorail operator at Disney World. She immediately began collecting evidence for what would quickly emerge as one of the most complicated and contentious cases OSHA had tackled in recent memory. According to many accounts, SeaWorld would go out of its way to make the Brancheau inquest extremely difficult for the feds.

Lara began by asking to interview a large number of SeaWorld employees, including orca trainers both present and absent when Dawn died. She requested trainer records from all SeaWorld parks, but only received some of them, while SeaWorld executives denied the existence of Animal Profiles and "aggression incident books," even though several trainers told OSHA they had personally seen them.[8] SeaWorld had also said nothing about a disturbance during the 12:30 p.m. "Believe" show that day, nor the impact it may have had on Tilikum in one of the back pools, Lara told John. He was not surprised by their lack of candor and neither was Lara, who told him that SeaWorld was not very forthcoming with information.

"I'm leery of SeaWorld's motives. They seem more interested in protecting themselves with mistruths than shedding light on this case," John told Lara, adding that SeaWorld put profit above safety. By portraying Dawn's death as an accidental drowning "they really downplayed the extent of her brutal death."

Lara Padgett was not only interviewing park staff and guests—including Todd and Suzanne Connell, who reached out to her—she was consulting with as many experts on wild and captive killer whales as she possibly could. It didn't take her long to find Naomi and allies such as Courtney Vail of WDCS and Bill Rossiter of CSI. They helped Lara and her investigation by providing her with published research on wild whales and their social behavior, with information on the stress of captivity in large, wide-ranging predators, and with contact information for cetacean specialists around the world.

Among the people on Naomi's suggested list were Graeme Ellis, John Ford, and Peter Olesiuk of the Department of Fisheries and Oceans' Pacific Biological Station in Nanaimo on Vancouver Island, Robin Baird of Cascadia Research in Olympia, Washington, and Dave Duffus of the University of Victoria, British Columbia, the professor who headed up the Keltie Byrne coroner's inquest back in 1991. Lara also contacted industry experts from outside SeaWorld, including members of the independent panel the company had assembled to review its killer whale safety procedures and make suggestions for any needed improvements.

Most of the wild whale experts responded enthusiastically and sent Lara their professional opinions on the matter. In contrast, most of the industry experts begged off the request to assist OSHA in its inquest, usually by citing a lack of firsthand knowledge of the actual events. Privately, Lara would learn, many of them were complaining to SeaWorld about her outreach to them.

Some of the wild orca experts, however, were reluctant to get involved. In an e-mail to one of the foot draggers, Lara laid out her reasons for seeking help.

OSHA had no regulations on working with animals, Lara explained. If the agency were to pursue a violation, it would have to issue a "general duty clause" citation. Under that clause, Congress ordered every employer to create a place of employment "free from recognized hazards that cause or are likely to cause death or serious physical harm to employees."

To prove such a violation, OSHA would have to show that "a condition or activity in the workplace presented a hazard to an employee," Lara said. It had to be a "recognized" hazard likely to cause death or serious physical harm, "with a feasible means of eliminating or greatly reducing the danger." Finally, the evidence had to show that the employer "knew, or with the exercise of reasonable diligence, could have known of the violative conditions," Lara wrote.

Given all the previous human injuries and deaths involving its killer whales, Lara said she was "looking at recommending that SeaWorld should keep their staff out of the water and in the near proximity to the pools." The company had already attacked her credibility, she added, "because I do not have expertise to express opinions on killer whales—they are right, all I have is a BA in Management and 17 years in employee safety—that is why I am reaching out to every orca expert that is willing to weigh in."

Back in 2007, SeaWorld had cowed Cal/OSHA into rescinding its own reprimand of the company and even issuing a public apology. Lara Padgett and her colleagues were painfully aware of that. They were determined to prevent a repeat of any kind of capitulation. Naomi was equally determined to help them.

SeaWorld would need all the PR assistance it could muster if park officials were going to convince the public that Dawn's death was a bizarre accident and that Tilikum was merely "playing" with her ponytail. That task became considerably more difficult on March 31, when Dawn Brancheau's autopsy was released.[9]

In the opinion of the medical examiner, the death was the result of "drowning and traumatic injuries," though there seemed to be far more evidence of trauma than of drowning.

According to the report, Dawn had sustained multiple blunt-force injuries of the head and neck, including the avulsion (ripping away) of the scalp and associated bleeding of the skull, lacerations of the right ear, abrasions of the left cheek, fracture of the mandible (lower jaw) with associated laceration and hemorrhaging of the oral cavity, fracture of a cervical vertebra, bleeding from the spinal-cord outer membrane, and softening of the spinal cord.

Blunt-force injuries of the torso included abrasions of the left upper back, fractures of three ribs, fracture of the sternum, lacerations of the liver, and blood in the abdominal cavity. The arms and legs also suffered abrasions, lacerations, and contusions (bruises), with a complete tearing away of the left arm and dislocation of the left elbow and left knee.

The evidence of drowning, listed at the end of the findings: approximately four milliliters (one-seventh of an ounce) of liquid in one sinus chamber.

Jeff Ventre and John Jett found the autopsy odd. To begin with, the evidence of drowning was listed almost as an afterthought, with nearly all of the evidence pertaining to traumatic injuries. Yet in the summary, which was picked up and repeated by media outlets around the world, the causes were put in reverse order.

"My gut reaction is that this is potentially misleading, though it may not have been written with that intention," Jeff, the MD, said to his friend. "But why was drowning listed before trauma?"

What's more, evidence of drowning was flimsy. Why was there no report of fluid in the lungs, which would have been a sure sign of asphyxiation by water? "The small amount of water in her sinuses could have been there from a regular show sequence," Jeff said. For example, when trainers went to the bottom to prepare for a hydro maneuver, water was routinely pushed into their sinuses. "That helps create the perception that this was a tragic drowning, not a brutal killing."

John agreed that four milliliters of liquid in one sinus was hardly enough to demonstrate that Dawn had drowned. "I probably use two milliliters of toothpaste on my toothbrush," he said. "I routinely had that much water in my sinuses following water work, and I still do after scuba diving or snorkeling."

Naomi was also disturbed by the autopsy. The attack was much more serious than she had thought. She mentioned her concerns to Lara Padgett at OSHA and also wrote them down for a small number of colleagues to consider.

"The number and type of injuries she had obviously set this incident well apart from the other two in which he was involved," Naomi contended. "Rather than merely preventing her from leaving the tank (inflicting only minor cuts and contusions) and eventually holding her under until she drowned, this time Tilikum shook and slammed her around, inflicting several serious traumas (severing her spine, breaking her neck, breaking other bones—these are signs of serious intent of some kind, although I'm still not convinced he meant to kill her)."

Whatever he meant to do, Tilikum was "extremely agitated," Naomi wrote.

"What snapped in his brain that led him to beat her up so badly, after 27 years of never showing QUITE this behavior before, and after 11 years of relative calm?"

That answer might never be known. Even so, something was "seriously wrong here," Naomi added. "Whether it was the result of her high energy behavior or whether he really has just had enough of small spaces—or whether he's just finally gone mad—I don't know."

Some circles speculated about the feeding habits of Icelandic orcas. Was Tilikum actually raised to eat marine mammals, and not fish, like a Transient whale? Naomi didn't think so. That sort of prey-handling behavior would have been learned in his first three years of life. If that was the case, it was "hard to explain why this would only come out now," she said.

During her investigation, Lara would hear from a killer whale expert from the UK who could help shed some light on Tilikum's natural diet and hunting instincts. Andrew Foote, PhD, of the University of Aberdeen, had studied Pacific and North Atlantic killer whales since 2000, including the population from which Tilikum had been taken, he told Lara in a letter to OSHA.

Foote wrote that fish-eating Resident-ecotype whales take marine mammals such as porpoises and "apparently 'play' " with them. This involved dragging the animals underwater, tossing them in the air, and blocking any escape with their bodies until they drowned.

The Icelandic population, like the Pacific Resident ecotype, was known to feed upon herring, Foote said. But recent research found that certain whales photographed near the area of Iceland where many orcas were captured also hunt seals around the Shetland Islands of Scotland. "During summer months they specialize in taking these mammals," he observed. Photo-identification and genetic testing suggested they were part of the same population seen feeding on fish. In other words, Tilikum might naturally have preyed upon herring *and* mammals.

Killer whales were "capable of causing injury or death to smaller mammals such as ourselves, either through predation or 'play' behavior," Foote concluded. "The conditions under which they are kept at SeaWorld represent a severely impoverished environment and may be an additional factor that could cause unpredictable behavior."

OSHA's investigation attracted a lot of attention around Orlando. Before long, Lara was being contacted by people on all sides of the captivity issue. Some of them were trainers, still working at SeaWorld, who did not want to be publicly identified. One of them chided Atchison for telling the media that Tilikum was never separated from the female killer whales (something that even Thad had contradicted). "To put it bluntly, that is a lie," the trainer wrote. Mostly, Tilikum was kept isolated from the females. "Jim Atchison knows this and yet he told the press the exact opposite."

The trainer went on to say that Tilikum became agitated "when the females are out doing the shows and he is left completely isolated," as was the case at midday on February 24. The writer also claimed that, up until a few years ago, trainers were never allowed in the water with Tilikum—even on a shallow watery ledge the way Dawn had done. All interactions with him had to be conducted from behind a wall in one of the back pools. They confirmed what Thad had told the media: Trainers never got that close when he was corporate vice president.

The trainer brought up another possible contributing factor for the attack, in addition to Tilly's isolation and Dawn's dangerous (and previously forbidden) proximity to him. "Potential corporate sponsors may have been in attendance at that Dine With Shamu show along with SeaWorld corporate staff," the writer speculated. While Dawn was not told to violate any safety protocols, "she may have been asked to 'get as intimate' as you can with Tilikum to impress our clients."

The trainer worried, "Our fear is that Dawn will get the blame for something when she was simply following orders."

34

Oversight

Given all the news about SeaWorld, much of it unflattering, Naomi thought it was time to bring in the big guns of Congress—that is, *if* members of a key committee in Washington could be convinced to hold a hearing. Within a week of the incident, she began contacting sympathetic Capitol Hill staffers to gauge their bosses' appetite for a hearing on the nation's lucrative and underregulated marine mammal industry.

Naomi contacted a friend who once worked for the International Fund for Animal Welfare (IFAW) and was now on the majority staff (Democrats controlled the House in 2010) handling marine mammal issues for the oceans and wildlife subcommittee of the House Committee on Natural Resources. The contact asked Naomi if she would like there to be an "oversight hearing."

That was the top item on Naomi's wish list.

Most congressional hearings are related to particular bills or ideas for bills. But Congress can also hold *oversight* hearings on existing laws and policies. They are typically called after some issue makes big news—steroid use among professional baseball players, say, or cars with faulty brakes. Because this was the oceans subcommittee, its oversight was limited to the industry's adherence to the Marine Mammal Protection Act, which since 1994 had been limited to the educational aspects of public display.

The display industry had never been subjected to such close scrutiny by Congress before. Most previous congressional hearings on marine mammal public display were called for and guided by industry. This would be the first time they would be testifying in front of a potentially hostile panel of lawmakers. For the anti-caps, it could be a once-in-a-lifetime chance to air an entire washroom of dirty laundry belonging to SeaWorld and its allies.

"We have succeeded in getting our hearing. It will be held April 27,"

Naomi wrote to colleagues. "The worst that will happen is that the impact of the hearing is kind of fizzly—we get some issues aired, but nothing terribly dramatic . . . the bad guys will certainly try to trip us up during the Q&A, but we have taken hits from all their ammo before and know how to keep our cool and respond rationally." And the best that could happen? "We blow them out of the water." The ideal outcome would be stronger regulation and "the beginning of the end of the 'laissez-faire' era of public display. This is OUR hearing—the relevant staff are looking to us for input and guidance."

HSUS was trying to convince CSPAN to cover the hearing, wrote Naomi, who said she was also drafting a two-page proposal for Tilikum's retirement for the Blackstone Group. Meanwhile, HSUS had issued an action alert directed at APHIS urging the agency to conduct a "thorough and unbiased investigation." The letter went directly to Agriculture Secretary Tom Vilsack, bypassing all bureaucratic filters.

For much of early April, Naomi was busy preparing a comprehensive memo and package of supporting documents to share with her allies on the Hill. Because the topic was limited to enforcement of the Marine Mammal Protection Act, whatever she presented to the committee had to be somehow related to the education and conservation provisions of that law.

One rich source of information was the industry itself. It was important to learn what these facilities were saying about their *own* education and conservation efforts, Naomi figured, in order to draft the most pointed lines of questioning possible. She worked with one of Chris's graduate students at George Mason University, Megan Draheim, who searched the websites of SeaWorld and other marine parks as well as the sites of the Alliance and the AZA. Naomi also dug up a written Q&A from her files that SeaWorld San Diego had completed in 2007 for a local news station, KGTV, after Ken Peters was injured by Kasatka.

The questions were tough but fair. SeaWorld began by admonishing the station for speaking with industry opponents. "We must express our disappointment regarding your sources," the unsigned document complained. "You have accepted as fact the propaganda of animal rights extremists whose sole objective is denying Americans the privilege of experiencing marine mammals in places like SeaWorld."

SeaWorld had put killer whales on display for nearly forty years and conducted more than 160,000 shows featuring millions of interactions, the letter stressed. "While we recognize that there is an element of risk in any animal interaction, there have been a remarkably small number of serious incidents and no fatalities." In addition, "virtually nothing" was known about killer whales before the orca display industry was born. "They were animals so feared and despised that they were routinely used by military pilots for aerial target practice."

That argument, Naomi thought, was so outdated it had become moot. No one today would seriously advocate a return to firing upon killer whales. The display industry may have turned public opinion squarely against such barbarity, but perpetual captivity was hardly required to prevent a return to those bad old days.

As Naomi prepared her subcommittee memo, most of the issues she addressed pertained to public education, orca research and conservation, and the debate over killer whale longevity:

Education: SeaWorld *did* post informative signage around the park with various factoids about the species in its collection, and preshow segments included information about the animals. In the KGTV Q&A, SeaWorld also boasted of its "extraordinary number of special programs." These included the SeaWorld/Busch Gardens Environmental Excellence Awards, which grant $80,000 to students and teachers working to preserve the environment; Shamu TV, an Emmy Award–winning education series available to more than 50 million viewers; a major animal information website filled with more than four thousand pages of materials; classroom curricula; conservation project "calls-to-action"; zoo career information; and "environmentally focused family resources."

Public opinion polls seemed to support SeaWorld's claims that people were being educated during their visits. A Harris Interactive poll commissioned by the Alliance found virtual unanimity (an almost-unheard-of 97 percent of respondents) that marine parks, aquariums, and zoos "play an important role in educating the public about marine mammals they might not otherwise have the chance to see." Nearly the same number, 96 percent, agreed that they "provide people with valuable information about the importance of oceans, waters and the animals that live there."[1]

But Naomi questioned how much of that education was being absorbed by guests, many of whom, frankly, came to SeaWorld for roller coasters and Shamu shows, and not lessons on the rich variety of the family Delphinidae. SeaWorld was not at fault if some people didn't want to be educated, but it was still obliged to offer information that was accurate. She was equally cynical about poll numbers. "Sometimes they're asking people for their opinions about SeaWorld as they are leaving the park," she said. "After they've just had a great day, most people are going to say, 'Sure, you guys are great. You betcha I learned a lot.' But what do they know after a visit? What do they do afterward? Public display doesn't actually teach people much. It just makes them feel good, which actually leads to less conservation action rather than more."

Indeed, signs were now informing guests that, because a small portion of their entry fee was earmarked for wildlife conservation, "Just by visiting SeaWorld today, you've helped save species and protect precious places."[2]

That SeaWorld did some research and some conservation was not in question, Naomi thought, but how many people actually left SeaWorld thinking, "Hey, we need to go save Puget Sound!" Sorely lacking was an objective, academic means to gauge the educational value of a day at SeaWorld. A controlled, systematic evaluation of SeaWorld's education programs and how effective they were among park guests would be a fascinating and worthy undertaking, Naomi believed. "But SeaWorld has never done it," she said.

Research and Conservation: "Visitors are exposed to marine mammals in an exhilarating and educational manner that is designed to instill an appreciation and respect for all living creatures and natural environments," the company said in its written Q&A with KGTV. Exhibits inspired visitors to "conserve our valuable natural resources by increasing awareness for the interrelationships of humans and the marine environment," which were the "first steps in conservation." In the Harris Interactive survey, 93 percent agreed that "visiting a marine life park, aquarium or zoo can inspire conservation action that can help marine mammals and their natural environment." The same number agreed that "people are more likely to be concerned about animals if they learn about them at marine life parks, aquariums and zoos."

The success of the company's rescue and rehabilitation, breeding, and educational programs were "unparalleled in the world," the company said. Since 1970, SeaWorld parks had rescued more than thirteen thousand animals, including endangered and threatened species. "That averages out to about one animal rescue every day for the past 34 years," it said. "Our financial commitment to our Animal Rescue and Rehabilitation Program is estimated at more than one million dollars per year." SeaWorld did not mention how much money it had received in federal, state, and other government grants for animal rescue and rehab, or what percentage of its annual revenue was represented by the million dollars.

As for research and conservation pertaining specifically to killer whales, SeaWorld noted in the Q&A that it had "funded countless studies of killer whale behavior, including conflicts between fishermen and wild whales in Alaska's Prince William Sound. We have sent teams to assist wild whales in distress, including a trip in the mid-1990s to rescue eight killer whales from Barnes Lake in Alaska and a trip just last month to stabilize and treat a newborn killer whale beached near Puerto Vallarta, Mexico." Knowledge gained through its breeding programs "has contributed to our understanding of killer whale biology, reproductive physiology, and behavior. This knowledge is important in assessing the status of wild populations."

Naomi felt all of these claims were inflated and hyperbolic. Yes, SeaWorld's often-trumpeted support for gestation research showing that killer whales were pregnant for seventeen months—and not twelve as previously thought—certainly qualified as valuable life-history information. As the company noted, it was "important to modeling expected population growth rates in the wild." SeaWorld had also "documented growth patterns" in killer whale calves, studied vocal development in killer whales, and demonstrated that calves learned vocalizations "in a manner similar to the way that human children learn language." But, Naomi knew, this work might not apply directly to wild whales (would growth rates in calves be the same in captivity versus in the wild?), and most of it would need to be replicated in the wild before its value could be assessed.

When KGTV pressed SeaWorld on what it was doing specifically "to aid the whale population in the Puget Sound, the pods that provided the first generation of whales for SeaWorld," the company seemed at a loss, and it reverted to its general support of orca research. But it did cite the "several" times it had been called upon to assist killer whales in the Sound, including the rescue of an orca named Sandy and rehabilitation of a young whale named Luna. For the record, Luna was never "rehabilitated"; he was not found in Puget Sound but far away on the west coast of Vancouver Island; and SeaWorld's main role in the saga of the young male orca appears to have been trying to send him to a marine park.

Longevity: Nothing was more central to the killer whale debate than the question of longevity. Solid proof that orcas lived significantly longer in the ocean than in tanks would clearly render captivity unjustifiable in the minds of many people. But if SeaWorld could show that orca life spans were no different in captivity than in the wild—or cast doubt on science's ability to measure such things—it would deflate much of the opposition's argument. In its written Q&A with the San Diego news station, SeaWorld said, "No one knows how long captive or wild killer whales live because no one has ever followed a group from birth to death. We have often said that 30 years is as good an estimate of average killer whale life span as we currently have. Clearly animals can exceed that age, as evidenced by one of ours, Corky. She is at least 40 and perhaps as old as 42."

SeaWorld noted that three of "the world's most respected marine mammal scientists," Peter Olesiuk, Graeme Ellis, and John Ford, had concluded that "female killer whales in the Northern Resident community had a mean life expectancy of 31 years and males just 19 years." Contextually speaking, this was true—but misleading. These values appeared in a 2005 Canadian Science Advisory Secretariat document.[3]

The researchers analyzed two study periods. The first, 1973–96, was a period of "unrestrained growth" in which the population steadily increased. Mean life expectancy was calculated at forty-six years for females and thirty-one years for males. Maximum life span was estimated at seventy to eighty for females and sixty to seventy for males. In the second period, 1996–2004, the population trend reversed and the number of whales began to decline. Mean life expectancy during this period was significantly downgraded, to thirty-one years for females and nineteen for males, as SeaWorld had correctly noted. The decline was due almost entirely to rising mortality rates.

The authors went back to look at what had changed between 1996 and 1997. They found that the drop was correlated with a major crash in the chinook salmon population, which in turn was due to dams, fish farms, pollution, and overfishing. El Niño conditions were also implicated.[4]

SeaWorld officials failed to note that they were using values from a period of population *decline*. They also ignored that, after 2001, the killer whale population began to rebound. By 2004, it was back to its pre-decline highs. Even though this three-year increase was not enough to elevate mean life expectancy for the 1996–2004 period overall, it was likely that the population would return to pre-decline levels. SeaWorld presented the lowest published life expectancy figures out of context, Naomi wrote, without regard to the recent reversal in the data.

"SeaWorld ignores the fact that the second set of life expectancies was calculated when the orca population was in decline," Naomi wrote to the subcommittee about the sleight of hand. "It also ignores that the population began to increase again post-2001." Study coauthor Graeme Ellis concurred. "The SeaWorld types picked the statement that met their needs," he wrote in an e-mail later that year.[5] "Since about 2001 the populations have reversed this declining trend, so life expectancies should return to what was previously published." It would take several more years of data to confirm if the improving trend continued, but there was one other positive sign. In 2006 the Canadian government changed the status of chinook salmon from "endangered" to "threatened." Serious problems remained, but the population trends for both predators and prey were looking better.

In her memo to subcommittee staff, Naomi asked why SeaWorld was portraying killer whale longevity as up for debate when the science was settled. "The industry's information is as close to lying as it gets," she said. "Acknowledging that killer whales don't live as long in captivity as they do in the wild is clearly not to their advantage." And she added this thought-provoking statement: "If captivity is safer and healthier, then animals should live longer—it's simple logic and actually holds true for many wildlife species in zoos. The deaths of captive orcas are ABERRANT—they aren't achieving the life spans they could and should. That's simply indisputable fact and no amount of spin or deflection can change that (although SW certainly has tried over the years)."

A far more detailed assessment of the display industry's education and conservation programs would be conducted at the hearing in DC. Meanwhile,

SeaWorld was now under investigation from OCSO, OSHA, APHIS, and the US Congress. As if that weren't enough, an experienced DC-based investigative journalist named Tim Zimmermann was now snooping around for a major piece commissioned by *Outside* magazine. Tim contacted John Jett and Jeff Ventre for interviews and drove out to Gaithersburg to spend a couple of hours talking to Naomi Rose. His resulting article, "Killer in the Pool," would change the course of the growing killer whale debate and finally help bring together such diverse parties—and former antagonists—as the seasoned marine biologist/animal protection advocate and the former orca trainers from SeaWorld.

Naomi had never seen so many people try so hard to unearth what she considered to be "Shamu's dark and deadly secrets." One night as she was finishing her memo at home with her husband and three cats, Naomi marveled at the rapid pace of change. "I really think that a lot of SeaWorld's fans will desert," she predicted, "once they see what is under those rocks that OSHA and the congressional hearing are going to overturn."

Naomi was now closely involved in three separate Tilikum investigations: the OSHA probe, Tim Zimmermann's article for *Outside,* and the House subcommittee hearing on oceans and wildlife. She was also in steady demand from the media for interviews on the attack and the factors that might make an orca lose control and take someone's life. The tragedy had led to a full-fledged national conversation. It was time for another trip to SeaWorld, Naomi figured, for a close-up look at the three-time killer.

In early April 2010 she flew to Orlando with an HSUS videographer. It was the first time Naomi had been to SeaWorld since 2005, when she visited the park in San Diego during a scientific conference. In Orlando, they went to see the "Believe" show twice.

After her return, Naomi wrote an account of her visit to a small group of allies, including Courtney Vail and Cathy Williamson from WDCS, Bill Rossiter of CSI, and OrcaLab's Paul Spong. "My brain is still mush from the experience," she wrote. During the 11:00 a.m. show, she could not clearly see Tilikum because he remained in a back pool and was not moving much. The show itself was "devoid of educational content. The narrative here was the dream of a young boy . . . who saw a whale in the wild one day [and] wanted to grow up to be a trainer. And he 'believed' hard enough and his dream came true."

The preshow orca Q&A on the JumboTron seemed to draw only a smattering of attention among the crowds, and the factoids were largely centered on general husbandry issues: How much fish did SeaWorld feed Shamu each day? How do you train a killer whale? How cold is the water in Shamu's enclosure? Only two questions were related to orca biology: Why are they black-and-white? How big do they grow? The show itself "had no factual information at all."

Afterward, Naomi climbed to the back row of the bleachers, which afforded a view into the backstage area. She could see Tilikum, logging motionlessly by himself. "At no time could we see any trainer(s) back there interacting with him, distracting him while the show went on." Naomi returned for the 2:30 show, and this time sat high in the bleachers. "In the end we were able to observe Tilikum (mostly not very well) for about 2.5 hours total. Of that time, he was alone for one hour, with another whale for a bit more than one hour, and with his trainers (at a distance, with a fire hose) for a bit less than a half hour (they were probably 'interacting' with him for a total of 30–40 minutes between the shows)."

Her bottom-line assessment: "Tilikum seems very depressed. He seems focused or fixated on the gate into the performance pool and/or the gate into the Dine with Shamu pool. He is spending a lot of time alone (with neither other whales nor trainers). It's very sad."

Even though it had been nearly a month since Dawn Brancheau had died, park guest John Kielty found that he couldn't leave the case behind. The more he researched what had happened at SeaWorld, and the more he learned about killer whales in captivity, the angrier he grew. John was unwittingly on his way to becoming a full-fledged anti-captivity campaigner.

In April 2010, John found an ally and kindred spirit in a St. Petersburg woman named Colleen Gorman. Vivacious, in her forties, with blond hair and blue eyes, Colleen was adamantly opposed to SeaWorld's keeping marine mammals in captivity. Colleen used to visit SeaWorld as a young girl, and even back then the place disturbed her tremendously. Guests were greeted by a live elephant seal plopped in the middle of a busy walkway. Everyone reached out to touch him as they walked by, and it appalled the nine-year-old, who shouted, "Stop it! Stop touching him!" Young Colleen felt sorry for the orcas in the Shamu show, and once, when she was chosen to be kissed by the whale, she adamantly refused to take part. She wanted to leave.

Colleen was at her computer at her condo in St. Pete Beach when she got word of the attack. She immediately googled Tilikum and began reading. "You better not kill that whale!" she cried out loud at the screen. When the county sheriff's spokesman said the trainer had fallen in, Colleen instinctively knew that was not the truth.

Within hours, Colleen had located a small group discussing the incident on Facebook. Mostly women, they began calling themselves the "mermaids," or simply the "merms." Their main focus, aside from Tilikum, was the effort to free Lolita down in Miami. Colleen also went to Jack Hanna's Facebook page and began posting indignant but reasoned comments to counter what she felt were the "lies" he was writing. She did the same thing at the SeaWorld fan page on Facebook.

Colleen, now somewhat obsessed with the Tilikum saga, read all she could on the attack, including the excellent daily reporting by Jason Garcia and others at the *Orlando Sentinel*. While commenting on one of Garcia's stories, Colleen came across the passionate anti-cap opinions of John Kielty. Soon they were friends on Facebook, and before long John had become a merm. So did Howard Garrett.

It soon became clear to Howie, Jeff, and others that they were becoming part of a small network of killer whale advocates around the country—and beyond—concerned about the well-being of whales in captivity (and in the wild). They were scientists, environmentalists, animal rights activists, former trainers, a journalist, and concerned citizens. But they lacked a place to come together, compare notes, get information, share ideas, post new studies, and discuss strategies for defending the orcas. If someone had a question, chances were good that someone else in this group had the answer.

A new, private Google group was formed. On July 8, 2010, Jeff started the Orca Aware list, which began with a handful of allies, including former trainers John Jett and Carol Ray, who had kept up with Jeff over the years, and John Kielty and Colleen Gorman, who were fighting SeaWorld in Florida. Other founding members were Jeff's sister, Kim Ventre; his girlfriend, Michelle "Chica" Duncan, a GIS specialist/computer cartographer; his old friend and Theta Chi fraternity brother at Florida State, attorney Todd Bricker; and Stefan Jacobs, an IT specialist and orca website operator from Germany who volunteered every summer with Ken Balcomb's killer whale survey.[6]

Orca Aware was the germinating kernel of an opposition force that would only grow larger and more organized.

A week before the showdown on Capitol Hill between the anti-caps and industry leaders, Naomi got a startling e-mail from a former SeaWorld employee. It would play a role in OSHA's legal case against the company and affect the ongoing debate over captivity as well.

The message arrived via the HSUS website. It concerned Loro Parque, the Canary Island facility housing four young surplus killer whales under a "financial arrangement" with SeaWorld. The writer, an American woman named Suzanne Allee, had been recruited while employed at SeaWorld Texas to work in the Loro Parque show's production crew. She stayed at Loro Parque until July 2009.

"I'm trying to raise awareness about the plight of the orcas at Loro Parque—as unfortunately the worst premonitions have come true," Suzanne wrote. Park trainers had sometimes forced the four SeaWorld whales to perform "while they were injured and bleeding." Suzanne also claimed to have seen trainers leave discarded plastic wrappers on the pool ledge just feet from the animals. Meanwhile, the four whales were stripping away shreds of material used for lining on the newly installed pools. "The animals have been subjected to numerous endoscopes because the Spanish park never wanted to invest money into surfacing the pools correctly."

Three and a half years of "negligence and mistraining" had resulted in the near-fatal attack on Claudia Vollhardt in October 2007 and the death of Alexis Martinez in December 2009, Suzanne alleged. Left in their wake were "four very messed up orcas."

Suzanne was hoping to find a reputable group "that will bring this issue to the US public and put pressure on SeaWorld to not abandon their orcas to the ineptitude of the Canarian trainers. These animals deserve the highest quality of life possible, and as the owner of these animals, it is SeaWorld's responsibility to insure this."

Suzanne wanted to speak with someone at HSUS "who could help me devise a media awareness campaign. Alexis Martinez already lost his life, and I have no doubt that if the orcas are abandoned by SW, it will be one of them next."

As bad as things sounded in Spain, Naomi viewed the e-mail as an unexpected opportunity. SeaWorld might now face another avenue of investigation. Naomi was swamped with preparations for the congressional hearing, among all her other duties, but quickly wrote back to the whistle-blower.

"What you describe is horrendous and even worse than what we were all concerned about when the orcas left for the Canaries," Naomi wrote. While she initially assumed that SeaWorld would be monitoring its whales more carefully, "now it sounds like I shouldn't have given them even that much credit!" Naomi wanted to help. "No one wants to see another death of a trainer or an orca."

Suzanne wrote back one day before the hearing. She had started at Loro Parque in February 2006 and moved up through the ranks to become A/V supervisor and department head. Now she felt that she had to speak up, despite her worries about repercussions from the industry. "The fact of the matter is that most everyone who was associated with Loro Parque is distressed by what a failure it is, but (as you well know) most of them are not in a position to freely voice their opinions." Suzanne said she would be willing to speak to US lawmakers and wanted to file a written brief for the House subcommittee hearing.

Naomi asked Suzanne to compile a brief detailing everything she had witnessed in Tenerife that had put the US whales at risk. She also inquired if

Suzanne would be willing to speak with OSHA's Lara Padgett, down in Tampa. "This may not directly or immediately improve the situation for the four whales at Loro Parque," she admitted, "but it could have the longest-term and widest benefits for whales in the future."

As for repercussions, Naomi said, "I have always been told that the truth is the best defense. If you are telling the truth, then there is no defamation."

Seventeen years had gone by since Naomi Rose had first appeared at a congressional hearing on the Byzantine captivity provisions of the Marine Mammal Protection Act and its complicated division of labor among federal agencies. Congress had required facilities to offer conservation or education programs that met industry-recognized standards, but NMFS did nothing to oversee or enforce whether facilities were meeting those standards.

Not much had changed since then.

To Naomi, it was scandalous that nobody in the federal government was making sure that SeaWorld and other parks were providing correct and up-to-date educational materials on marine science and conservation. On the morning of April 27, 2010, just three months after the death at SeaWorld, the House Subcommittee on Fisheries, Wildlife, Oceans, and Insular Affairs would hold its hearing on the matter.

Naomi was invited to testify as an expert witness at the hearing, "Marine Mammals in Captivity: What Constitutes Meaningful Public Education?" She had worked hard to brief the Democratic majority members' staff. The Democrats were generally more supportive of greater regulations for captive marine mammals than their GOP counterparts, with a few notable exceptions—including freshman Democratic representative Alan Grayson, a darling of the American left from Florida's Eighth District, which includes SeaWorld and Orlando.

Joining Naomi on the same side would be Lori Marino, PhD, an old friend and colleague and veteran of the anti-captivity wars. A senior lecturer in neuroscience and behavioral biology at Emory University and faculty member in the Emory Center for Ethics, Lori had also served as a research associate at the Smithsonian Institution National Museum of Natural History.

Lori, in her early fifties with long, dark hair and a soft face, was a leader in cetacean neuroscience who had published more than eighty papers on animal behavior, neuroscience, and human-animal interactions—roughly half of them peer-reviewed studies on dolphin and whale brains, biology, intelligence, and cognition. She had studied dolphins both in the ocean and in captivity and, working with colleague Diana Reiss, produced the first definitive data showing that bottlenose dolphins could recognize themselves in a mirror—a rare sign of self-awareness among animals and something even human children do not typically display until they are two.[7]

Lori had also written or cowritten several peer-reviewed papers presenting data unfavorable to dolphin-assisted therapy and swim-with-the-dolphins programs, and she had worked on some highly critical reviews of educational claims by the zoo and aquarium community.

The scientist had also declared her opposition to captivity for all marine mammals, making her more "radical" than Naomi and HSUS. Lori opposed taking marine mammals even for research purposes, saying that all necessary studies could be conducted in the wild, especially with so-called *solitary sociable cetaceans* (free-ranging animals that for some reason become friendly with people). That position set Lori quite a bit apart from the majority of her marine mammal colleagues. Most people in the industry disliked her.

The House inquest was shaping up to be a major event. Naomi could not recall a time that industry leaders had been summoned to an oversight hearing in Washington under such unfriendly conditions. "It's time Congress and the agencies took a critical look at the conventional wisdom that marine mammal live displays serve the animals' best interests," Naomi said in a written statement to the media.

Subcommittee chairwoman Madeleine Bordallo, a Democrat and nonvoting representative from the US territory of Guam, chaired the proceedings. A decent number of committee members actually showed up, and the hearing room was filled with industry supporters and critics, and a sprinkling of media, including CSPAN and the *Orlando Sentinel*.

"As everyone well knows, this topic is not without controversy and can become very emotional," said Bordallo, a soft-spoken, older woman with carefully coiffed hair. The controversy and emotions, she added, had only intensified after the death of Dawn Brancheau.

Even so, for many of the millions of people who visited US marine mammal parks each year, they were the only source of information about "these iconic creatures and their life history and the many threats that they face," the chairwoman said. "It is imperative we ensure that conservation and education programs at all captive display facilities meet all of the highest professional standards." Naomi had evidently done a good job with the briefing paper.

There was one "special appearance" before the first panel of witnesses: Representative Alan Grayson, while not on the committee, reminded his colleagues that central Florida was home to world-famous zoos, aquariums, and amusement parks. "SeaWorld is a shining jewel among them," he read from a prepared statement. "The economic impact of SeaWorld Parks & Entertainment is tremendous. SeaWorld Orlando alone attracts six million visitors a year, including my five children."[8]

As for the Brancheau tragedy, the congressman downplayed its significance: "One unfortunate incident that made us all sad doesn't prove that SeaWorld acted in any way irresponsibly. In the end, we can't expect that the rules are going to make people good. In the end, people have to be good themselves."

The first panel consisted of Lori Marino; Eric Schwaab, assistant administrator at NMFS; Peter Corkeron, PhD, a whale biologist serving as a visiting fellow in the Bioacoustics Research Program at the Cornell Lab of Ornithology; and Paul Boyle, PhD, senior vice president for conservation and education at the Association of Zoos and Aquariums.

Schwaab acknowledged that education and conservation standards were being "set by the industry that is being regulated" and stated the 1994 changes to the MMPA (the ones supported by industry and adamantly opposed by HSUS and others) explicitly said that the secretary of commerce, who had ultimate authority over NMFS, lacked the authority to regulate marine park education and conservation programs. There was no need, therefore, for further "clarification" of NMFS's role—something that industry opponents such as Naomi wanted. "Congress and NMFS have acknowledged that standards used by AZA and the Alliance represent an acceptable industry standard," Schwaab explained.

Dr. Marino was next. She could not have disagreed more with Schwaab's sanguine assessment of the industry's self-imposed standards. All information offered on marine mammal natural history, biology, behavior, and conservation was required to meet two criteria: It had to be *accurate;* and it had to be based on *valid outcome measures* demonstrating that a facility was truly serving an educational or conservation purpose.

Industry was not meeting those criteria, she alleged. For example, the Alliance claimed that belugas and killer whales not only lived as long or longer in its facilities than they would in the ocean, but they were also living "happy lives." But as Lori pointed out, "these two species live much shorter lives in captivity." And the "emotional" claim that the Alliance's animals led "happy lives" was pure speculation.

Lori also disputed industry educational materials claiming that captive marine mammals showed no signs of elevated stress. All one need do was open the Marine Mammal Inventory Report to find lists of numerous animals dead from "stress-related disorders" such as ulcerative gastritis, perforating ulcer, and heart failure, "strongly indicating that stress is an important component of captive display."

That non-peer-reviewed, industry-sponsored research had detected more markers of stress in stranded animals than animals living in accredited display facilities was, essentially, meaningless, she charged. "Stranded animals would be *expected* to have higher rates of pathologies, including ulcers, as a matter of course." But the pro-captivity authors had failed to include a control group—a basic tenet of all scientific experiments—therefore rendering their conclusion "uninterpretable at best."

Lori had coauthored a policy paper on "dolphin-human interaction programs" that reviewed the evidence for stress in captive cetaceans. She said the study had concluded that many captive dolphins showed markers of stress, such as "elevated adrenocortical hormones, stereotypies, self-destruction, self-mutilation, and excessive aggressiveness." Contrary to Alliance claims in its public education materials that captive animals did not show signs of stress, there was "ample scientific evidence to the contrary," Lori asserted, adding that marine mammals were "often very seriously affected by it."

Another area where display industry materials shaded the truth was the question of cetacean intelligence, she continued. The industry clearly thought the animals were intelligent enough to take part in human activities, but it simultaneously downplayed that same intelligence "so as to undermine concerns about keeping these intelligent animals in captivity. Like the last bowl of porridge in the Goldilocks fairy tale, dolphin intelligence is *just right*."

As for SeaWorld, its materials were "littered with inaccuracies," deliberately designed to bias the perception of cetaceans as "interesting but rather ordinary animals" and therefore suitable for captivity. SeaWorld claimed that dolphins were "large" animals with "proportionately sized brains," a statement that was "patently false." In science, brain size was evaluated by taking body size into account. Many species' brains are proportionate to body size. But a few have much larger brains compared to their bodies. Human brains are seven times larger than they should be for our body size. Dolphin brains, "contrary to industry assertions," are proportionately large as well, Lori said. Many have brains three to five times larger than normally expected.

Species with larger-than-expected brains "tend to show exceptional intelligence in many ways," Lori said. The relative size of a dolphin brain is second only to that of people. The average killer whale brain, at more than twelve pounds, is about four times heavier than the average human brain and second only to that of sperm whales, at seventeen pounds. "The Alliance apparently wishes to hide this similarity along with any concerns that dolphin sensitivities may be too similar to that of humans for them to be in captivity." The Alliance claimed that brain size did not equate with intelligence, and that comparing intelligence between species was "impossible and inappropriate." As for dolphin brains being associated with high intellect, the Alliance called this an "untested and disputed" hypothesis and said there was no "consistent intelligence test" for humans.

All of those statements "range from false to misleading," Lori countered. "It is neither impossible nor inappropriate to compare different aspects of intelligence, such as learning, memory, problem solving, or behavioral flexibility, across species." There were also, of course, several reliable and cross-cultural intelligence tests for humans.

After challenging the accuracy in some of the industry's education materials, Lori tackled the much trickier and subjective second criteria: Was there evidence based on valid scientific measures to show that people truly learned something—or were moved to take action for marine conservation—after visiting a marine park? The industry said there was, but Lori wasn't buying it.

"It is not proper to simply ask students *whether* they have learned or what they *think* they have learned, or how much they enjoyed the class," Lori said. Direct testing of acquired knowledge, and not feel-good public opinion polling, was the only way to properly assess an education program. Industry-sponsored papers on the topic—few or none of which were published in peer-reviewed journals—"typically involve asking zoo and aquarium visitors whether they *think* they have been educated. But they do not actually test knowledge."

One of the largest and most recent papers, *Why Zoos Matter,* funded by the National Science Foundation and published online by the AZA, surveyed more than fifty-five hundred visitors and twelve zoos and aquariums. It concluded that zoos and aquariums have a "measurable positive impact" on conservation attitudes among adults. AZA president and CEO Jim Maddy called it the "first time we have reliable data validating the positive impact zoos and aquariums have in changing visitors' feelings and attitudes about conservation." It was the "Holy Grail" the industry had been searching for, Lori said, "to validate their educational and conservation claims."

But in a paper published in the journal *Society & Animals,* Lori and her coauthors determined "these conclusions are entirely unfounded." Flaws in the AZA paper included lack of control over confounding factors that could bias the results, such as a nonrandom study sample. The AZA assessed neither attitudes nor knowledge, but simply evaluated "what visitors *believed* they felt or learned." Surprisingly, even with those flaws, the authors still found "no significant gains in general knowledge" resulting from a zoo visit. The AZA was stretching the truth, Lori said. "It is, frankly, surprising that the authors based such strong claims on these flawed findings. There is no compelling or even strongly suggestive current evidence that visits to zoos and aquariums promote positive attitude change, learning, or conservation actions." She then urged the reluctant NMFS "to exert greater control over this important issue and its very serious consequences."

Lori was followed by the AZA's Paul Boyle. He looked somewhat dyspeptic but kept his cool. "With 175 million annual visitors to our 221 accredited zoos and aquariums, AZA's focus on connecting people and animals provides a critical link to helping animals in their native habitats," he began. "Far-reaching conservation programs at AZA institutions have provided over ninety million dollars per year over the past five years to support over four thousand field conservation and research projects in more than a hundred countries."

As centers for "conservation volunteerism," AZA members offered the public a "great way to discover connections to their environment and to learn how they can make a difference in conservation." Each year, more than fifty-eight thousand volunteers donated some 3 million hours of their time, Boyle said, "supporting virtually every aspect of zoo and aquarium operations."

AZA members communicated "science-based educational messages about the importance of conserving marine mammals and their habitats" to their multitude of visitors, Boyle continued, 50 million of whom were children. AZA-accredited zoos and aquariums had trained more than four hundred thousand teachers in informal science education methods and provided over $85 million annually in educational support in the prior decade. Members also generated an impressive $8.4 billion annually in economic activity and supported more than 126,000 jobs.

Lori Marino's testimony aside, Paul Boyle said that the "enormous value of informal education is proven" because research showed that effective learning can take place in an informal environment such as zoos or aquariums, which were "rich with educationally framed real-world phenomena." They were places where people "can pursue and develop science interests, engage in science inquiry, and reflect on their experiences through conversation."

He cited a 2009 study showing that visitors to dolphin performances displayed an "increase in conservation-related knowledge, attitudes, and behavioral intentions" right after their experience. Three months later, they reported "engaging in more conservation-related behavior." It showed that people *did* retain knowledge gained during the "marine mammal experience."

Not surprisingly, Boyle opposed greater regulations on display permits under the MMPA. He had "great concern" that AZA standards would become "undermined by those who deny the now significant body of evidence" showing positive conservation and education outcomes.

"Some people speaking to you during this hearing claim that zoos and aquariums are not educational facilities. I challenge these claims as an unsubstantiated academic hypothesis that is unsupported by empirical research and represents a deluded view of what we know about how people learn," said Boyle, now clearly on the offensive. "These naysayers are locked in the past, suggesting that the methods developed under rigorous scrutiny of an international community of scholars devoted to the study of informal learning experiences are somehow inadequate—nothing could be farther from the truth."

The paper by Dr. Marino and colleagues was simply a "veiled attempt to discredit a team of respected researchers" and to rip apart their good names in order to claim that zoos and aquariums "can't be educational environments." But that criticism did not meet "even the minimum criteria for scholarly ethics, nor the basic rules of evidence for scientific research." Marino et al. were attacking a "summary document." They did not refer to any original research journals or any publicly available data. "I cannot imagine why these scholars chose to play so fast and loose with the facts. Dr. Marino and her colleagues clearly have no knowledge of the education standards for evaluation."

The second panel was composed of Naomi; SeaWorld Parks & Entertainment curator Julie Scardina; Louie Psihoyos, executive director of the Oceanic Preservation Society and director of *The Cove;* and Dr. Rae Stone, of Hawaii's Dolphin Quest interactive experience, a marine mammal vet and past president of the Alliance, which she was representing.

Julie Scardina went first. To Naomi, the curator looked tense, as if she would have been more comfortable back at SeaWorld, speaking to adoring schoolkids, rather than answering questions from potentially ill-disposed members of the US Congress. Where was Chuck Tompkins, Naomi wondered, where was Jim Atchison? SeaWorld threw this poor woman to the wolves, Naomi thought, with a bit more relish than sympathy.

Scardina echoed the AZA in stating flatly that the current standards followed by its members were perfectly adequate because they required "exceptional public education programs." More federal regulation was simply not needed.

Meanwhile, SeaWorld parks had a "multifaceted" mission of being centers of both entertainment *and* education. "We work hard to ensure that our patrons leave our facilities having had an enjoyable experience, and with greater knowledge of and appreciation for animals and the natural world," she said in her written remarks. In 2009 alone, SeaWorld staff conducted more than half a million hours of "structured teaching" with more than half a million guests. Since 2003, SeaWorld & Busch Gardens Conservation Fund had granted more than $6 million to four hundred projects in sixty countries for "species research, habitat protection, animal rescue, and conservation education."

As for killer whales, SeaWorld research was having an "immediate impact on wild populations," Scardina said. "Many recent examples" illustrated how venues such as SeaWorld were uniquely suited to conduct research that would not be possible in the ocean. Researching dialects in wild whales, for instance, had been difficult because "behavioral interactions cannot be observed underwater." But thanks to the "unique characteristics of Shamu Stadium," such as hydrophones and underwater viewing areas, investigators from Hubbs-SeaWorld Research Institute made "significant progress in developing tools to identify calling whales." Other studies on the development of dialect in killer whales "may cast light on the evolution of human language," Scardina added, without explaining how this might produce an "immediate impact on wild populations."

Another study, supported by the SeaWorld & Busch Gardens Conservation Fund, was working to determine "how whales have solved the complex relationships of energy intake and expenditure and what their food requirements are. Knowing how much energy a whale requires is fundamental to resolving a number of questions about the role whales play in their ecosystem." Scientists were now able to assess the relationship between the metabolic and heart rates of killer whales at SeaWorld and use that knowledge to "evaluate metabolic rates of free-ranging whales and ultimately apply this information to the conservation and protection of wild whales." A worthy goal, Naomi thought. But to her knowledge, this work had not yet been published in the scientific literature, where it might benefit research in the wild.

Killer whales at SeaWorld were also being studied by scientists at the University of Central Florida, who were measuring "physiological parameters" including heart rate, metabolic rate, and thermal balance, under controlled conditions, to evaluate how they might be "affected by age or body mass." Naomi questioned how valuable physiological data from "controlled conditions" would be for studying whales in the open ocean.

Naomi was up next. She began by noting that a proposed rule to implement or amend public display regulations in the MMPA had been unveiled back in July 2001, but the rule was never finalized. Because of that, NMFS was not able to sanction any facility that failed to meet "professionally recognized standards" of the public display community. "In short, the public display community . . . has been largely self-regulating for sixteen years, to the detriment of the educational experience of those who visit these facilities and use their online resources."

Among other criticisms, Naomi assailed the industry's claim that the 2005 Harris poll proved that places like SeaWorld "educate the public to care about conservation." The poll only asked "what they think or believe, not what they know." As a counterweight to the study offered by Paul Boyle, Naomi cited a European analysis comparing the knowledge of students who received a marine mammal lesson in an aquarium setting, and students who were taught in a traditional classroom. The results suggested that, after several months, the aquarium group did not retain lesson materials any better than the classroom students. Informal settings might be fine for learning, but classrooms were just as good.

She also challenged that captive breeding programs were "a positive aspect of SeaWorld's work with endangered and protected animals." The company's own breeding efforts were "not consistent with the generally accepted concept of conservation-based captive breeding" because that concept included "the final step of releasing captive-bred progeny into the wild."

In the question-and-answer session following the panel, Naomi spoke from the heart, without a prepared script. Representative Bordallo asked her how the display industry could better champion the natural world, but Naomi was skeptical the industry even *wanted* to champion the natural world. Nature was not SeaWorld's companion; nature was SeaWorld's rival.

"I have heard again and again that people cannot go scuba diving or whale watching—it is inaccessible to the vast majority of people living in an urban environment," Naomi said. "The fact is a lot of people can't afford to go to SeaWorld either."

SeaWorld was competing against the great outdoors for people's dwindling dollars. From a purely business perspective, promoting nature was akin to promoting the new Harry Potter attraction at Universal Orlando that was denting SeaWorld's revenue. "There are certain things that they are simply not going to tell people," Naomi stated. "They don't really have a vested interest in telling you how wonderful nature is either, because that's their competitor, even though they are supposed to be educational facilities."

Naomi said she was not alone in that belief. "I think that there are more marine mammal scientists who would agree with that statement than anything about [animal] welfare or other issues that are of great concern to me."

When SeaWorld wasn't ignoring the competition, it was misleading customers about the virtues of its rival, Mother Nature. "These are very large, brilliantly colored animals—and when you see them leap high in the air in a SeaWorld show, it *is* spectacular. But it's not natural," Naomi explained. "When you get entertained by a show that has not one scrap of natural history information in it—the preshow does and the signage does, but the show does not—and that's what you came to see, that's what you paid your seventy-nine dollars to see . . . to be entertained [and] I just don't want people to be misled that they are also being educated."

Visitors to SeaWorld were not learning "what really happens with these animals in the wild, how they really live. The fact is they portray nature as a scary place, full of hazards both natural and human-caused, and yet they equate them—'predators are scary.' Well, predators are like all sorts of things we have to deal with every day. . . . And that is what these animals have to deal with out in the wild. They have a job to do. And it's challenging and it's engrossing and, yes, it has hazards. But that's natural, that's normal, that's nature. And they should be celebrating that instead of telling young kids, 'SeaWorld is going to protect you from that.' I feel very strongly that young kids get a sense that the animals are safe at SeaWorld, but it's really scary *'out there.'* I don't think that's a conservation message. I think that's an anti-conservation message."

Julie Scardina rejected Naomi's assertions and held up a thick stack of SeaWorld educational materials, most of which were "geared towards children," she said. She also brandished a sixty-six-page document listing published studies conducted by SeaWorld staff, including "peer-reviewed and also other information that comes out from people that work at SeaWorld." She didn't say what percentage of the research papers was peer-reviewed.

"We are making sure that we are putting out the highest standards of available information that not only our scientists and our educators find out and research, but also outside researchers through AZA and the Alliance as well as people in the scientific community—basically the greatest and most in-depth experience that there is on conservation and education in the world, if you are going to put it all together," Scardina said.

The curator also defended the use of killer whales in performances for the public. "What you see there are the capabilities of these animals—I mean, why do we have the Olympics? Because people get to the greatest height of what they can do, and we love watching that. Well, our animals have been trained to do something that they are capable of. We can't train them to fly because they can't fly, but we can train them to do all these magnificent things that people are inspired by."

Then Scardina took refuge in the familiar fold of the "ocean as scary place" defense, without grasping the irony, Naomi thought, of providing bad education at a hearing on education.

"You know, Dr. Rose talks about nature being a place where there are only natural dangers. And if that were the case, we really wouldn't have too much of a need for additional education, but that hasn't been the case for many, many years. You add [all] the changes that humans have made over the last couple hundred years, and that's why we have so many threatened and endangered animals right now. There's pollution in the oceans that's greater than it has been certainly at any time in history—and today, that we shouldn't inspire people to care enough to change their behavior, to make sure that we take care of the oceans, just doesn't make a whole lot of sense to me."

Naomi thought Scardina was getting rattled. Before she finished her answer, the curator got personal; never a good idea at an oversight hearing.

"She talks about us and our efforts as we show off these animals and inspire people in our parks that, at the same time, we are making money," Scardina said of the HSUS scientist sitting by her side. "But I would like to point out that HSUS makes money when they try to convince people that, that we are the bad guys and that there's nothing going on outside in the world, that they want people to give money to them. So just remember that it's not Dr. Rose coming from a purely unbiased standing and that we are not the bad guys here."

Bordallo asked Scardina if the benefits of killer whale shows outweighed the risks.

"Yes," she answered. "They do."

Naomi did not agree. She noted that fewer than two hundred killer whales had ever been held in captivity, and of those, about two dozen had injured or killed their trainers. "That's more than a ten percent accident rate. If a brand of can opener had that kind of accident rate, it would be pulled from the shelves. So I think the risks *absolutely* outweigh the benefits."

The chairwoman then announced "probably the sixty-four-thousand-dollar question to round out this hearing, and that goes to Ms. Scardina: What happens now to Tilikum?"

Julie Scardina mustered a pleasant smile. "We are definitely committed to taking care of that animal for the rest of his life. He didn't do anything wrong. We are caring for him in the social pod that he knows, with the people that he knows, with the excellent care that he receives."

Once again, Naomi was itching to take issue with the SeaWorld official—and her rosy depiction of Tilikum's life. "I am a killer whale biologist and I spent hundreds of hours on the water watching these animals and their social interactions. And in fact, I specialized in male behavior," Naomi began, reminding everyone that her knowledge of killer whale social behavior likely exceeded that of anyone else in the room (if not on Capitol Hill altogether). Tilikum was no longer being touched in the wake of Dawn's death, she noted, "because they are keeping trainers back to a safe distance." He was still allowed to interact with other whales, "and that of course is good. But he must know that something is different, perhaps even that something is wrong, because he is not being touched by his trainers. . . . It must be very confusing for him, at the very least, and potentially very depressing for him. And I am not trying to be anthropomorphic: These animals are intelligent, they are social, and they can feel things like depression."

Finally, as for the "scary ocean" rhetoric, it was true: Pollution and other human-caused hazards were critical problems that demanded solutions. But, Naomi said, "I think the self-congratulatory tone of the public display representatives at this hearing is discordant with that."

This was not to excuse the environmental protection community, Naomi assured the subcommittee. All sides were "missing the boat in getting people to adequately change their behavior and wake up and recognize how serious this situation is, and how much their individual behavior matters and counts. To be perfectly blunt, as someone who works in regulatory and legislative endeavors around the world to try to protect these animals, I don't see this great education that is occurring, apparently, at these facilities."

Naomi noted in closing that Britain had one of the "most staunchly conservation-oriented publics . . . they are strongly antiwhaling . . . they are an island nation who feel they must protect the marine environment." Yet there were *zero* cetacean displays left in the UK. "Clearly they are getting their marine education, their marine ethic, from some other source."

When the hearing adjourned, Naomi was told by the HSUS public relations representative who attended the proceedings that she had overheard an industry representative on his cell phone out in the hall. "Yeah," the official had said, "that woman who testified for the Humane Society . . . she was pretty articulate."

Naomi was energized by the event. The opposition was not used to being under a microscope. A few years earlier, this inquiry would never have taken place: The display industry would have brought out its big guns and killed the very idea of an oversight hearing long before it gained a foothold on the Hill.

A few days later Naomi sent an e-mail message of hopefulness to some HSUS colleagues. The majority staff wanted to introduce a bill to address all issues related to public display—including educational oversight. "I am working with my colleague on public display lobbying, Courtney Vail of WDCS, on an outline of a possible bill," Naomi wrote. "I don't want to be overly optimistic, but the momentum on this issue seems to be in our favor for the first time in a long time . . . maybe ever." At the very least, she said, "We may finally convince the public to question the value of orcas on display, which might affect the marine theme parks' bottom line."

Six months later, voters would sweep the Republicans into power in the House in a stinging rebuke to President Obama's economic and health-care policies. With the environmentally friendly Democrats now banished to the minority and the pro-business, anti-regulation GOP in control, Naomi's hopes of amending the MMPA would have to wait at least two years—if not more.

35

Citation

After the congressional hearing in late April 2010, the Dawn Brancheau controversy shifted onto the back burners of American conversation and media scrutiny. But by no means did it fade away.

As the spring and summer swept by, combatants on *all* sides of the reinvigorated anti-captivity issue braced themselves for the findings of the OSHA investigation. The stakes could not have been higher. If OSHA found that SeaWorld had willfully violated federal worker safety laws by allowing Dawn to assume such a dangerous position with a whale already involved in the deaths of two people, it would be a huge blow to SeaWorld, and to Blackstone's investors. The publicity would be bad enough. But a high-level citation from the feds would carry significant consequences for the theme park company and its new owner.

A "willful" violation, as opposed to, say, a "serious" violation, would expose SeaWorld to a high level of liability, especially in regard to Dawn's husband. Scott Brancheau was keeping a low profile in his private mourning and not speaking to the media—at *all*. Second, such a black mark on the safety reputation of SeaWorld might send the premiums on its liability insurance up into prohibitively expensive territory.

But mostly, SeaWorld was worried about whatever type of "abatements" OSHA might mandate to resolve the safety issues in its workplace. In the worst-case scenario, from the company's point of view, the federal government would order all trainers out of the water during Shamu shows *permanently*.

Lara Padgett and OSHA, meanwhile, had a statutory deadline of six months from the date of the incident to finalize their report and possibly issue a citation for any safety violations. Everything had to be wrapped up by

August 23, 2010. As Naomi, Jeff, Howie, and other anti-captivity advocates learned, SeaWorld officials would drag their feet in the OSHA investigation by withholding materials or denying they even existed, trying to block interviews with park employees, complaining about the agency contacting activists, marine mammalogists, and officials at other marine parks, and doing everything to prevent the issuance of any citation at all through SeaWorld's considerably powerful network of lobbyists, lawyers, and loyal lawmakers on Capitol Hill.

OSHA was going to need every day available to pull the final report together.

Even as Naomi Rose toiled hard to furnish OSHA with contacts, research papers, and other information, Jeff Ventre and John Jett—along with Howard Garrett of the Orca Network—were working with DC-based journalist Tim Zimmermann as he completed his eighty-seven-hundred-word investigative article, "The Killer in the Pool," for *Outside* magazine. Tim's research had been exhaustive and highly productive. In addition to Naomi, the former trainers Jeff and John, Ken Balcomb, and other industry critics, Tim had secured an interview with Kelly Flaherty Clark and Brad Andrews at SeaWorld.

On a trip to Florida, Tim also spoke with former SeaWorld employee Thad Lacinak, as well as John Jett and Mark Simmons. John and Jeff suggested that Tim speak with Mark. The three of them were still friends, despite the tough language Mark had used against John in his unpublished blog for *AC-360*.

On Mother's Day weekend, when Jeff flew into town from New Orleans to spend time with his parents at their Lake Mills home, he invited both of his old buddies out to the place for some beers and fond memories. It was a good day. The guys agreed to disagree about killer whales and captivity and left it at that. Instead they spent the warm afternoon catching up and discussing Mark's company, Ocean Embassy, which had reportedly been involved in the capture and sale of wild dolphins. Mark bragged about sustaining zero dolphin fatalities in his operations. Jeff and John found the discussion surreal. It was difficult for them to fathom how their dear friend could be pulling wild dolphins from the ocean. The détente would not last long.

Life for Tilikum had radically changed after he killed his trainer of six years. The massive bull—twice the size of any other killer whale at SeaWorld Orlando—now swam alone in his own artificial "habitat." As Naomi had testified in Congress, no one would touch him—quite literally. Tilikum must have felt like a pariah.

Did Tilikum realize what he had done to Dawn? Surely he was aware that the friendly blond woman with bright eyes and warm smile no longer came around to feed, train, or play with him. Instead of rubdowns and other forms of touching by other trainers, Tilly's tactile stimulation mostly came from a brush fastened to a pole, or a fire hose blasting against his skin—pointed from a safe distance away.

The notorious whale was now approaching thirty, the mean life expectancy for wild male orcas, at least among Northern and Southern Residents. In nature, he might live another ten or twenty years—or perhaps even more. By SeaWorld's standards, however, Tilikum was geriatric. Indeed, captivity defenders routinely maintained that Tilikum was too "old" to be suitable for release or retirement to a sea pen, even though male killer whales his age were thriving in the sea.

According to a reliable source with access to backstage information, Tilikum was now separated from the other whales more than ever before—with one exception: the high-spirited Taima. The clever and mischievous female, now twenty, was still adored, despite her moods and occasional bouts of aggression. But around 2000 (the exact date is not clear) SeaWorld trainers stopped doing water work with her.

After February 24, 2010, Taima and Tilikum spent a lot of time together in the same pool. With the harsh spotlight now turned on SeaWorld, park officials were desperate that Tilikum be seen and photographed with a companion, observers speculated. Whatever the reason, the two whales grew closer.

Like Tilikum's, Taima's life had hardly been rosy. Her younger half sister Nyar suffered from physical and cognitive ailments and passed away before age three. Their mother Gudrun died horribly after a stillbirth. Taima then turned violently against her first calf, Sumar, a son sired by Tilikum in 1998, who was transferred to San Diego for his own safety. Two years later Taima and Tilikum had a second calf, another male, named Tekoa. Taima started out as a better mother than before, nursing and bonding with Tekoa, but soon turned against the youngster, who had to be separated to another tank. (In 2006, Tekoa was shipped to Loro Parque, where, in 2007, he repeatedly slammed trainer Claudia Vollhardt, seriously injuring her.)

In March of 2007, Taima had given birth to another calf, a female, also sired by Tilikum, and given the name Malia. Taima was occasionally rough with this calf as well, sometimes displacing her onto slide-out areas and blocking her return to the water. Other times, however, Taima was affectionate with her daughter.

In December of 2008, when Malia was not even two, Taima was impregnated—once again by Tilikum.

Pregnancy seemed to calm Taima, though she reportedly devised a devilish way to startle park visitors during gestation. She would use her considerable charisma to engage tourists and beckon them to approach the viewing window, the source said, then slam the glass with a violent *thwack* of her head. Taima seemed to sincerely delight in scaring her audience. "SeaWorld is still trying to stop the other orcas from doing this, after they learned it from Taima," the insider said.

During the last show of the day on June 5, 2010, Taima went into labor. The show was canceled. She remained in labor overnight and into the following morning. Eventually, she delivered the placenta, but not the calf, which was dead inside her and in an unusual position in the birth canal. The calf was turned in such a way that Taima could not deliver it naturally. In the frenetic attempt to extract the dead calf, Taima herself passed away. Her death, reminiscent of her mother's, was supposedly gruesome.

The "Believe" show resumed the very next morning. SeaWorld "put on a show about their bond and love for these animals," the source said. "There will be no service to remember her, no memorial, and no consideration for Taima's memory. How can those closest to Taima shrug off her terrible death and go on with the show flashing a cheery smile after watching their friend die hours before?"

Given its proximity to the Brancheau tragedy, Taima's death was big news—much bigger than it would otherwise have been. Dr. Chris Dold, SeaWorld's vice president of veterinary services, said stillbirths were more common in the wild than at SeaWorld.

Naomi blasted that claim in a news release: "SeaWorld has been quoted as saying that successful birth rates for orcas are about 50 percent in the wild and 85 percent at SeaWorld; that stillbirth rates in the wild are significantly higher than at SeaWorld; and that no orca has died while giving birth at SeaWorld in twenty-five years. None of these statements is true."

She cited one study estimating that the survival rate, up to six months of age, in the wild was 60 percent. Survival at birth "would of course be higher," Naomi said. Actual stillborn rates in nature could not be determined. "Births are rarely observed in the wild, let alone stillbirths." It was "simply nonsense to state that the wild can be compared unfavorably to SeaWorld in this regard."

At least four orcas had now died while pregnant or in labor at SeaWorld. As for Taima, "I have a weird idea," Naomi wrote privately to some colleagues. "What if the whales noticed something was off with her pregnancy back in February, and THAT'S what was causing the ruckus in the performance tank that day and got Tilikum all riled up? If Taima didn't feel right, or the calf was sitting wrong, she might have been quite testy."

As for Tilikum, his solitude was now more manifest. In late February, he had lost his longtime trainer. Now, in early June, his closest orca companion had vanished as well.

In late June of 2010, SeaWorld went on the counteroffensive, in a fourth-quarter attempt to head off a damaging final OSHA report and citation. On

June 28, Jason Garcia reported in the *Orlando Sentinel* that the company had "reached out to OSHA about the possibility of negotiating a settlement even before the safety probe is complete." Such "pre-citation settlements" were rare, Garcia wrote, and SeaWorld would pay a price if it managed to wrangle one: It would have to agree to "worker safety changes" at its parks.

Still, settling with the feds would herald a welcome respite for the beleaguered company. A preemptive agreement "could allow SeaWorld to effectively blunt any fallout from the closely watched probe by ensuring that it ends as quickly as possible," Garcia wrote, "and by avoiding the kind of scathing indictment that investigators issued three years ago after a separate incident at a SeaWorld in San Diego, when they declared it was 'only a matter of time' before a killer whale killed a trainer."

SeaWorld had yet another motive. If OSHA slapped it with a significant violation, SeaWorld would have no choice but to appeal the citation before a judge, and the public, in an open trial that would entail "both expensive litigation costs and continued negative publicity," as Jason Garcia noted.

It was not clear whether OSHA was inclined to even consider an advance settlement, though Lara Padgett confided in Naomi she was worried about the prospects. And Representative Alan Grayson, the Democrat from Orlando and stalwart SeaWorld booster who spoke at the congressional hearing, Naomi learned, was beginning to cast about for support in Washington. He was allegedly lobbying for a watered-down report, if not an entire dismissal of the case.

Averting federal action was not going to be easy. OSHA officials were loath to repeat the 2007 reversal that SeaWorld had extracted from Cal/OSHA. Indeed, the feds had recently "signaled a more aggressive enforcement approach for theme parks and other entertainment venues," Jason Garcia reported in the *Sentinel*. Assistant Labor Secretary David Michaels, the head of OSHA, had put the entertainment industry "on notice" in May following a number of worker deaths, including at SeaWorld Orlando and Disney World.

But SeaWorld insisted it had merely "communicated with OSHA" and not entered into any negotiations "at this time," Garcia wrote. The company had, however, retained attorney Carla Gunnin, formerly with the Labor Department and now at the Atlanta-based firm Constangy, Brooks & Smith, whose OSHA Practice Group was led by Patrick Tyson, a former top official at OSHA. SeaWorld was clearly going to fight this every step of the way.

As if SeaWorld officials didn't have enough to deal with, on June 27, Tim Zimmermann's article was published in *Outside* magazine.[1] It was a blockbuster and garnered a lot of attention. Tim covered everything from the life of Tilikum to the death of Dawn, from the history of early captures to the bitter fight over captivity and the majestic sight of Southern Resident whales meandering up Haro Strait and "huffing spumes of mist into the salty, spruce-scented air."

Tim wrote that the incident with Dawn had come as "a shock to Americans accustomed to thinking of Shamu as a lovable national icon, with an extensive line of plush dolls and a relentlessly cheerful Twitter account." SeaWorld was hoping to avoid exactly this kind of high-profile coverage, and it may have prompted Director of Animal Training Kelly Flaherty Clark to speak out and help soften and humanize the company's mien.

"Every safety protocol that we have failed," she told Tim one month after Dawn died, her voice "still tight with emotion," he wrote. "That's why we don't have our friend anymore, and that's why we are taking a step back."

As for the key question about what *triggered* the attack, Tim turned to Jeff Ventre. "Some trainers believe that killer whales are acutely aware of what they're doing," Tim wrote. He then quoted Jeff: "I've seen animals put trainers in their mouths and know exactly what the breaking point of a rib cage is. And how long to hold a trainer on the bottom. . . . Tilly was a good guy that got beat down by the women. . . . So there are a lot of reasons he might be unhappy."

John Jett told Tim that sometimes Tilikum "might be 'off' for days, 'splitting' from his trainer to swim at high speed around the pool, acting agitated around the females, or opening his eyes wide and emitting distress vocals if asked to get into a vulnerable position (like rolling over on his back). 'It's extremely sad if you think about being in Tilly's situation,' says Jett. 'The poor guy just has no place to run.' " Ken Balcomb had a slightly different take for Tim: "Tilikum is basically psychotic. He has been maintained in a situation where I think he is psychologically unrecoverable in terms of being a wild whale."

John Kielty returned to Florida in the summer of 2010. He could not get Tilikum out of his mind. John began working alongside Colleen Gorman at her condo in St. Pete Beach on a number of captive-orca issues, especially the situation with Tilikum. Colleen let John move into the spare bedroom overlooking the lagoon, and they turned the living room into command central for fighting SeaWorld. They were only roommates, but they felt like soul mates when it came to Tilikum.

By July of 2010, John and Colleen began going to SeaWorld regularly. They would drive the two hours or so—much more with traffic—to Orlando and use their annual passes to get in and park for free. "We keep going and going," Colleen told Howie Garrett, who was part of the merm group started by Colleen, "and it's the same story, every time we see him. He's always alone."

Colleen and John had spent hours watching Tilikum in the back pools from high up in the stadium and from the underwater viewing area at G Pool. Most of the time he was floating motionlessly at the surface. She said to Howie, "Yes, trainers come up for ten minutes, twelve minutes, sometimes only seven minutes. Sometimes they only come up for three."

The lives of John and Colleen had forever been changed because of the events of February 24. Colleen, who considered herself quite spiritual, believed some force in the universe had brought her and John to the SeaWorld war together. "This was all because of Tilikum, this was his cry for help," she said. "I believe that everything happens for a reason, and he was speaking out that day, from a more cosmic point of view. He'd killed people before, yes. But this time he made a conscious decision to attack her, so that he could be *heard*. I believe he really needed and wanted change."

Meanwhile, in Tampa, Lara Padgett's inquest into the incident doggedly continued. On July 3 she reached out to Howard Garrett. "Would you care to weigh in on the SeaWorld incident?" Lara asked. OSHA was investigating the occupational exposure of working in the water or close to orcas, she said, "and I was hoping that you might consider writing a statement." Howard said he would be pleased to comply. He also urged Lara to contact two former SeaWorld trainers who could provide invaluable insight about life at Shamu Stadium: Dr. Jett, and Dr. Ventre. She did.

The first reply came from Jeff. Tilikum had "made (at least) three decisions," he wrote. "The first was to pull Dawn into the water. The second was to displace and prevent her escape. The third was to possess-at-all-cost, leading to her death, scalping, and dismemberment." The "call-back safety-spotter system" had failed as it had before, with examples publicly available on YouTube. "Including the recent death of trainer Alexis Martinez, we've seen morbidity and mortality generated by *both* captured and theme park bred animals. Incidents have been perpetrated by *both* male & female orcas."

Jeff commended SeaWorld for its study of orca gestation periods, for its manatee and other rescue programs, and even its artificial insemination methods "that could one-day help restore a threatened species." Even so, he said, "Those accomplishments may or may not justify the 45-year history of orca captivity, and its associated *cause of human injury and suffering*. From a scientific perspective, there is a notable lack of peer-reviewed literature to justify the perpetuation of this *clearly unsafe* practice."

John wrote to Lara about a week later. He told her that SeaWorld believed it needed to maintain water work in order to protect its bottom line even though the evidence suggested that "this activity puts human life at unnecessary risk." It was now obvious to him that working with killer whales was inherently dangerous, "both in and out of the water." Trainers would continue to be hurt or killed because of their "close contact with these immensely strong, easily-bored, social, and contemplative animals," he warned. John had personally witnessed whales "coming out" at trainers in incidents that were neither investigated nor listed in any report.

Of special concern were the frequent times he had seen whales ignore callback stimuli (water slaps, stage slaps, underwater callback tones, etc.). Callbacks were the most fundamental means of controlling whales during "emergent situations," he told the OSHA investigator. All trainers had experienced callback failures. It was part of their daily lives. "Anyone cued into observing these stimuli and their responses can discover numerous examples of whales simply choosing to ignore them (often during emergencies)."

Other hazards lurked around the orca tank. One day John accidentally slipped and fell into C Pool when Tilikum was in it. He was able to scuttle out before Tilikum came racing over to investigate the splash. John accidentally fell into a pool with Gudrun once, although she didn't react at all. "These incidents went unreported, yet either could have escalated into a major event." John also mentioned the time he slammed his hip into the stage when doing a foot push with Katina. "I was never able to fully understand why I hit the stage, and again the incident went unreported."

John said in closing, "Accidents happen in many professional settings and certain professions are evidently more dangerous than others. However, few professions are as unnecessary as training killer whales for purely entertainment purposes."

Tim Zimmermann's groundbreaking article on the industry helped move the Tilikum saga back into the spotlight as the deadline for an OSHA ruling approached. On July 21, Tom Ashbrook, of the popular NPR-syndicated public affairs program *On Point,* out of WBUR-FM Boston, dedicated his hour-long show to "the whole disturbing history of killer whales in captivity." Tim was one of three scheduled guests, along with Thad Lacinak and Ken Balcomb. During the show, Suzanne Connell called in to give her account of what happened at the "Dine with Shamu" event that fateful day, and John Jett phoned in to describe what he had witnessed working at SeaWorld.[2]

Thad, when asked for a response to John's allegations, assailed the accuser rather than the accusations. John had "never even made it to trainer level," and he "probably" had quit after learning he was being transferred to Whale and Dolphin "because he was not cutting it at Shamu Stadium, he was not up to par with the other trainers."

"Will we see more deaths like this?" Ashbrook asked.

"No, we're *not* going to see more deaths like this." Thad sounded exasperated. "I mean, SeaWorld is going to address this, they are addressing it right now."

After the show, John's and Jeff's old friend Mark Simmons went on the attack. Clearly they were no longer friends. In a comment on the *On Point* website, Mark accused the two former trainers of being "disgruntled past employees of SeaWorld that have an axe to grind." John had "no more than three years [*sic*] experience [John had more than four years at SeaWorld], never was approved to do show water work with killer whales, was not Tilikum's 'team leader.' Ventre was fired. Balcomb has never worked with whales in a zoological setting."

Tim had been one-sided in his reporting, Mark charged. "You have given very little weight or 'air time' to those with relevant and suffiecient [*sic*] experience to comment. If you give your audience 10% of the picture—their conclusions are dictated by ignorance. If this was meant to be a useful investigative journalism peice [*sic*] then try representing a balanced picture. . . . You cannot be an objective writer when your conclusion was preconceived."[3]

Mark wasn't done. He also posted a comment on Tim's website, which Tim has since deleted: "It seems I've lost two friends and gained a detractor." He added that the only reason people such as Tim, John, and Jeff believed that killer whales were "worth fighting for" was because "SeaWorld taught you they were. Why do children from Iowa know what a killer whale even is or for that matter how it's social, cares for its young or learns complex behavior? Because of SeaWorld. . . . Get out of your armchair and roll up your sleeves before you point your pen at me."[4]

Tim was undeterred. He defended his sources, John and Jeff, against the condemnations of Mark and Thad. "I'd like to provide some more context to make clear that the effort to dismiss them now as disgruntled, substandard, former trainers who should not be listened to is absurd," Tim wrote on his own blog, "and in fact says more about the critics than Ventre and Jett."

The former trainers had not become "go-to sources for any muckraker that wanted to poke a stick in SeaWorld's eye," Tim continued. "I had to seek them out, and even then it took many conversations and a level of trust before they felt comfortable going on the record."

For the record, Tim added, despite what Thad and Mark had claimed about John Jett, "he left SeaWorld at the level of full trainer, he certainly was cleared for Shamu show water work (and has the videos to prove it), and he was indeed Tilikum's team leader for many months before he left, writing the daily training goals up on a whiteboard in the trainer area."

Suzanne Allee, the American video and production specialist who had worked at Loro Parque in the Canary Islands, took Naomi's advice and put to paper her recollections and allegations about conditions for the four SeaWorld-owned killer whales on Spanish soil, and the trainers who worked with them. According to Suzanne, safety lapses were everywhere.

For example, one young woman hired by chief trainer Miguel Diaz had panicked during a diving accident and tried to scramble up a gate that separated her from the orcas. Two of the whales grabbed her scuba gear through the bars and began to "thrash her about," Suzanne said. The woman screamed. "There's no doubt if she hadn't been able to reach the top of the gate she would've been killed."

Another time, a member of Suzanne's crew, David Quintero, was in the pool when the orca gate unexpectedly opened. Miguel Diaz had demanded the gate key from the "safety spotter," who handed it over, a direct violation of protocol. Diaz opened the gate, unaware that David was in the pool. "The four orcas immediately took off towards David," Suzanne said. "His frantic efforts to exit before they reached him only increased their curiosity and speed." He was yanked out when they were just a few feet away.

Suzanne was equally worried about what she saw as glaring violations of the US Animal Welfare Act, which arguably applied to these killer whales no matter where in the world they were. "Loro Parque management was unable to live up to the standards SeaWorld maintains for its trainers, and unwilling to provide the quality of life that SeaWorld claims to provide for its orcas. It became increasingly disillusioning to see how little support the on-site SeaWorld trainers actually received from their US superiors."

Less than five weeks after the orcas moved in, the main pool had to be shut down because the whales were tearing the coating off the walls and eating it, Suzanne contended. The park's owner had selected a new material, Metflex, which was never before used in orca pools. "In the rush to open the show, they ignored the MetFlex supplier and sprayed the coating onto walls and floors while the cement was still damp from rain."

Just before the first show, in March 2006, all four orcas appeared backstage with Metflex strips hanging from their teeth and paint smeared on their heads. "The trainers were so desperate to go on with the show that they actually tried to clean the paint using isopropyl alcohol," Suzanne said. The show was canceled, mostly because the whales wanted to eat the lining rather than obey commands.

The "bio balls," plastic canisters in water filters to reduce bubbles, also had problems. "They purchased the cheapest bio balls possible." They were breaking up and seeping into the pools. One morning the whales were found "swimming in a stew of plastic shards." The whales got sick and had to endure multiple endoscopies due to GI ailments, likely caused by the toxic chemicals they ate. Only then did SeaWorld tell Loro Parque to fix the pools for good. But after the repairs, the orcas continued to pick at the epoxy lining. It had become a learned behavior.

Suzanne provided a few details on Tekoa's 2007 attack on Claudia Vollhardt and, more important, on the Christmas Eve 2009 killing of Alexis Martinez by Keto. Unlike Tilikum, who never did water work, Keto was the "most reliable and oft-used orca in both dry and wet shows," she said, "yet he killed his trainer." Alexis's body had been riddled with "various fractures, including every bone in his anterior rib cage, lesions on his vital organs, and the bite marks of an orca."

After two deaths in two months, it was time to confront SeaWorld on worker safety and animal well-being, Suzanne said. The problems in Tenerife were caused by a "grossly inadequate permit process and inadequate regulations under the MMPA concerning the exportation of captive marine mammals." The Loro Parque "experiment" had been "an unmitigated disaster," Suzanne wrote, "and for Alexis, his loved ones, and the four orcas, it is a tragedy, and it should never be allowed to happen again."

Meanwhile, the four US whales, "whose quality of life, quality of care, and quality of training are not comparable to SeaWorld standards," should be removed from the venue at once. "There are simply too many people in charge" at Loro Parque, she alleged, "who don't and can't put the well-being of the animals as the priority."

Details of the Alexis Martinez disaster continued to trickle in. Naomi learned from OSHA's Lara Padgett that footage of the attack had been sent to SeaWorld officials, who reviewed the video. Brian Rokeach from SeaWorld San Diego, who was involved in the 2006 incident with Orkid, was the visiting SeaWorld trainer at the time. He was acting as control trainer on the evening of December 24, 2009, during a rehearsal session, when Martinez was in the water with Keto. SeaWorld was not allowing Lara to interview Rokeach.

"Apparently there was some sort of problem with Keto and Alexis," Lara informed Naomi. "Brian recalled Keto to the stage but didn't reinforce Keto to stay there. Then he motioned for Alexis to swim to stage. That's when Keto broke from control with Brian and went after Alexis. He struck Alexis underwater, to the chest." Alexis stayed under for about two and a half minutes. Staff members finally managed to move Keto out of the pool and dived down to recover their colleague. He reportedly had blood seeping from every orifice.

One week before the deadline for OSHA's anxiously anticipated ruling, representatives from SeaWorld flew up to Atlanta to meet face-to-face with Cindy Coe, a leading official in OSHA's Southeastern US region, Jason Garcia reported in the *Orlando Sentinel.* Neither party would discuss details, though SeaWorld said in a written statement it had "cooperated fully" with OSHA throughout the inquest. "Because the process has not yet reached its conclusion, it would be inappropriate to comment further."

Naomi learned two competing drafts of the OSHA findings were in circulation among Department of Labor attorneys in DC, Atlanta, and Tampa. The more lenient version was being championed by none other than Representative Alan Grayson (who would be defeated in the November 2010 election). The *Orlando Sentinel* reported that the liberal congressman had "intervened" in the probe on behalf of SeaWorld. Grayson confirmed to Jason Garcia that he had personally contacted OSHA to "discuss" the case, but refused to answer any other questions on the matter. Later, his chief of staff sent an e-mail to the *Sentinel* explaining that Grayson simply wanted to learn about the status of the investigation "first hand," and to "share his own views and impressions regarding the matter."

Grayson also issued a prepared statement that gave SeaWorld's safety record a thumbs-up: "As everyone knows, working with sharks and whales is inherently dangerous. But SeaWorld has done what it could to make that work as safe as possible. Beyond that, SeaWorld has raised people's knowledge and understanding of cetaceans enormously, and contributed greatly to the well-being of Central Florida and our community." Neither SeaWorld nor OSHA would comment on the matter, but captivity opponents cried foul.

"None of this smells very good," Naomi told Jason Garcia, who noted that SeaWorld had a "sizable presence" in DC and spent nearly $600,000 on Capitol Hill lobbyists in the first six months of 2010. Nearly two-thirds of that, $380,000, "occurred in the April-through-June quarter, amid the federal scrutiny triggered by the SeaWorld trainer's death," Garcia wrote.

Howard Garrett, who shared Grayson's progressive political values, was appalled by what he considered the lawmaker's cloak-and-dagger machinations to save SeaWorld's corporate hide. Howie was also deeply conflicted. "Yes, we want Grayson to be re-elected many times over, but this blatant, corrupt interference in due process is an assault on democratic values," he wrote in an August 19 blog on the left-wing website Daily Kos. "Grayson is acting as a perfect example of the corporate stranglehold on the legislative process that he otherwise so effectively battles." Then, in a "personal note" to Grayson, he let loose. "You have seriously soiled your reputation by trashing OSHA at a time when it is trying to perform its legislative mandate . . . in the face of a corporate bully. You have lined up alongside the bully this time, and that won't be forgotten."[5]

Howie, Naomi, Jeff Ventre, and other industry foes may have been troubled that Grayson was trying to derail action against SeaWorld, but in the end they needn't have been.

On August 23, 2010—OSHA's deadline to act in the Brancheau case—the agency slapped SeaWorld with the maximum penalties and sanctions allowed under federal law.[6] It was, by far, the greatest legal blow ever delivered to SeaWorld, and it inflamed people within the company. Opponents were amazed, but gratified.

OSHA had issued three citations against SeaWorld, including one that classified their violation as "willful," something the company's opponents were hoping for. "SeaWorld recognized the inherent risk of allowing trainers to interact with potentially dangerous animals," said Cindy Coe, OSHA's regional administrator in Atlanta, in a news release. "Nonetheless, it required its employees to work within the pool walls, on ledges and on shelves where they were subject to dangerous behavior by the animals."

Video footage in OSHA's possession showed Tilikum "repeatedly striking and thrashing the trainer, and pulling her under water even as she attempted to escape." SeaWorld had "forbidden trainers from swimming with this whale because of his dangerous past," but still allowed them to interact with him, including touching him, while "lying on the pool edge in shallow water."

OSHA's investigation had uncovered "an extensive history of unexpected and potentially dangerous incidents involving killer whales at its various facilities," the agency said in atypically harsh language. "Despite this record, management failed to make meaningful changes to improve the safety of the work environment for its employees."

The "willful" citation was issued against SeaWorld for "failing to provide a workplace free from recognized hazards that were causing or likely to cause death or serious physical harm to employees," the OSHA document said, citing two reasons.

First, trainers were exposed to "struck-by and drowning hazards in that they were allowed unprotected contact with Tilikum while conducting 'drywork' performances on pool ledges, slideouts, and platforms." The agency noted that a willful violation was "one committed with plain indifference to or intentional disregard for employee safety and health."

One "feasible and acceptable" means of abatement for this violation, OSHA wrote, "would be to not allow animal trainers to have any contact with Tilikum unless they are protected by a physical barrier."

The second reason cited by OSHA concerned the other whales at Shamu Stadium, who placed trainers at risk for "struck-by and drowning hazards" because they were allowed to "engage in other work and dry work performances without adequate protection."

Feasible means of abatement on this count would include a ban on water work and dry work with the whales, unless trainers were protected by physical barriers, decking systems, oxygen supply systems, or other forms of control.

The fine for the violation was $70,000, a pittance to SeaWorld Parks & Entertainment but nonetheless the maximum that OSHA could impose under federal law.

OSHA also issued a "serious" violation and fine for $5,000 for the lack of a stairway railing on one side of a ten-foot-high bridge at the "Believe" stage in Shamu Stadium. OSHA issues a serious citation when "death or serious physical harm is likely to result from a hazard about which the employer knew or should have known," the news release said. A third "other-than-serious" violation was issued for failing to equip electrical outlets at Shamu Stadium with weatherproof enclosures.

Under federal law, SeaWorld had fifteen business days to comply with the citation by paying the fines and enacting the means of abatement. Alternatively, it could request an informal conference with OSHA's area director or "contest the citations and penalties before the independent Occupational Safety and Health Review Commission."

Lara Padgett personally delivered the citation papers to SeaWorld Orlando. By the time she had driven back to Tampa, the company was already proclaiming its innocence—and declaring plans to fight the citations.

"SeaWorld disagrees with the unfounded allegations made by OSHA today and we have already informed the agency that we will contest this citation," the company announced.[7] "OSHA's allegations in this citation are unsupported by any evidence or precedent and reflect a fundamental lack of understanding of the safety requirements associated with marine mammal care."

This same tactic had been employed by Thad Lacinak and other industry insiders: Accuse the accusers. As for SeaWorld's contention of no evidence to support the citation, that fight would have to be taken up later, by a federal judge. To the average person, however, a severed arm, a torn-away scalp, and two previous deaths in Tilikum's pool likely qualified as evidence enough.

But SeaWorld insisted it was innocent and would be vindicated. The company noted that an independent review conducted by "some of the world's most respected marine mammal experts" had arrived at conclusions "in stark contrast to OSHA's." (In other words, why was the government persecuting SeaWorld when it had already been cleared by its own industry?)

"The safety of SeaWorld's killer whale program was already a model for marine zoological facilities around the world," the statement said. "The changes we are now undertaking in personal safety, facility design and communication will make the display of killer whales at SeaWorld parks safer still."

Trainers would remain out of the water until those new safety measures were put in place.

"SeaWorld trainers are among the most skilled, trained and committed zoological professionals in the world today," the statement continued. "The fact that there have been so few incidents over more than two million separate interactions with killer whales is evidence not just of SeaWorld's commitment to safety, but to the success of that training and the skill and professionalism of our staff.

"We look forward to challenging OSHA's unfounded allegations," it concluded, "and are confident that we will prevail."

Naomi Rose and HSUS had a decidedly different take on the citations. Naomi was presenting a talk at a zoological conference in China, but before she left, she and the PR department had drafted two press releases—one to address a "bad" outcome at OSHA, and one for a "good" outcome. This was a good outcome.

HSUS was "guardedly optimistic" about the citation, Naomi's statement said, adding that if orcas decide to act dangerously, "there is nothing anyone—including SeaWorld—can do about it. Trainers should never be in a vulnerable position around the water," she said, "and they certainly shouldn't get in the water for the show."

The media ran with the story. SeaWorld's publicity department was flooded with calls from reporters as company officials braced themselves for yet another bruising storm of controversy. For some, it must have felt like late February all over again.

This was turning out to be the worst year ever for SeaWorld. In early 2010, park officials were still dealing with the fatal attack by Keto on Alexis Martinez when their own employee was killed by Tilikum. That triggered an avalanche of negative media; investigations by the local sheriff's office and the not-so-friendly OSHA; condemnation by celebrities and animal protection groups; calls for Tilikum's extermination; a stinging investigative exposé in a popular magazine; heavy media coverage of Taima's death; and now, a citation for a willful violation and high-profile fines from OSHA that could, if not overturned, herald the end of water work during orca shows.

But now things were about to get really ugly: The lawyers had arrived. Lawsuits—and threats of lawsuits—quickly began circulating.

In the wake of Dawn's death, a number of current and former SeaWorld employees stepped forward—some privately, some quite publicly—to launch a startling array of extremely serious accusations against the company. If even one of the claims was proven true, it would be devastating for SeaWorld. But the mere airing of some of the allegations, whether fair or not, further tarnished the company's already dented and scratched reputation.

Early on the morning of August 23, a few hours before the OSHA citation was released to SeaWorld and the media, a matronly woman and her attorney appeared on ABC's *Good Morning America* to unleash a firestorm of accusations against SeaWorld and its handling of the OSHA investigation.[8]

The whistle-blower's name was Linda Simons. Just one week before Dawn died, Simons was hired as the health and safety director for SeaWorld Orlando, Discovery Cove, and Aquatica park, responsible for safety, health services, risk management, and related issues. Two months after the disaster, she was abruptly terminated by the company.

Linda Simons, backed by more than twenty years in the safety and security business, was now "speaking out on what she calls questionable or even dangerous safety practices at the Florida park that could result in another tragedy," ABC News reported. Simons, accompanied by her attorney, Maurice Arcadier, told *Good Morning America* that she was speaking out so that the safety of "the team members that remain is not jeopardized. If they're put into that close proximity, it could easily happen again."

According to Simons, everyone at SeaWorld was fully aware of the damage that Tilikum could do: "They talk to you about going into the water with Tilly. That if you go into the water with Tilly, you would come out as a corpse."

Many of Simons's allegations were described on the website of her attorney's law firm, Arcadier and Associates in Melbourne, Florida. "From the very beginning of OSHA's investigation, the SeaWorld executives involved were not cooperating," Maurice Arcadier wrote. Company brass, he alleged, "wanted to block them [OSHA] from coming onsite" and "obstructed the disclosure of relevant documents, videos, and/or information that was damaging to SeaWorld."

According to Arcadier, Simons tried to explain to her superiors that "OSHA's goal is to investigate the incident so that no other team members will be killed," and that cooperating with an OSHA investigation was mandated by law. "However as the investigation continued Mrs. Simons began to doubt that safety was SeaWorld's first priority. When OSHA requested specific trainers to be interviewed, SeaWorld changed them to different trainers for fear that the requested trainers may provide damaging information to OSHA; if necessary, schedules would be altered, claiming they were on vacation, or unavailable for other reasons. Mrs. Simons opposed this unscrupulous practice."

Arcadier added, "SeaWorld instructed Mrs. Simons to change and withhold documents from OSHA, which she refused to do."

When she was fired on April 20 for what she claimed was "attempting to cooperate with the OSHA investigation," Simons was told that she wasn't "a good fit" for the corporation. But she had never received any negative feedback "verbal or written," she said in the statement. In fact, she claimed to have received "repeated praise" from many executives at the park.

Simons filed a whistle-blower protection claim with OSHA and was also pursuing a wrongful-termination claim against SeaWorld, and "other claims available under law," Arcadier said. "SeaWorld aggressively threatened to pursue legal action against Mrs. Simons if she went public with her information. Mrs. Simons would not be bullyied [*sic*] and has been brave to bring forward the reprehensible conduct of SeaWorld."

Then Arcadier turned up the heat: "SeaWorld goes through great lengths to keep the public and media blinded. They use threats of lawsuits and their big time lawyers to keep the truth out." SeaWorld required that all employment disputes "be privately arbitrated in a confidential manner." Therefore, there was "so little information of all of SeaWorld's improprieties."

Linda Simons, in a signed affidavit that Arcadier faxed to SeaWorld's law firm, offered a day-by-day account of her recollections of what happened within the executive offices in Orlando between February 24 and April 20, 2010—the day of her firing. A copy was sent to Lara Padgett at OSHA's Tampa office.

Linda listed the many documents that Padgett had requested via e-mail on March 2, including any investigation report prepared by or for SeaWorld or the Blackstone Group; all witness statements; video recordings and still images; training records for Dawn, Jan Topoleski, and Lynne Schaber; all internal accident investigation reports regarding orcas in Orlando; and all "documents, permits and animal records for Tilikum," such as "historical records received when purchased from SeaLand of the Pacific."

A week later, OSHA issued a *second* round of requested documents, according to Simons. This time, Lara Padgett wanted such items as technical information and blueprint drawings for Shamu Stadium; all Behavior Review Committee safety reviews for all SeaWorld killer whales; the past five performance reviews for Dawn, Jan, and Lynne; all "Animal Profiles and Aggression Incident Report Notebooks" on orcas in Orlando; and a copy of the company video used for training purposes "compiled from previous incidents involving killer whales."

Padgett was thorough, if anything. She was not a popular figure around the park.

SeaWorld officials were indignant at the scope of the requests, Simons indicated. On March 16, for example, two executives in particular "were very vocal on not providing the documents and videos concerning any of the killer whales. They repeatedly said that the documents and videos would be released to the media/public."

Tilikum's documents were a particular concern, Simons recalled. On March 31, park curator Bill Hughes pointed out that the whale's vet records showed he was "on injections and was worried this would be viewed as a concern." Other executives urged that Tilikum's "meds list" be withheld from the feds.

Kelly Flaherty Clark, meanwhile, was worried because, a bit more than three years earlier, Dawn had been "written up for a safety violation (diving into one of the pools without the gates being shut and locked)," Simons wrote. That information would be withheld, she alleged, because it was more than three years old and attorney Carla Gunnin said SeaWorld would only turn over trainer records covering the prior three years.

The resistance to produce requested documents continued: "OSHA was told that they would have to subpoena Carla for any documents that they were not given [or] allowed to review."

When Simons went public, SeaWorld officials wasted no time fighting back; and they did so with full force. The company excoriated its former safety officer in a harsh statement brimming with denunciations:

> *The safety of our staff, guests, and animals is SeaWorld's highest priority. And we have cooperated fully in OSHA's inspection of the February 24th accident. We're not at all surprised to hear that Ms. Simons has reached out to the media with these unfounded charges. Since her termination several months ago, her representatives have used the threat of negative publicity to seek a sizable monetary payment from SeaWorld in exchange for her not going public with these false allegations. Linda Simons worked for SeaWorld for only a few weeks and was fired not for the reasons she cites, but rather for poor performance during the OSHA inspection of Dawn Brancheau's death. During those critical weeks, Ms. Simons repeatedly demonstrated an inability to conduct herself to the acceptable standards of competence, transparency, integrity or professionalism demanded of an inspection of this magnitude. Any claim to the contrary is simply false.*

Privately, Mark A. Hanley, an attorney representing SeaWorld, had already written to Maurice Arcadier telling him to back off. He said that an investigation of Simons's allegations found they were "without merit and apparently have been raised solely to deflect criticism of her own inadequate job performance."

Linda Simons had vehemently objected to working with outside counsel, Carla Gunnin, Hanley wrote, and her "subsequent job performance reflected her unreasonable resentment." Simons had not been "privy to all the information that was provided to OSHA" because she had been "essentially cut out of the loop."

Hanley said, "It is SeaWorld's very strong position that neither the law nor the facts support any claim under OSHA" or Florida whistle-blower laws. SeaWorld had offered Arcadier's client a "generous severance package" and a neutral job reference, which she rejected and instead sought $300,000 in damages—a demand that SeaWorld had now "categorically rejected. Should Ms. Simons, nonetheless, choose to pursue legal action, we have been instructed to vigorously defend all claims and seek all appropriate relief upon behalf of SeaWorld."

Simons chose to pursue legal action.

On the evening of August 23, Simons and Arcadier appeared on *Larry*

King Live in a segment that included yet another on-air dustup between Ric O'Barry and Thad Lacinak, who attacked Linda Simons as well.[9]

"I find it funny that we have someone that works for SeaWorld for—what did you say, a couple weeks or a week before the incident?—and then she gets fired. And then she's all of a sudden a credible witness on TV," Thad said. "I worked for SeaWorld for thirty-five years. I never saw any of the accusations that she said. SeaWorld cares about their animals. They aren't going to do things to hurt their trainers. This thing from OSHA is wrong. They should never have put this statement out, recommending that they do not do anything with the killer whales."

Larry King asked Simons for her response.

"I am a safety professional with over twenty years' experience," she said with full composure. "And my claim is that they were not cooperating with OSHA's investigation. I am not claiming anything to do with how the whales were treated. He was a dangerous whale. And Thad knew about the Tilly talk. That is not my claim, though. . . . They blocked documents that were requested from a federal agency because they had damaging information in them. And because I wanted to cooperate, they terminated me."

"Maurice, is Linda planning a lawsuit?" King asked of Arcadier.

"Right now we filed an OSHA retaliation complaint," the lawyer said.

"But no lawsuit? Will there be a lawsuit?"

"We'll see. SeaWorld prevents employees from filing lawsuits because they have an arbitration clause," Arcadier said. "We have to go through arbitration. And they have all these kinds of confidentialities to keep the media out."

36

Superpod

Samantha Berg, the sharp-witted New Yorker who worked at SeaWorld with Jeff, John, and Carol Ray, was the last of the ex-trainers to join the Orca Aware group. Carol and Sam had kept in contact, visiting each other in Seattle or Anchorage, where Sam had moved in 1994 and married her boyfriend, Kevin Meddleton. In 2003 Sam got her master's in acupuncture from Tai Sophia Institute in Maryland and returned to Palmer, Alaska, to open an acupuncture clinic with Kevin.

After Tim Zimmermann's article "The Killer in the Pool" was published, Sam wrote to Carol to ask if she had read it. Many of the things he mentioned in *Outside* magazine surprised her. Sam had left SeaWorld with a mostly positive opinion of the place.

Carol, who had been in contact with Jeff since Dawn died, knew about the article before it was published. She also told Sam about the Orca Aware group and how they had assisted Tim in his research. The group had been growing, Carol said. Members now included Howie's wife, Susan Berta, the other half of the Orca Network, and Astrid van Ginneken, the Dutch doctor who still flew to Washington every summer to help with Ken Balcomb's orca census. Yet another member, Wendy Cooke, was a blogger from Northern California who tracked the genealogy of captive orcas and ran a website, Without Me There Is No You, to raise awareness about whales. Its name came from the idea that "Without the patrons of the marine parks, there would be no show," the site explained.

The former trainers realized they were in a unique position to influence the captivity debate *and* the OSHA legal battle that was already taking shape. One "feasible" abatement OSHA had listed was an "oxygen supply system," better known as "spare air." A modified scuba system, it allowed trainers to carry compact amounts of air somewhere on their person as they worked in the water. Jeff learned that SeaWorld was vigorously exploring spare-air technologies as part of the safety review it had launched after Dawn died.

The ex-trainers dismissed it as a worthless idea. Jeff and John suggested they write to OSHA. Lara Padgett was preparing the battle against SeaWorld's legal appeal. All four wrote to Padgett. Sam's letter spelled out the main counterarguments well: "I don't see how 'spare-air' would have been any use to Dawn or any of the other trainers in the past who have been injured or killed by killer whales."

To begin with, killer whales "move FAST," Sam wrote. "Imagine trying to breathe normally or remember to slowly exhale with your arm or leg in a killer whale's jaws, while you are being forcibly dragged all over the pool. This is completely unrealistic and, frankly, preposterous." She mentioned the danger of severe lung damage if a trainer took a breath of spare air while at the bottom of the pool if a whale bolted him to the surface.

Some systems might also offer the whales "something else to grab," Sam said. "All marine parks" had trouble keeping things out of the mouths of their cetaceans. Sam stated with New York bluntness, "A marginally conscious, hypothermic, badly injured trainer in shock in the jaws of a 12,000 lb whale would be just as dead if he or she were carrying extra air or not."

When OSHA issued its citations, SeaWorld's legal troubles multiplied. One might forgive company officials for thinking Tilikum had become more trouble than he was worth.

Linda Simons was now filing a defamation suit against SeaWorld, in addition to her wrongful termination suit and OSHA complaint. SeaWorld's response to her allegations had been libelous, she claimed, by stating she lacked "acceptable standards of competence, transparency, integrity or professionalism" and that she tried to extort money in exchange for her silence.

Then, on August 24, 2010, the day after OSHA acted, the *Orlando Sentinel*'s Jason Garcia reported that Scott Brancheau had retained a Chicago law firm with expertise in wrongful-death litigation, O'Connor & Nakos. "Any potential lawsuit gained ammunition Monday when [OSHA] cited SeaWorld," Garcia noted. Brancheau's lawyer, Dan O'Connor, declined to say if his client would sue, though he did state, "It's not every day that OSHA issues a willful citation for plain indifference or intentional disregard for human life." He added intriguingly, "It is clear, after reviewing the willful finding, that more of the true facts will be brought out." Dawn did not violate any safety protocol; she was "working within the customs, practices, and procedures" of SeaWorld, O'Connor said.

It was "the first public sign of a fissure between the Brancheau family and

Orlando-based SeaWorld Parks & Entertainment," Garcia wrote in the *Sentinel*. The "Brancheau family" did not include Dawn's siblings and their mother, Marion Loverde, however, who were not party to any potential suit. SeaWorld was helping them to establish a foundation in Dawn's honor. Scott did join with SeaWorld and the Loverde family to block videos and photos of the incident from being released to the press and public.

Meanwhile the Connells, the New Hampshire family who were filming at "Dine with Shamu" moments before the carnage unfolded, filed suit on behalf of their son, Bobby, for negligence and intentional infliction of emotional distress, after OSHA issued its citations. Bobby had been profoundly traumatized by what he witnessed, the complaint alleged. The boy had looked directly into the victim's eyes and "saw the look of horror and desperation on Dawn's face as she was swimming for her life," the lawsuit charged.

Now he was haunted by nightmares, crying a lot, not eating well, and prone to bursts of anger. "The school's had some counseling for him," Mrs. Connell said. "It affected all of us. I'll start crying while driving. . . . I see that face every night before I go to bed."

Most legal analysts pooh-poohed the suit. Eyewitnesses to even the most unspeakable tragedies had no legal standing in court unless they were closely related to the victim. On the Internet, commentators eviscerated the parents. "These people are just disgusting greedy money grubbing low life pigs" was a typical remark. Another critic asked, "When the planes hit the towers on September 11, did all of those people sue the plane companies?"

As the summer of 2010 morphed into the fall, another former employee stepped forward with a legal complaint. She also claimed to possess troublesome secrets about SeaWorld, including its alleged obstruction of the federal Brancheau probe. Some of her allegations supported much of what Simons contended.

The plaintiff, Michelle Dillard, began working in 2007 as a human resources manager for SeaWorld, Discovery Cove, and Aquatica and rose to director of human resources before quitting in anguish in 2010, five months after Dawn's death. After she quit, Dillard retained Simons's attorney, Maurice Arcadier, and signed a sworn affidavit outlining her allegations. The document was sent to OSHA and copied to SeaWorld.

Dillard's first task in 2007 was organizing the chaotic applicant and employee records in advance of a Labor Department audit on possible discrimination in hiring. The records were in "an abominable state," Dillard said. Her boss, human resources vice president Christine O'Neal, allegedly told her to relax. The feds were "complete idiots" and had "no idea what they are looking for," O'Neal said, according to the affidavit. Dillard also alleged to have witnessed O'Neal repeatedly "stonewalling the investigation." The park failed its paper audit and federal inspectors descended on-site. "We created what we wanted them to have," Dillard charged. "We shredded some papers and stored others."

Even as the stressful inspection continued, Dillard said she was hit with yet another HR scandal. A senior male trainer at Shamu Stadium entered her office and unburdened himself of a dark secret: "There's a lot of quid pro quo sexual harassment going on around here. It makes me extremely uncomfortable." Some male senior trainers were demanding sexual favors from junior female trainers. They would allegedly ask for sex "in return for being given more 'water time' with the whales." Often, they would play the adult make-out game Dirty Jenga, pulling small blocks from a tower with written instructions to do things like spend "5 minutes in heaven" in another room.

According to Dillard, she began an investigation, grilling dozens of staff, and confirming that "indeed, there were severe, pervasive sexual improprieties committed against the junior trainers." As a result, "two male senior trainers/ supervisors were terminated," she said, though they left with "hefty severance."

Dillard also learned about safety issues. "August Busch III felt that trainers should make more money because they work with dangerous animals," she wrote to OSHA. Busch also wanted greater safety measures. "They looked at such things as an underwater breathing apparatus, but trainers felt they were cumbersome and wouldn't really help." Trainers also lobbied for false-bottom floors, which could rapidly be raised to beach a rampaging whale. But these were deemed too costly by the "bean counters," Dillard alleged.

By 2009, Dillard said, she was suffering from emotional anxiety on the job. "I repeatedly witnessed behavior that was morally reprehensible to me. It was proving to be detrimental to me in terms of the unrelenting stress I suffered." By Christmas of 2009, when Keto killed Alexis Martinez, she was seeking medical help for chronic anxiety and depression.

Then Dawn Brancheau was attacked.

Dillard had sprinted to Shamu Stadium to help with the rescue but was too late. "I was within 30 feet of the whale's mouth with Dawn's dismembered body in it," she wrote, recalling that one of the curators on-site had grumbled, "If we had put in those false-bottom floors, this never would've happened."

In the aftermath of the disaster, during the OSHA investigation, Dillard denounced what she called SeaWorld's "historical culture" among management of "outright lying to OSHA [and] using misinformation and cover-up, using intentional delay tactics to stonewall the investigation. She wrote that her boss "hid documents, pretended to not know that documents existed and obstructed OSHA's investigation."

Dillard also backed up Linda Simons, saying Simons "did everything in her power" to cooperate with OSHA. Simons had urged management to cooperate "so that another park employee might avoid Ms. Brancheau's fate." But Dillard feared that Simons would face retaliation.

Two months after Dawn died, the Department of Labor cited SeaWorld parks for employment records that showed a "disparate impact" on African-American and Hispanic applicants. As a result, the company agreed to try to contact hundreds of qualified applicants who had been denied employment, and some were offered either a job or a modest financial settlement.

Meanwhile, Dawn's death "began haunting me," Dillard wrote. She sought stronger antianxiety meds. Angst over her job grew, rendering her more physically and mentally unfit. In June, Dillard's doctor ordered an immediate leave of absence. When she returned in mid-July, nothing had changed. When she asked O'Neal about the probe, the reply was "If Dawn had just reacted like a trainer that day she'd still be alive," Dillard alleged. She was appalled by the reply. That same day, Dillard tendered her two-week notice. Told by her boss that she was free to go right then, she walked out for good.

Dillard filed a complaint for "constructive discharge"—a term used when working conditions become so unbearable one had to quit. The case was mandated to confidential, binding arbitration: Dillard had signed a standard agreement handed to all new employees requiring them to arbitrate personnel disputes and barring them from seeking damages in court. Nothing about the arbitration, including its outcome, was to be discussed in public.

SeaWorld's vice president of communications, Fred Jacobs, said neither Dillard nor Simons should be believed. "Ms. Dillard was not involved in the OSHA inspection of Dawn Brancheau's death," he said. "SeaWorld cooperated fully with the agency and, since the OSHA investigation continued for months after [their] employment with SeaWorld ended, they have no knowledge as to what documents or witnesses were provided to the agency." Both Dillard and Simons were suing SeaWorld and both were represented by Arcadier, who, Jacobs said, "threatened that Simons would go to the press with these unfounded allegations unless SeaWorld agreed to pay her hundreds of thousands of dollars."

Simons and Arcadier flatly rejected those charges. Meanwhile Fred Jacobs did not deny there were charges of discrimination in hiring at SeaWorld or sexual harassment at Shamu Stadium.

Colleen Gorman and John Kielty still lived together in Colleen's two-bedroom condo in St. Pete Beach. Every day, they worked at their computer stations in the living room overlooking a shallow lagoon frequented by wild dolphin pods. ("There's no need to go to SeaWorld when you can see them right from here!" Colleen liked to say.) The two were central members of the Orca Aware group. They shared with their growing network intelligence they gathered from marine mammal blogs and websites, and even contacts they had established inside the gates of SeaWorld.

John and Colleen kept busy tracking captive-orca issues at parks around the country and the world—Canada, Japan, Spain, France, Holland, and Argentina still had captive killer whales on display—but mostly they were focused on Florida, in particular Lolita at the Miami Seaquarium and Tilikum at SeaWorld. They made frequent drives up to Orlando to perch high in the bleachers of Shamu Stadium to observe Tilikum, often floating alone, in a backstage pool during the "Believe" show.

The show, designed for hotdogging and other types of water work, had already banished all trainers from the water for safety. For "Believe" segments where trainers previously rode whales or launched from their rostrums, now they had little else to do but dance silly jigs on dry land to the blasting music.

Between shows, Colleen would spend long periods gazing through the gate at Tilikum in G Pool. More than once, SeaWorld security asked her to move along.

In September of 2010, John and Colleen decided to formalize their efforts. They began creating a 501(c)(3) nonprofit organization to advance their anticaptivity agenda. They dubbed themselves The Orca Project (TOP) and launched a news and opinion blog by the same name. TOP rapidly grew into an influential and well-trafficked site among whale advocates.

John and Colleen shared writing duties. In early September TOP's first entry, "Seeing Is Believing: Tilikum's Lonely Life After Dawn," was posted by Colleen. For the past six months, as SeaWorld's internal review was under way, she wrote, Tilikum had been kept out of his segment in the "Believe" show. On one typical day, he "barely moved" from the time Colleen arrived at 9:00 a.m. He "simply rested all day long until it came time to be fed, once at 1 pm, 3 pm and 6 pm when they came with a bucket of ice and a few fish."

Such prolonged isolation, Colleen wrote, "begs the questions: what does he do with his time? Why does SeaWorld hold on to him versus retiring him to a sea pen? . . . What does he do for stimulation?" She had asked around among SeaWorld staff about Tilikum. These were social animals, she said. Shouldn't he be with the other whales? She was told that Tilikum was "an Icelandic 'transient,' therefore implying that transients are drifters and spend time alone."

Colleen continued, "Any sentient being would go out of their mind in such a situation." If SeaWorld thought that Tilikum was dangerous before, she asked, "What will happen to this poor soul if they continue to keep him separated, isolated, ignored [and] neglected?"

SeaWorld did not respond to the blog. But as Colleen reported, company officials were telling the media that Tilikum was receiving "much interaction, play, stimulation from the trainers and is doing very well."

On Monday, September 6, trainers at SeaWorld San Diego noticed that Sumar, the twelve-year-old son of Tilikum and Taima (who died during a stillbirth in Orlando), was not feeling well. The young male, separated years earlier from his mother after she tried to drown him, began to show signs of "slowness" during the afternoon. Staff veterinarians took blood samples and quickly initiated a full regimen of antibiotics.

On Tuesday, his condition deteriorated. The 12:30 Shamu show was canceled to allow trainers and veterinarians to devote their full attention to Sumar. "Despite these efforts," SeaWorld said in a statement, "Sumar did not survive."[1]

Sumar had no history of health problems and his passing "shocked" park employees, the *San Diego Union-Tribune* reported. Shamu Stadium was closed, although the "Believe" show resumed on Wednesday. "All of us, especially his trainers, will miss Sumar more than you know. He was a truly wonderful whale and was very deeply loved," the company said on its Facebook page. The cause of death was unknown and necropsy results were not expected for several weeks. It left SeaWorld with six killer whales in San Diego and eighteen in total.

Sumar's death "complicated challenges for SeaWorld," which relied on the breeding of whales such as Sumar, instead of capturing them in the ocean, the *Union-Tribune* noted. But that wasn't the only thing SeaWorld relied upon. It also needed continued public support for keeping killer whales in captivity, "something that animal rights groups said is eroding after a string of high-profile incidents," the paper observed.

Despite Sumar's youth, SeaWorld spokesman Dave Koontz said captive orcas "routinely live into their thirties or forties." Colleen and John went on the attack. "They DO NOT routinely live into their 30s and 40s," TOP wrote. Of the 41 captive orcas still alive, only 2 had reached age forty and 3 had reached thirty. More than 150 died before those ages. Put another way, of the more than 190 whales held in captivity since the 1960s, only 5 so far had survived past the age of thirty—the mean age of longevity for males in the oceans. The figures TOP used had been compiled by Wendy Cooke, although both HSUS and Stefan Jacobs's Orcahome website had also reported much of the data.

"To give you a sense of how young orcas in captivity mostly are when they die," TOP wrote, "here is a list" (SeaWorld whales only, and some calves who died before age one, excluding stillbirths and miscarriages):

Shamu (F), *lived 6 years*
Ramu (M), *lived 15 years*
Kilroy (M), *lived 11.5 years*
Kandu (F), *lived 4 years*
Orky 2 (M), *lived 20 years*
Nootka (F), *lived 20 years*
Winston (M), *lived 15.5 years*
Kandu 3 (F), *lived 4 years*
Sandy (F), *lived 4.5 years*
Kona (F), *lived 6 years*
Canuck (M), *lived 2.5 years*
Frankie (M), *lived 5 months*
Kanduke (M), *lived 15 years*
Kenau (F), *lived 15 years*
Gudrun (F), *lived 19.5 years*
Canuck 2 (M), *lived 4 years*
Kona 2 (F), *lived 10 years*
Kandu 5 (F), *lived 12 years*
Winnie (F), *lived 24.5 years*
Kotar (M), *lived 16.5 years*
Shawn (F), *lived 1 year*
Kahana (F), *lived 12.5 years*
Nootka 4 (F), *lived 12 years*
Haida 2 (F), *lived 19 years*
Samoa (F), *lived 8.5 years*
Bjossa (F), *lived 21 years*
Katerina (F), *lived 10.5 years*
Splash (M), *lived 15.5 years*
Taku (M), *lived 14 years*
Nyar (F), *lived 2 years*
Baby, *lived 38 days* (Haida 2)
Halyn (F), *lived 2.5 years*
Taima (F), *lived 21 years*
Baby Shamu 2, *lived 11 days*
Sumar (M), *lived 12 years*

Local news accounts stated that parts of Sumar's corpse were donated to research, some may have been sold to pet-food processors, and the rest went to landfill.[2]

By the end of the summer, SeaWorld had still not put orca trainers back in the water—though park officials made no secret of their desire to do so. If they could deploy the right technologies, they might be able to return to water work safely and still be in compliance with OSHA's "feasible means of abatement."

Tim Zimmermann and the Orca Aware members knew that SeaWorld was in the early phases of developing a spare-air system. Tim, who was busy with another assignment, collected the four former trainers' OSHA statements against spare air and forwarded them to the *Orlando Sentinel*'s Jason Garcia. Garcia ran with it.

The "SeaWorld Four" were now united and going public in a coordinated campaign against their former employer. On September 13, Garcia published his story, "Former Whale Trainers Criticize SeaWorld Safety Proposal." John called spare air a public-relations ploy and Jeff denounced it as "a phony" solution. It might be useful "when a killer whale is holding a trainer underwater but not otherwise brutalizing the trainer," he said. "But it's not a real solution."

Spokesman Fred Jacobs said the spare-air plan was "very early" in the research and development stage and the company still had to establish its feasibility.

That same morning, Sam appeared on the popular *Fox & Friends* show.

Sam was "not alone" in her accusations, noted anchorwoman Gretchen Carlson. "That's right," Sam said, "a group of trainers now believe that, for the safety of *all* trainers, they shouldn't go back in the water." Former staff speaking out against SeaWorld was "unprecedented," Sam said, adding that she hoped others would "come forward who actually witnessed incidents in the past."

Sam's appearance triggered a flurry of comments on her Facebook page, some from trainers still at SeaWorld. "I took that interview personally. I wish I didn't," one wrote. "Never in all the years we have known each other have you ever expressed the slightest concern for the trainers. . . . How long were you at Shamu? Months? I'm sorry but that hardly qualifies you as an expert."

Sam wrote back quickly: "In order for me to live in integrity, I absolutely have to speak on behalf of not only Dawn and all the other injured or deceased trainers, but also on behalf of all the whales who have died prematurely."

Another trainer was more circumspect: "Ultimately, Dawn made a mistake, but I believe 'the system' allowed the environment for that kind of mistake with Tilikum to occur. It could have been any one of us." SeaWorld had eighteen years to make Tilikum safer through training him how to respond appropriately. And though trainers knew there was a risk, "upper, upper, management have created an 'umbrella' of training methods that have lulled people into believing these animals are safer than they really are. They have been marketed and showcased that way with the Shamu brand over the years."

Sam wished the conflicted trainer peace and predicted, "'Something' good will come out of it all." The contact agreed—something positive *had* to result from Dawn's death "because that's how I sleep at night."

Less than a month after Sumar perished of "mysterious causes" in San Diego (his necropsy later revealed a twisted intestine),[3] Kalina, the original Baby Shamu and daughter of the venerable Katina, died in Orlando, on Monday, October 4. She was twenty-five. Like Sumar, Kalina died suddenly and with little warning. She had exhibited no signs of illness as recently as Friday the first, and was eating well over the weekend. But by Monday afternoon, she began showing signs of discomfort and quickly died early that evening.[4]

The celebrated Kalina, the first successfully bred killer whale in captivity, was the third to die at SeaWorld in just four months and the twenty-fourth in twenty-five years. "The trio of deaths comes amid what may be the most challenging period in Orlando-based SeaWorld Entertainment's nearly 50-year corporate history," Jason Garcia wrote in the *Sentinel*.

SeaWorld pushed back. Spokeswoman Becca Bides told Garcia there was "no credible evidence" to suggest that killer whales at SeaWorld were stressed. The company had achieved twenty-six successful births over twenty-five years, and seventeen of those whales were still alive. They were all "content and healthy," Bides contended, "a fact supported by medical, scientific, and behavioral evidence."

Naomi Rose appeared on *Good Morning America* to discuss the loss of both Kalina and Sumar. "When they're confined, I think they're under stress, I think they're at the very least very bored; and both stress and boredom, and depression, can kill you." She added that the three whales who died were twenty-one, twelve, and twenty-five, which was "not even middle-aged for wild orcas." Orcas were just too big for life in a tank. "SeaWorld is telling you otherwise, but they are misleading the public."

ABC coanchor George Stephanopoulos, who delivered the story live from Orlando, read a SeaWorld statement on air. "Animals in our care live and die just as they do in the wild," it began. Naomi had heard all this before, but maybe, she thought, this time it might ring hollow to the public. In terms of longevity, foraging, social behaviors, natural ranges, and "activity budgets," the comparison was patently disingenuous, if not borderline Orwellian.

Fortunately for her, the lead reporter on the story, Matt Gutman, told Stephanopoulos that in the wild, orcas "live forty years; in captivity, twenty years." Some wild whales lived up to eighty years, he added, "and clearly that's not happening in these parks."

"It's not even middle age," Stephanopoulos concurred, echoing Naomi's words. More than seven months had passed since Dawn had died, and Naomi was *still* struck by how radically the national media had altered its once-fawning posture toward orca display.

Later that day, Naomi issued an uncompromising statement in an HSUS press release on the spate of deaths. Kalina's passing "should be the final word: orcas do not belong in tanks. With the tragic death of trainer Dawn Brancheau and these three orca deaths, it is difficult to escape the conclusion that a 45-year experiment has resulted in a spectacular failure."

Naomi had not forgotten about the troubles allegedly plaguing Loro Parque. She marshaled her considerable resources to see that Suzanne Allee's disturbing allegations about SeaWorld's orcas at the park's Orca Ocean stadium were heard by the proper US authorities.

Allee had refined her written eyewitness account into a fourteen-page white paper titled "SeaWorld's 'Excess Orcas' at Loro Parque." In early October 2010, Naomi delivered copies to her contacts at NMFS, APHIS, the federal Marine Mammal Commission, and the House Committee on Natural Resources. Allee's accusations about death, danger, and toxic pool-lining material being peeled away by whales had not yet been made public.

Naomi, joined by Courtney Vail at WDCS and Susan Millward at the Animal Welfare Institute (AWI), arranged a series of meetings to introduce Allee to the federal regulators, present them with the evidence, and request that they investigate all potential violations of the Marine Mammal Protection Act. Naomi wanted NMFS to confiscate the animals if Loro Parque did not improve conditions or if SeaWorld refused to repatriate the animals. She realized the request was a long shot: It would involve complex application of US and international law, not to mention potential diplomatic turmoil between Washington and Madrid.

A few weeks later, SeaWorld responded to Allee's accusations in a six-page letter to NMFS written by Brad Andrews, chief zoological officer of SeaWorld Parks & Entertainment, and copied to the deputy administrator of APHIS.

Andrews rejected nearly every point in Allee's paper: "At no time during Ms. Allee's employment did she express concerns involving killer whales. These baseless allegations are troubling only because they appear to reflect the bias of a disgruntled former employee which was used by HSUS, WDCS, and AWI to advance a political agenda."

Park operations and zoological practices at both SeaWorld and Loro Parque were "widely acknowledged by industry experts as exceeding high standards, and we stand by them." Andrews did acknowledge that the SeaWorld whales had been exposed to "bio balls" from the broken filtration system (though he insisted the animals did not ingest the material) and had torn and swallowed pieces of the Metflex lining in the pools. "Some ingestion and regurgitation of MetFlex was documented," Andrews admitted, but the coating was ultimately replaced.

"In any event, NMFS has no authority to confiscate the killer whales as the HSUS requests," Andrews told the agency. Nor could NMFS revoke the killer whale display permit now that the animals were at Loro Parque, which, as a foreign facility, had no MMPA permit. "There is nothing for the Secretary [of Commerce] to revoke," he concluded.

Naomi was hardly surprised by the rebuke from SeaWorld. But she remained hopeful that NMFS might intervene and inspect Loro Parque. She would wait many months for an answer.

John Jett, the scientist and PhD, wanted to publish a paper on the effect of captivity on orcas. John and Jeff Ventre had been bandying the idea around for months. They had asked Mark Simmons during their lakeside reunion, back in May, if he wanted to coauthor a manuscript and submit it to a peer-reviewed journal for consideration. Mark didn't respond. Now, in early November of 2010, John and Jeff decided it was time to get on their computers and start their research in earnest. If they were going to make claims against their former employer, they would need hard science to get published and advance their case.

As Jeff and John organized the structure of their paper, John Kielty was tirelessly filing FOIA requests to help reconstruct the entire Marine Mammal Inventory Report, and searching online for media accounts of captive-orca deaths. Part of the paper was an extension of the "Five Medical Issues with Captivity" Jeff had supplied to Tom Ashbrook for the *On Point* segment he did back in July. Ashbrook mentioned Jeff and read the list on the air: (1) chronic tooth decay; (2) retinal damage from looking up at the sun; (3) collapsed dorsal fins; (4) forced social reorganization; and (5) pathological behaviors from "living in a blue box."

The physician and the biology professor fully intended to write a paper suitable for peer review, but the rush of outside events quickly overtook them. In November, the federal Occupational Safety and Health Review Commission set a time and venue for the big trial—*Secretary of Labor v. SeaWorld of Florida, LLC*—in which the company would fight to overturn the potentially crippling citation issued by OSHA. The trial—actually, a legal hearing presided over by an administrative law judge employed by the Labor Department—would begin on February 14, 2011, in Orlando.

Jeff and John wanted to have a document in circulation by then, but with just three months to write it had no time for a peer-reviewed article.

The weird and woeful saga of Tilikum—taken from his mother, sent to shore in Iceland and from there to Canada and Florida, forced to live with whales who abused him, involved in three human deaths—had filled the pages of newspapers for decades. But no stories about Tilikum's bizarre life—even in supermarket tabloids—were as lurid as the day that rocker Tommy Lee took on SeaWorld's vaunted artificial insemination program for orcas.

Lee, the bad-boy drummer from the heavy metal band Mötley Crüe, was a PETA supporter. He blasted SeaWorld for refusing "to release this frustrated whale because he is your chief sperm bank," he said in a letter to Terry Prather, who became president of SeaWorld Orlando in October.[5]

Lee said he knew how SeaWorld staff extracted the whale's semen: "Having someone get into the pool and masturbate him with a cow's vagina filled with hot water." Even during his most frenetic days, he said, "I never could've imagined something so sick and twisted." Lee asked, how could SeaWorld say that trainers did not get close to Tilikum anymore "when they are jacking him off?" That, he said, was "about as 'direct' as it gets."

Lee's letter was posted online by celebrity-tracker *TMZ,* and it went viral. The tale caused a sensation on the Internet. But SeaWorld went on the counteroffensive. The following day Fred Jacobs fired off a sneering letter to *TMZ:* "If Mr. Lee's information on Tilikum and SeaWorld's artificial insemination program for killer whales comes from PETA we're not surprised that it's wrong. PETA is as careless with facts as they are extreme in their views." Jacobs asserted that "collecting semen for whale artificial insemination is similar to techniques employed in managing livestock or other species for zoological display." Contrary to the rocker's assertions, trainers "do not now nor have they ever entered the water with Tilikum for this purpose. Whatever his views on SeaWorld, Mr. Lee would be wise to spend more time checking his facts."

Lee was also incorrect about the cow vaginas, although SeaWorld did at one point use artificial vaginas (AVs), which were considered standard practice.

Lee's letter and SeaWorld's response—which provoked "tweets" (and titters) around the world and T-shirts demanding FREE TILLY'S WILLY!—were written up by Colleen and John at The Orca Project. One of their self-proclaimed SeaWorld insiders explained the semen-extraction procedure to them, and they published it.[6]

These days, trainers masturbated the bulls with latex gloves and K-Y jelly, TOP's source said. "The animals already know how to roll over. It's a basic requirement for accessing their tails and genital slits, and other husbandry. It's a well-rewarded behavior, and the whales are usually cooperative." Sometimes, other whales were brought in to "stimulate the whale undergoing the AI session. Interestingly, it was usually another male." Any bulge or erection was rewarded with fish and manual stimulation. "The whale eventually figures out you're stimulating it in an erotic way. The ejaculate is stored in a bag or Nalgene plastic bottle."

According to John and Colleen, Tilikum had sired ten living calves and was grandfather to three more. SeaWorld owned nineteen captive-born whales, and thirteen of them were his blood relatives. "This means that the gene pool is dangerously lacking diversity," TOP reported.

Sam Berg, who had studied artificial insemination and animal genetics at Cornell, discussed SeaWorld's breeding programs with her former colleagues, the ex-trainers. "They are just trying to make more whales, which jeopardizes the health of their population," she said in an e-mail to Jeff, John, and Carol. "I don't know if there's any regard for the health of the whales that they are creating. I think you could say that SW has created more of an 'inbreeding program.' "

By the end of 2010, Jeff and John had made considerable progress on their captivity paper. The OSHA hearing was looming and they rushed to complete it. John was coming up with some shocking numbers on survivorship, and Jeff was discovering revealing information, including on captive-orca tooth damage and pulpotomy (drilling) and orca husbandry. Jeff also had a copy of the inspection report compiled by APHIS when animal health inspectors went to SeaWorld after Dawn died to investigate Tilikum and look at his husbandry records. John Kielty had obtained the APHIS report through the Freedom of Information Act and posted it to the Orca Aware list.

Jeff shared his progress with a few close allies, including Howard Garrett. Howie forwarded a draft of the paper to Ingrid Visser, PhD, head of the Orca Research Trust, a New Zealand nonprofit, and an authority on wild orcas in the waters of New Zealand, Antarctica, Argentina, and Papua New Guinea. Dr. Visser had worked on the Keiko project and produced several scientific papers and documentary films on orcas. Visser shared a copy with Dr. Lori Marino, the friend and colleague of Naomi Rose's who testified at the House subcommittee oversight hearing in April.

Lori, who forwarded a copy to Naomi, wrote back to Jeff in December, saying, "This is the beginning of what could be a major paper. I would be happy and enthusiastic to help you turn it into something that could be submitted to a major journal." Ingrid also urged them to publish.

Naomi sent Jeff an e-mail in late January that was also supportive, but noted that one of their statistics was also calculated in Small and DeMaster's landmark paper in 1995. She urged them to acknowledge the previous work and to complete the calculations that would update the annual survival rates (ASRs) for captive orcas, a necessary endeavor given how many years had passed since 1995.

Jeff and John did not welcome Naomi's opinion on their paper, though the interjection was softened somewhat by her adding: "The information about CAUSES of this higher mortality that you discuss in your paper, particularly the issue of poor dental health, is new (at least outside the high walls of SeaWorld!) and deserves publication."

Naomi Rose had finally made the acquaintance of Jeff and John—albeit electronically, and not on the best of footings. Naomi was wary of animal trainers—even former trainers who had turned against the industry—but she had to admit that Jeff and John had gotten a good start on an important paper. Perhaps most important, she wrote, "Your voices, as former trainers with a science background, carry enormous weight in this debate, which has gained such tremendous momentum since Dawn Brancheau's tragic death." She added, "I hope together all of us who have been working to change the public's mind about the acceptability of orcas in captivity can work to push society past the tipping point."

The paper, "Keto and Tilikum Express the Stress of Orca Captivity,"[7] was ready to go in February. SeaWorld, however, was not ready to go to trial. The company requested—and received—a postponement of the hearing until April 24. Had Jeff and John known the trial would be delayed, they would have held off on releasing the paper and aimed for publication in a peer-reviewed journal instead.

"The practice of keeping killer whales in captivity has proven to be detrimental to the health and safety of animals and trainers alike," John and Jeff's paper began. They described the killings of Alexis Martinez and Dawn Brancheau, exactly two months apart.

Tilikum, they stated, was "representative of the many social and health issues plaguing captive orcas." The APHIS report on Tilikum that John Kielty had obtained, for example, showed he was "on antibiotic and antifungal drugs for an 'inflammatory issue' with an elevated white blood cell count beginning 'about February 11' (2010)," and lasting until at least the day he attacked Dawn: Tilikum was fighting off an infection when he attacked.

The authors said captive orcas were subject to boredom, social strife and aggression, and acting out. They discussed the large range of wild orcas versus the cramped quarters of captivity, and they talked about dorsal fin collapse. They claimed "poor dentition can lead to a host of diseases including valvular heart disease, gingivitis, pneumonia, stroke, and heart attack." The open bore holes left behind by "pulpotomies" created a direct route for pathogens to enter the blood, "where they can then be deposited into the tissue of various organs throughout the body, such as the heart or kidney." Were bore holes contributing to poor health at SeaWorld? We may never know. "Unfortunately, orca necropsies are mostly done in-house, by park personnel, and under a relative cloak of secrecy. Could the cause of pneumonia be bacteria carried to the lungs from rotting food plugs or tooth decay? This is unclear due to insufficient research and lack of scrutiny."

Even less understood were the "consequences from increased ultraviolet (UV) radiation exposure to the skin, eyes, and immune systems as animals float motionless at the surface." Marine-park orcas sometimes got sunburn and required coats of black zinc oxide to protect their backs. "Furthermore, at least one serious trainer injury (John Sillick) has been linked to a whale's poor visual acuity, possibly secondary to cataract formation. It is known that UV radiation exposure is a factor in the development of cataracts." It can also lead to immunosuppression and retinal damage. Despite these risks, APHIS—charged with enforcement of Animal Welfare Act (AWA) provisions such as protection from direct sunlight—had "historically been ineffective" in doing anything about them.

Turning to survival issues, the authors needed to provide some metrics for comparison, and they came up with two. The first value they calculated was the "mean duration of captivity" (MDC), which took into account the 152 killer whales that had lived and died in captivity.

The results were shocking. The mean duration of captivity among the 152 animals was 2,413 days—a brief 6.6 years. But the MDC had a weakness: It only included animals who had died. John and Jeff needed a statistic accounting for living animals as well. John consulted a colleague at Stetson University, who suggested the Kaplan-Meier statistical formula, often used in large-scale drug trials, to best represent the "central tendency" of the data. This allowed "credit" to be given for living animals, including the older whales Corky, Lolita, Ulises, Katina, Kasatka, and Tilikum. After factoring them in, however, the figure rose to just 8.5 years.

Privately, Naomi was concerned that these calculations repeated some of the same mistakes she herself had made back in 1993—such as including calves that had died before they were one year old—and she didn't know if the Kaplan-Meier method was applicable to this situation. John and his colleague at Stetson insisted it was, and John's calculations certainly suggested that captive orca survivorship was no better, and probably worse, now than in 1995.

"Their life spans are decreased and their behaviors altered from the stressors associated with confinement," Jeff and John asserted. They also called for an end to orca shows. One solution would be phasing out captive populations via attrition. "In short, stop breeding the animals and let those already in captivity live out their lives."

John and Jeff still hoped to publish their work in a journal (*Zoo Biology* encouraged them to submit a formal manuscript). But now it was posted at TOP for all to see. "As it turned out, many more people read the paper online than would have in some journal," John told his orca colleagues. Traffic data by the end of 2011 showed that the paper was read some forty thousand times at TOP and downloaded nearly two thousand times. Jeff's own post of the paper was hit five thousand times, and a Spanish version (translated by Ester Quintana-Rizzo, PhD, a marine scientist at the University of South Florida) was read about fifteen hundred times. The paper was also posted at the website of the British Natural History Museum.

Howie sent the paper to the MARMAM list, allowing thousands more to read it. "Ventre and Jett have kept their focus on the evidence, on what's happening to the orcas," he wrote, "a key element that is seldom seen in the controversies over orca captivity."

• • •

In early 2011, Florida attorney Maurice Arcadier, who represented the two SeaWorld whistle-blowers, braced himself for a day of depositions. He was scheduled to meet with client Michelle Dillard, who had filed a "constructive discharge" claim after witnessing a sensational array of misdeeds at work.

Arcadier entered the conference room and sat across the table from a battery of unsmiling lawyers. His client walked in and sat down. After she was sworn in, SeaWorld attorneys handed her the affidavit she had signed. Dillard read the document aloud, line by line.

After each sentence, he said she declared, "I retract that statement in full."

Arcadier's stomach sank. His client had turned. She had not even told him in advance. Somehow, they got to her, he thought. Either with money or with threats, but SeaWorld had found a way to make this go away.

Arcadier looked across at the opposing attorneys. They looked back, no longer unsmiling.

On February 23, 2011, the eve of the first anniversary of the Dawn Brancheau killing, the anti-caps anticipated a wave of publicity. It came, but not as they expected. Instead of thoughtful discussions on policy changes in killer whale captivity, the news cycle was dominated by SeaWorld's own carefully crafted story: Orca trainers were preparing a return to the water.

Park officials told *Sentinel* reporter Jason Garcia that trainers at all three parks would initiate water work within a few months, with interactions "initially restricted to small medical pools equipped with false-bottom floors that can be lifted out of the water."

SeaWorld was investing tens of millions of dollars on lifesaving technologies such as fast-rising pool floors and "underwater vehicles that could be used to distract an out-of-control killer whale with pulsing lights and whale vocalizations." The officials had not decided when, or even if, water work during shows would resume, Garcia reported, but it was not dependent on the outcome of the OSHA hearing.

Opponents were unimpressed. Both John and Jeff appeared on CBS's *Early Show,* where John called the plan "a recipe for disaster." The Orca Project ripped into the proposed measures, saying they came "on top of other inadequate safety improvements SeaWorld has made in recent months," citing removable guardrails around stage ledges and net boxes that reportedly deployed quickly.

SeaWorld's Julie Scardina responded that the reforms were essential to resume water work, which was the "best way to not only showcase those animals but to care for the animals, as well" because it allowed close access to the whales. Meanwhile, trainers were attending seminars "on topics such as whale aggression," Jason Garcia reported. Perhaps anticipating further mishaps despite these precautions, executives were investigating another technological innovation: a machine to quickly pry open a whale's jaws.

It was time to put Tilikum back in the show.

On March 29, 2011, Jason Garcia reported in the *Orlando Sentinel* that the 12,500-pound, three-time killer would appear in that morning's "Believe" show. It was important for Tilikum's health and husbandry, SeaWorld insisted. "Participating in shows is just a portion of Tilikum's day, but we feel it is an important component of his physical, social and mental enrichment," noted Kelly Flaherty Clark in a prepared statement.

The house was packed that morning. "After the audience's trademark tribal chant of 'Shamu, Shamu,' [Tilikum] answered by splashing icy water into the first several rows of the stadium," *Sentinel* reporter Dewayne Bevil wrote. The fans loved it; not so the demonstrators outside the park hoisting placards that said LIFE IN A CONCRETE BOX IS NO LIFE and IT WILL HAPPEN AGAIN. Inside SeaWorld, Bevil caught up with Colleen Gorman and John Kielty, who had driven up from St. Pete Beach to see Tilikum perform for the first time in thirteen months.

They were "unhappy after the show," Bevil wrote. Tilikum had "moved a little slow today. . . . He looked a little lethargic," according to Colleen. She was afraid he would kill again, she said. Tilikum should be retired to a sea pen, where caregivers could teach him to catch live fish "and just teach him how to be a whale again."

In Alaska, former trainer Samantha Berg was asked to speak on CBS's *Early Show* about Tilikum's return. Having him perform again was good for him, she told coanchor Erica Hill, but "you're risking that another tragedy could happen. It might not be today, it might not be tomorrow, but somebody could make a mistake and there could be a fourth death." Sam's own awareness of Tilikum's dark past had been "very limited" while working there, she said. "In fact, there had been thirty incidents between killer whales and trainers prior to my being hired at the park. I didn't know about any of them until I left SeaWorld."

Tilikum's return to show business was given mostly positive reviews—though nobody could report the story without mentioning that three people had perished in his tanks.

On other fronts, things were shaping up to SeaWorld's advantage. On April 14, the administrative law judge overseeing *Secretary of Labor v. SeaWorld of Florida, LLC*, Ken Welsch, agreed to postpone the hearing, yet again, from

April 24 until September 19. SeaWorld claimed that most if not all of the discovery documents it had turned over were proprietary trade secrets—maintaining that revealing them in open court would harm SeaWorld's business and aid its competitors. It would take more time to sort through this issue.

SeaWorld was reportedly trying to use its muscle to close some or all of the proceedings to the public. The idea was fiercely opposed by HSUS. Naomi wrote a protest letter to Labor Secretary Hilda Solis and issued an action alert to members asking them to do the same. Other anti-captivity advocates, including TOP and the former trainers, also decried any proposed closing to the public. Whether SeaWorld would win that battle remained to be seen.

Two weeks after the hearing was postponed, another legal difficulty for SeaWorld suddenly fizzled away.

Linda Simons, just like Michelle Dillard, recanted.

Citing what it said was a sworn statement from Simons, SeaWorld quoted the ex-employee as saying that she had reviewed new materials since bringing her claim against the company and that "I now realize SeaWorld acted appropriately under the law."

"As we have said throughout this process, Ms. Simons' allegations were unfounded and we are grateful that she has fully recanted those charges in a sworn statement," SeaWorld said in a prepared statement. In a separate communiqué Fred Jacobs noted that Simons admitted that her "earlier opinions and allegations were formed and made without the necessary facts and were not accurate." Jacobs likewise said that Michelle Dillard "admitted that she had no factual basis to support her allegations and dropped her claims as well."

Word of Simons's turnaround swept through the anti-cap community within seconds. Most people assumed she was paid off. Perhaps so. But attorney Maurice Arcadier had previously alleged that SeaWorld was applying intimidation tactics against his client, including a threat to make her reimburse them for all legal fees if she lost. After Simons folded, Arcadier would no longer comment, except to note that Simons had disavowed her allegations about SeaWorld's behavior during the Brancheau probe, leaving open the question of whether she had agreed to recant and drop her wrongful termination complaint in exchange for a settlement of her defamation suit.

In the spring of 2011, Naomi decided to write her own white paper on orcas in captivity. One of her main goals was to update the annual survival rates (ASRs) of captive killer whales and compare them to the rates reported by Small and DeMaster in 1995. If SeaWorld's claims were accurate, and animal husbandry practices had improved over time, then the ASR of captive whales should have improved as well.

Naomi hired a graduate student consultant with expertise in statistics and had her communicate with Colleen and John at TOP. John's database of all known captive orcas worldwide was far more complete than the government's Marine Mammal Inventory Report, with all its holes and omissions.

When Naomi completed a first draft, she shared copies with Lori Marino, Paul Spong, and Tim Zimmermann, to solicit their comments. She wanted to have the white paper finished by September, when SeaWorld and the feds fought out their differences in court. It was sure to be a magnet for the media, and a fine venue for unveiling the report.

It was getting hard to keep track of all the current and former SeaWorld employees who had stepped forward—some quietly, some quite vociferously—and challenged the company in the wake of the Brancheau tragedy. In May of 2011, their ranks grew by one.

Yet another current trainer who did not want to be named sent a message to Colleen via TOP. Colleen sent a copy to Jeff. "I would proceed with a bit of caution just to make sure this person is not pulling some BS," she said. "But if this guy/girl is for real, it's wonderful news indeed!"

Jeff was suspicious, though he was certain the contact worked at Shamu Stadium—or was close to someone who did.

"After giving this a lot of thought and consideration I have decided that I want to anonymously give information I have," the person had written to Colleen. "Why? Because I thought that SeaWorld would have truly learned something from what happened to Dawn but we continue to go the wrong direction and down the wrong path, and I am impressed with the investigative reporting I have seen from you."

The source, who claimed to have years of experience at SeaWorld, asked for Jeff's e-mail address. "I know the hearings with OSHA have been pushed to September now and I want to help Jeff who I know will be called as an expert witness for OSHA. Some information I can provide may be helpful in those proceedings. . . . The focus of my intention, and I believe your work as well, is to provide the best life possible for these animals who have given so much."

Jeff decided it was worth responding. "Hello Sir," he wrote, not knowing the person's gender or park location. "Thank you for reaching out. It must be tough. I'm not sure that anything provided anonymously can be of use. However please feel free to give your accurate and current perspective. I agree that the animals have given a lot."

The trainer confessed to an initial sense of betrayal when Jeff first spoke out against SeaWorld, "but then I actually read your and John's paper and watched your interviews and I had an entirely different view. . . . How can you get angry when it's the truth?"

The trainer shared that sentiment with others at work, adding, "I need to be careful before I start getting pulled in behind those closed door meetings because I'm singing your praises and applauding the Orca Project . . . definitely a slippery slope there. I just want to say thank you. And if there is anything that you may feel I can answer and that could help you, feel free to ask. I want the truth to come out in the OSHA hearings and the public able to witness it."

As for the trainers at Shamu, "we have all been kind of just hangin' in there waiting to see what happens with these hearings," the e-mail said. "Will we swim again? Will we get a higher pay grade? If we get the answers to these that we feel we will get, I believe several trainers at Shamu from each park will jump ship."

Then the person suggested something that Jeff had suspected: Dawn did not let her hair "drift" into Tilikum's mouth; instead, the whale most likely grabbed her by the arm, a seemingly small but important distinction.

"One interesting response to the ponytail issue and how he pulled her in," the source wrote, "a very high up individual over training conceded that Tilikum may have very well pulled her in by her arm and not the ponytail and noted that they just didn't know for sure. But they come across as sure to the public. . . . Good luck to you Jeff and for having the balls to advocate for the whales and us trainers," the person signed off. "Something's gotta give."

In a follow-up e-mail, the trainer complained about construction under way for the new orca show, "One Ocean," which would soon replace "Believe." The new production was costing each park millions, "with absolutely no benefit or reward being given to the whales. This new, fancy, and ridiculously expensive set does not enrich the lives of the whales at all, and they are the ones who suffered throughout construction, most often being confined together in a single pool with all the sound of jack hammering and concrete cutting."

Jeff asked the trainer how SeaWorld had reacted when Keto killed Alexis. The three parks went "dry" for only a day or two before returning to water work, the trainer said, "without even knowing the specific details on the incident." There was no hard-copy report, and "details didn't even start to come out until after curator Mike Scarpuzzi flew to Spain because Brian Rokeach was so distraught. They were obviously not overly concerned about our safety at that time, to put us back in the water."

As for Dawn and Tilikum, the insider said it was "infuriating" that blame was being placed on her for being too complacent around the giant whale. "The truth is, those who were on Tilly's close contact team had interacted with him that same way for years; a trainer even had her profile pic on Facebook of her laying face to face with Tilly the same way."

The truth was, being *anywhere* near Tilikum—or "over the wall," as the trainer put it—put anyone in danger. If someone were to be grabbed while irrigating his teeth, for example, the person would "ultimately be blamed for being too complacent," for taking an eye off the whale, perhaps, or getting a forearm too close. Whatever happened, it was always the trainer's fault.

"I believe you are a perfect spokesman for the cause and I hope that you guys have something good up your sleeve in preparation for this case," the source told Jeff. "I am not looking to ever ask you any questions regarding that because I don't want you to question my intentions. As I said before, we all need to be cautious."

In another e-mail, the insider complained about poor worker morale and low pay. After Dawn died, some trainers requested a transfer out; others were planning to do the same. But new safety protocols required that more trainers be on-site at all times, so all transfer requests were allegedly denied. "People seem to be getting more disgruntled by the minute." Nobody was getting rich: Senior pay, including a $5-an-hour "water work bonus," was $23 per hour in Texas and $26 in California, the insider wrote. That was better than it had been before Kasatka attacked Ken Peters in 2006, but was "still disgusting considering the responsibility and abuse your body takes."

Despite the bruising work, there was "huge pressure to suck it up and not to report injuries . . . or else. And there are definitely situations with the whales that are not written up—and for a variety of reasons" they tended to get "swept under the rug."

Jeff asked his source if he could share their e-mail exchanges with John Black, the Department of Labor attorney who would lead the courtroom battle against SeaWorld in September. The trainer agreed.

Jeff passed the information along to Black. Coming from an anonymous source, it was of no use at trial, but it certainly provided insight behind the scenes at SeaWorld. To Jeff, it seemed as if little had changed since he worked there.

"You are right on the money with your assessment," the trainer wrote back. "Regardless if it's 1995 or 2011, it's the same."

Naomi Rose was headed back to British Columbia. She'd been to Vancouver on business over the years, but not to her field site since 1994. In May of 2011, she and Chris were on a plane to Victoria, for the Society for Conservation Biology's International Marine Conservation Congress. Chris was president of the society's Marine Section and was serving as the conference program cochair.

Naomi was scheduled to make a presentation on science and policy in Victoria, but Chris and his colleague Leslie Cornick wanted to turn it into something more. Leslie tried to organize a debate on marine mammal captivity as an evening event but, despite a widespread search, could not find anyone willing to take the pro side. It was both gratifying and infuriating.

The debate became a discussion. Naomi shared the stage with Paul Spong, who delighted some one hundred conferees with his stories about Skana, Hyak, Springer, and other orcas. One audience member, an older researcher who strongly supported cetacean display, spoke up during the Q&A, and the one-sided discussion briefly took on the liveliness of a real debate. Alexandra Morton, the naturalist and environmental activist, gave a conference keynote address, speaking of salmon farms throughout the killer whales' habitat, where salmon fry were infected with farm-generated diseases that in turn infected wild salmon. She spoke passionately about her efforts to curb the aquaculture industry in British Columbia, which was indirectly devastating the whales.

After Victoria, Naomi and Chris headed up the island to Telegraph Cove. The place had changed: The cove itself was now choked with pleasure-boat slips, awaiting the summer crush. An unsightly two-story hotel crowded one side of the inlet; custom-built homes were going up on the bluff overlooking the strait. The couple stayed in one of the original wooden bungalows built on stilts over the water.

Because it was still May, the Northern Residents had not yet made their appearance, though that hardly put a damper on the trip. Naomi got to show Chris her old stomping grounds and catch up with friends from her field-study days. Telegraph Cove might have changed, but the people she knew were still the same.

It was wonderful to spend time with Jim Borrowman and his second wife, Mary. Jim took Naomi out on a highly nostalgic spin on the *Gikumi*. They called at Alert Bay and picked up Paul Spong and Helena Symonds before setting out for Hanson Island, OrcaLab, and Blackfish Sound. Dall's porpoises played at the bow at one point, thrilling all aboard.

Naomi and Chris also drove over to Hidden Cove Lodge (a road had finally been cleared) and paid a surprise visit to Dan and Sandra Kirby. It felt great to be back. One morning, Bill MacKay and his wife, Donna, took the couple out on their state-of-the-art tour boat, the *Naiad Explorer,* which was fast enough to take them all the way down to West Cracroft and back before noon. The island had not changed. Naomi pointed out where Cliff Camp and Boat Camp used to be. Across Johnstone Strait, the Mike Bigg–Robson Bight reserve sat in its isolated glory.

In May, there would normally have been at least some salmon running, but today none were to be seen. "These damn Norwegian fish farms," Bill said as he navigated the vessel. "They're wiping out the wild salmon," he added bitterly.

If the salmon farms could somehow be stopped, the whales would have a fighting chance.

• • •

Naomi had worked to convince federal regulators that the situation at Loro Parque was dire enough to warrant intervention by US officials. She had spearheaded the lobbying efforts by HSUS, WDCS, and the Animal Welfare Institute to bring official attention to Suzanne Allee's report on the SeaWorld whales in Spain.

It had been months. Finally, on June 2, 2011, she got an answer.

A NMFS official, Michael Payne, wrote to Naomi advising her that the agency was in receipt of Naomi's letter, Suzanne's statement, and SeaWorld's response denying any serious problems at Loro Parque.

NMFS and the Marine Mammal Commission had met with SeaWorld executives to review HSUS's concern. "SeaWorld explained their oversight role over the care and maintenance of the killer whales," Payne said, and maintained "direct oversight" over their care through regular conference calls, site visits, and staff exchanges.

Based on that, NMFS was "satisfied that the maintenance of the killer whales at Loro Parque is in compliance" with MMPA provisions on public display.

"We note that HSUS has not visited Loro Parque," Payne wrote with a whiff of bureaucratic condescension. "Should you choose to do so, we would encourage you to report back your findings to both NMFS and the Marine Mammal Commission for consideration."

Many lives were altered by the terrible events that occurred that gray February day at SeaWorld Orlando. That of DC-based journalist Tim Zimmermann was no exception. The attack and its aftermath sent him on a dogged investigation of SeaWorld, captive killer whales, and the entire display industry. After his "Killer in the Pool" article appeared in *Outside* magazine, Tim launched a series of hard-hitting exposés, called "Diary of a Killer Whale," on his own website.

In July of 2011, Tim completed another article on the industry, this time digging deep into the disaster at Loro Parque that left young Spanish trainer Alexis Martinez dead in the orca pool. The article, "Blood in the Water," was published online by *Outside* magazine.[8] Tim had obtained and translated the corporate incident report and the Canary Islands government inquest report and brought new details of the death to the English-speaking world.

On Christmas Eve of 2009 Brian Rokeach, the senior trainer from San Diego who was on rotation at Loro Parque, was rehearsing a new show with Loro Parque staff, including trainer Alexis Martinez. Rokeach was working as the control trainer from the main stage as Martinez joined Keto in the pool.

Martinez initiated a stand-on behavior (balancing atop the whale's rostrum as the animal rises vertically out of the water). Keto drove Martinez with proper force, but as he rose, he leaned slightly to one side. Martinez toppled off, and Rokeach decided not to bridge Keto because of the error.

Martinez tried the stunt again, but Keto began twisting as he rose toward the surface. Martinez stopped the behavior and gave Keto an LRS (least reinforcing scenario)—or a three-second neutral response, which is usually followed up with an opportunity for the whale to remain calm and attentive. That in turn is reinforced with a reward. It worked: Keto obeyed a call to go to a shallow ledge across from the stage, where another trainer gave him some fish.

Martinez said he was going to perform a haul-out, in which Keto would propel him down into the water and then up onto the stage. But Keto went too deep, and Martinez aborted the behavior. He held out his hand, a signal for Keto to follow him, and they floated to the surface in unison. Martinez again delivered an LRS to Keto for having failed. But Keto did not respond properly this time around.

The whale remained calm, but he moved in between Martinez and the stage. Martinez requested an underwater tone to call Keto over to Rokeach at the stage. Keto responded. But according to the incident report (translated from Spanish), Rokeach noticed that Keto was "not committed to remaining under control" and growing "big-eyed." Rokeach used a hand target to get Keto to focus on him. Martinez treaded water patiently in the middle of the tank, waiting for the right moment to swim toward the slide-over (a ramp that links the main pool with those backstage) and exit the water. As Keto calmed down, Martinez slowly made his move.

But Keto noticed the trainer. He started to lean toward Martinez. Rokeach knew he was about to lose control of the whale and gave him another hand target. The effort failed. Keto charged. The orca reached the young man in seconds, then "rammed him and violently played with his body," a supervisor told investigators.

Keto forced Martinez to the tank's bottom. Rokeach slapped the water—with no response. After several agonizing moments, Keto finally obeyed the call-to-stage signal. Martinez lay motionless down in the depths. "Rokeach sounded the emergency alarm," Tim wrote in the *Outside* article. "Keto took a quick breath, returned to Martinez, and then came back to the surface carrying Martinez' limp body across his rostrum. Rokeach called for the team to get a net while others raced to corral the other three killer whales into one of the back pools."

Keto submerged with the trainer again. Martinez had been underwater for nearly three minutes. Rokeach and another trainer dove in to retrieve him. Martinez was unconscious. Blood spewed from his nose and mouth. "A distraught Rokeach immediately initiated CPR. A defibrillator was brought out, and Loro Parque called for an ambulance," Tim wrote. "But Martinez was never revived."

Loro Parque called the killing an "unfortunate accident" in a statement released in Spanish. An "unexpected reaction of the animal" had led Keto to knock into Martinez, compressing his chest and causing asphyxiation. Keto's behavior "did not correspond to the way in which these marine mammals attack their prey in the wild." It was instead merely a "shifting of position." But Martinez's autopsy told another story. His was a "violent death," it said. He had sustained multiple cuts and abrasions, a lacerated liver, badly damaged internal organs, a collapse of both lungs, fractures of ribs and his sternum, and puncture wounds "consistent with the teeth" of killer whales.

"The corporate incident report," Tim wrote, "in effect acknowledges the imperfect understanding between man and whale." It concluded that Keto's behavior was "unforeseen" and "incorrect," Tim noted. "'Incorrect' is a wholly inadequate description of what Keto did to Alexis Martinez. But it's the 'unforeseen' part that should make any trainer nervous."

SeaWorld had sent four young orcas to Loro Parque and three of them—Keto, Tekoa, and Skyla—now had a "history of incidents," Tim wrote. The fourth whale, Kohana, gave birth to her first calf, Adán, in 2010—she rejected him immediately and never nursed him. He was raised by hand. Now five young whales were living without the guidance and supervision of more mature members of their species.

The Canary Islands labor ministry, meanwhile, reached a similar conclusion to OSHA's findings from Orlando. The danger of water work lay "precisely in the interaction with an animal that weighs more than [sixty-six hundred pounds] and is in its natural environment." The only way to eliminate the risk to workers was "prohibition of the activity."

As for SeaWorld, whether water work should be ended or "severely constrained by safety measures," Tim wrote, "will be decided by the OSHA proceedings."

Officials at NMFS may have felt little compulsion to investigate SeaWorld's "loaner" orcas in Spain, but for many US activists that didn't excuse the agency from ignoring what was happening at home.

When Kalina died the previous October, SeaWorld told ABC News, "Animals in our care live and die just as they do in the wild." But members of the Orca Aware confederacy were discovering more and more evidence to show that some killer whales in SeaWorld's care neither lived nor died "just" as they do in nature. Wendy Cooke of the Without Me There Is No You site found a paper in a scientific journal that revealed the true cause of death for Kanduke, the big male Transient who died in 1990. Wendy sent the study to John Kielty at The Orca Project, who had obtained Kanduke's necropsy report via FOIA. John then sent all the information to Courtney Vail, who wrote about Kanduke on the WDCS blog:

"WDCS is shocked to learn of another way that captivity kills." Kanduke had died "suddenly and unexpectedly after deteriorating very quickly." In one of the scientific studies actually conducted on a SeaWorld orca, company vets were puzzled by his rapid decline and sent tissue samples to Yale University for assessment. In 1993, a study funded in part by SeaWorld and published in *Clinical and Diagnostic Virology* reported finding St. Louis encephalitis (SLE) virus, an avian pathogen transmitted by mosquitoes, in Kanduke.[9] No marine mammal had ever before been found with SLE.

The necropsy and article suggested "a direct causal link between orca confinement and premature death," Courtney wrote. "There was an outbreak of St. Louis Virus in Florida during the summer of 1990, when the whale was bitten by mosquitoes that were infected with this virus." She said former trainers Jeff Ventre and Carol Ray witnessed mosquitoes on Kanduke's back and the sunburned and peeling condition of his skin. In contrast, wild whales "would not be stationary and suspended with their backs at the water's surface for hours each night, thereby exposed to Florida mosquitoes carrying the avian virus."

Kanduke's case, it turned out, was not an isolated incident. Another study coauthored by SeaWorld researchers, published in *Emerging Infectious Diseases*, described the death of fourteen-year-old Taku, a male orca who died suddenly in San Antonio in 2007. The cause was West Nile virus. Equally alarming, all six of the other orcas in San Antonio tested positive for the virus, which can be transmitted to people via mosquitoes. In 2007, Jeff discovered, Texas was among the five states with the highest number of West Nile cases.

Again, a whale in British Columbia or Iceland was highly unlikely to contract a tropical disease such as West Nile virus. "From what I can tell, and after browsing several papers and articles, these types of mosquitoes are limited to coastal marshes and mangrove swamps," Jeff wrote to the group. "Wild killer whales would not be exposed to such vectors of disease transmission. And thus it could be stated that the exotic infections that have killed at least two captive SW orcas . . . were directly attributed to their confinement in unnatural spaces."

Throughout 2011, the Orca Aware group kept growing. Several new people, including scientists Naomi Rose, Lori Marino, and Ingrid Visser, had joined the discussion. The alliance had spilled out of the confines of its

Google home and morphed into an informal e-mail chain, with people writing back and forth several times a day. The exchange of news, gossip, and inside intelligence collected from a clandestine cadre of well-placed informants was voluminous.

Jeff Ventre was amazed by the aggregation of individuals who had been drawn together by their passion for killer whales. They came from distant corners of the country and the world, each with unique insights, experiences, and talents.

It was time for a powwow. Jeff had begun planning a mass gathering earlier in the year, and now, in mid-July, it was actually happening. Nearly the entire group was headed for San Juan Island for several days of eating, drinking, kayaking, and watching and talking about whales. Naomi decided not to go: She had just been to British Columbia, had been traveling a lot since then, and had run out of personal leave.

Most people stayed in some cozy cabins at Snug Harbor, on the island's west side near Ken Balcomb's Center for Whale Research. Jeff and his girlfriend, Chica, were among the first to arrive, along with Jeff's old friend Todd Bricker and his wife, Jamie. They were soon joined by Samantha Berg and her husband, Kevin Meddleton; Carol Ray and her husband, Eric Peterson; Jeff's sister, Kim, and her partner, Marty Neese; and Dean Gomersall, who used to work at Whale and Dolphin Stadium with Jeff and the others and was now anti-SeaWorld, and his friend Breanne Antonius. Dr. Lori Marino flew in from Atlanta, joined by her significant other, Michael Mountain, cofounder of Best Friends Animal Society and author of a blog on animal rights and other issues called Zoe Nature.

Tim Zimmermann also traveled to the island, along with a film crew that had retained him to work on a feature documentary about the entire Tilikum/captive-orca saga; Gabriela Cowperthwaite was the director and Manny Oteyza the producer. Already on San Juan Island were Howie Garrett and Susan Berta, Ken Balcomb, Astrid van Ginneken from Holland, and Stefan Jacobs from Germany.

Days were spent whale watching from chartered boats, or from the cliffs of Lime Kiln Point or American Camp. In the evenings, everyone would gather for dinner, drinks, and whale talk. They often socialized at the ocean-view home of Candace Calloway Whiting, one of many people who'd moved to the island "to be close to the orcas, and the water," she said. Candace, who had studied and trained dolphins, seals, and orcas, volunteered at the Center for Whale Research and authored a blog about cetaceans for the *Seattle Post-Intelligencer*.

One evening after dinner, the black fins of Resident whales were seen from Candace's home, rolling along the island's edge in the lingering summer dusk. A researcher at NMFS's National Marine Mammal Laboratory in Seattle, Dr.

John Durban, presented recent data on orca ecotypes in Antarctica, and Howie spoke about the importance of symbols in the formation of orca culture. Lori Marino followed up with an engaging talk on orca intelligence, followed by the screening of a documentary, *The Whale,* produced by Michael Parfit and Suzanne Chisholm, a couple from the Canadian side of Haro Strait, who boated over for the evening.

The Whale told the emotional saga of Luna, the young killer whale who was separated from his family deep within a fjord on Vancouver Island. With no other orcas to bond with, Luna began bonding with boaters and other people. The film's executive producers were Ryan Reynolds and Scarlett Johansson, and the narration was provided by Reynolds, who grew up in British Columbia. *The Whale* was a *New York Times* Critics' Pick and *The Seattle Times* called it "a gorgeous and provocative film."

One morning, some of the group set out into Haro Strait on a boat captained by Spencer Damico to see the Southern Residents. Michael Mountain described the outing on the Zoe Nature site: "I'm hearing crackling voices coming in on the radio saying that the K and L pods are heading south in our direction from Canada. 'That means we could be really lucky,' Spencer says. 'They may all be joining up for a superpod.' "

Superpods, Michael wrote, were "like orca conventions, with lots of excitement among the whales as they share information, learn about each other and perhaps do things like learning how to live together peaceably." Younger whales got to meet peers from other pods, "and mothers encourage their sons to 'go talk to that nice girl over there.' "

Spencer shut off the engines. Members of K and L pods drew near. " 'There's one at seven o'clock,' someone says as we start seeing the orcas surfacing, breathing, then diving again," Michael wrote. " 'Three o'clock.' . . . 'Five o'clock.' We soon stop pointing them out—there are just so many of them." Suddenly, J2, or "Granny," the matriarch of J pod, now estimated at a hundred years of age, surfaced a few meters from the boat. It was "like getting to meet your favorite celebrity on the red carpet at the Oscars. We've actually met the most remarkable orca on record."

As the reunion wound down, Jeff remarked how the group was like a bunch of orcas, coming together for a gathering. Sam, after returning home, wrote to the gang that Jeff's comment had made her think about their group: "We created a whole new version of the Superpod consisting of scientists, researchers, activists, journalists, writers, film-makers, and ex-trainers. We're a Superpod now, even when we're not together, but we need a name."

Michael Mountain had an idea. "How about 'Superpod'?" he suggested. "As in 'The Superpod is meeting at Candace's.' If it sounds at all self-aggrandizing, it's also obviously slightly tongue-in-cheek, so it would work. It's catchy. And a bit alarming if you're SeaWorld."

• • •

OSHA's legal team was almost ready for trial. It had been difficult to secure an expert witness. Even most scientists who had provided written statements to OSHA did not wish to be further dragged into a showdown between SeaWorld and the feds. Dr. David Duffus, the University of Victoria professor who had acted as foreman in the 1991 Keltie Byrne inquest, agreed to testify on behalf of OSHA.

The government's list of other potential witnesses was long, including many current SeaWorld employees. Chuck Tompkins and Kelly Flaherty Clark were on it, as were trainers, including Jay Topoleski and Lynne Schaber, who were at G Pool on February 24, and from San Diego, Ken Peters, who was attacked by Kasatka in 2006, and Brian Rokeach, who was attacked by Orkid the same year and had been at Loro Parque when Keto killed Alexis Martinez.

Three former trainers were also on the list: Jeff Ventre, John Jett, and Kim Ashdown, who had submitted her resignation to SeaWorld Orlando in December of 2009, shortly before Dawn died.

According to Ashdown, after the Ken Peters incident in 2006, animal trainers were told that only those who wanted to work with killer whales would be sent to Shamu Stadium. "I voiced my preference to not ever move to Shamu Stadium," Ashdown wrote, but was transferred there anyway. "It was a very intimidating pool to me, not only to be in it with such large animals but to know that I was responsible for other trainers' lives when an incident occurred. There were several trainers that were asking for a transfer out of Shamu. We were put in life and death situations every moment we were there."

Two months before Dawn died, Ashdown once again asked for a transfer. The request was denied. She said higher-ups informed her that "due to admitting my concerns and my anxiety, my performance would be highly scrutinized, and one mistake was grounds for termination." She informed her bosses that she was neither sleeping nor eating and was "consistently stressed out being at Shamu Stadium," a situation that could affect safety, she said. "And they told me they were watching." Ashdown said she could no longer "work for a company that was waiting for me to make a mistake in an environment where life and death hangs in the balance on a daily basis."

But even as new legal headaches surfaced for SeaWorld, old ones faded away. On September 7, 2011, SeaWorld attorneys crushed the Connell family of New Hampshire in court. The family had sought damages for their son Bobby's emotional distress. Judge Julie O'Kane dismissed their suit with prejudice, meaning they could never refile their claim. O'Kane could not find a single case where "a complete stranger" to the injured party was allowed to proceed for emotional distress, she said, calling the family's legal quest "futile."

Naomi finished her white paper, "Killer Controversy: Why Orcas Should No Longer Be Kept in Captivity," in time for the big OSHA trial set for September.[10] It was, in many ways, the net product of her twenty years of fighting the industry. "The science is in," she said in an HSUS statement issued with the report, "and we should realize that nothing—not profit, not education, not conservation—can justify keeping this large, social, intelligent predator in a small box."

HSUS said the paper highlighted "the growing body of scientific evidence showing that orcas do not adapt to captivity," including:

- Data collected up until 1992 showed captive orcas had higher mortality rates than wild ones; now, new analyses of data up until 2010 had confirmed that the situation had not improved in the past eighteen years "and in fact has worsened."
- Captive female orcas were giving birth too young and too often, leading to a higher death rate among both adults and infants.
- The most common cause of death for captive orcas was infection. Chronic stress might also be an important factor in weakening the animals' immune response.
- Captive orcas had poor dental health compared to wild whales, "which may be another factor in their susceptibility to fatal infections."
- Orcas in captivity were more aggressive toward each other, and females behaved abnormally toward their calves more often than in the wild.
- Captive orcas had "seriously threatened the lives and safety of dozens of people," four of whom had died.

"The predicted improvement in survivorship [hasn't] materialized," Naomi wrote in the report, "despite supposedly continued improvement in husbandry techniques." The most "parsimonious explanation" was that "orcas are *inherently* unsuited to confinement." Kalina, who died at age twenty-five, "set the current upper limit for captive-born orca longevity. No improvements or advances in training, nutrition, veterinary care, husbandry, or transport can 'fix' this poor survivorship."

The paper called for the gradual phasing out of killer whale displays. "The population of captive orcas can be eliminated through attrition, with the animals currently alive evaluated for continued display, retirement to sea pens, or rehabilitation and possible release to the wild if appropriate." HSUS was not calling for a "blanket closure of oceanaria," but a phasing out of orca shows "taking as long as three decades, giving oceanaria sufficient time to repurpose their orca enclosures." It was simply no longer justifiable to perpetuate orca shows for entertainment, or even for education, "especially when that education is biased toward information that supports a corporate narrative rather than good science."

The report concluded, "No more orcas should have to die prematurely. No more trainers should be put at risk. It is time to accept that we have been wrong in our assumptions. The orcas deserve no less."

September in Orlando can be blindingly hot. Naomi Rose, dressed in shorts and a sun hat, strode through the clinging humidity across the wooden footbridge that led to Shamu Stadium with a friend, Nancy Yates, who lived near Orlando. In town for the OSHA trial, Naomi figured she would visit SeaWorld once more to take in the new "One Ocean" show that had replaced "Believe." Later that day, they would meet up with Colleen Gorman at the park.

The show seemed unimpressive, Naomi thought, with trainers still doing dry work and bopping around onstage in their sleek ocean-blue wet suits, as if they might dive in with the whales at any moment. It ended with the big splash segment, and Tilikum performed splendidly. He looked strong and healthy, at least from the bleachers.

Afterward Naomi and Nancy tried to peer through the fence of G Pool, which was undergoing renovation. What appeared to be air bags the size of Volkswagens were being installed in the drained tank. In an emergency, they would inflate and lift the bottom within sixty seconds. If a whale had a trainer in its mouth, rescuers could reach the scene instantly, rather than in the twenty minutes it took to corral Tilikum into the medical pool when he attacked Dawn.

"This must've cost millions," Nancy said with a grimace. "SeaWorld is definitely going to put their trainers back in the water. They wouldn't be doing this otherwise."

The OSHA hearing was set for the following morning at the modern Seminole County Courthouse, thirty miles north of Orlando in the nondescript suburb of Sanford. OSHA had a difficult time securing a venue: The sensational Casey Anthony murder trial had ended in July, and courthouses were wary of overflow crowds and media-circus disruptions.

The crowds did not materialize, but the media showed up en masse. TV trucks from local stations fed regular dispatches via satellite to the networks in New York. CNN and Fox News Channel sent crews; AP and Reuters sent reporters. Jason Garcia from the *Sentinel* also attended. Naomi handed them copies of her orca white paper and conducted several interviews on the courthouse grounds, under the furnace of tropical heat.

Colleen Gorman and Courtney Vail of WDCS were there, as was a PETA lawyer named Jared Goodman. PETA was about to file a lawsuit against SeaWorld on behalf of its five wild-caught orcas, Tilikum, Katina, Kasatka, Corky, and Ulises, seeking their retirement to a sea pen. Sam Berg, Carol Ray, Howard Garrett, Ric O'Barry, and New Zealand's Dr. Ingrid Visser were representing the whales as "next friends."

PETA's argumentation was novel, and highly controversial. The group contended that the five whales had been taken from their home and held in "involuntary servitude" in violation of the Thirteenth Amendment to the US Constitution, which prohibited slavery but did not specifically mention humans. SeaWorld had moved to dismiss the case. Some legal scholars and African-American leaders blasted PETA, but a few experts said the suit had merit. "People may well look back on this lawsuit and see in it a perceptive glimpse into a future of greater compassion for species other than our own," opined Laurence Tribe, the famous Harvard law professor and constitutional scholar.[11] Whatever the outcome, PETA claimed, the hearing on SeaWorld's motion would mark the first time the constitutional rights of animals had been argued in a US court.

Naomi finished her last interview and gratefully walked into the blast of cold air inside the building. She entered the courtroom and sat down on a hard oak bench. It was going to be a long week.

The marine biologist nodded hello to OSHA investigator Lara Padgett and waited in silence for the trial to begin. What a momentous year and a half it had been, she thought. The media glare, the various investigations, the congressional oversight hearing, Tim Zimmermann's exposé articles, the radio and TV debates, the ex-trainers and whistle-blowers stepping forth, the lawsuits and the general public awareness about killer whales had reached critical mass. None of this would have transpired before Dawn died. It was hard to believe it had transpired at all.

Now an extraordinary courtroom drama was about to unfold. It had the potential to forever change the way SeaWorld did business. Captivity was not on trial here. Important as the hearing might be, it was limited to the safety of SeaWorld trainers who performed with orcas in the water. While an end to water work might chip away at SeaWorld's revenue, it would hardly deliver a fatal blow; trainers had been out of the water for nineteen months and Shamu Stadium was still packing them in.

Naomi did not hate SeaWorld, nor did she wish to see the place closed down. Her opposition centered on cetaceans in captivity, but other things about SeaWorld were worthy. The manatee exhibit was educational, for instance, and the little-known aquarium was exemplary: Beautifully designed and filled with exotic species, it also had well-lit signage explaining interesting facts about each type of fish in the tank.

Naomi had no doubt that some children were inspired by their SeaWorld visits to learn more about marine mammals and conservation, or even to become marine biologists. She knew about SeaWorld's animal rescues and grants for habitat restoration. She knew about the educational materials, some of them flawed, including the Ask Shamu web page that had kid-friendly information on orca diet, coloration, and size. But did the rescue of manatees and production of high-tech materials actually justify the confinement of killer whales to watery cell blocks?

SeaWorld supporters implied that because of human activity the ocean was now too dangerous; it was no longer a fitting place for killer whales to live. Captivity was thus an important undertaking. SeaWorld argued that whales living in pools helped whales in the wild by raising awareness, boosting conservation, and allowing research on captive animals in their corporate collection.

But how much of those efforts were really benefiting wild killer whales? Why were the Northern Residents still threatened and the Southern Residents endangered? How much did people actually learn about saving the whales, and what, exactly, was SeaWorld doing to help J, K, and L pods recover?

At the company's website, a few general conservation tips were given, such as "never flush medications or put harmful chemicals down the drain" and "always recycle plastic." Another way to salvage the sea was to "throw an ocean party [or] make some sea-themed jewelry." SeaWorld also suggested that people "meet a marine animal" because it was easier to care about animals "once you've felt a personal connection to them." Meanwhile, a literature search Naomi directed revealed just under thirty papers on SeaWorld orcas. At least half of them pertained to husbandry issues, with little bearing on wild whales.

In contrast, conservation materials at the Vancouver Aquarium (which, ironically, no longer housed orcas) were hard-hitting and thorough. They provided detailed science on the threat to orcas from declining food supplies, pollution, vessel traffic, and human-caused noise. Yes, the ocean *had* grown more perilous for killer whales, the aquarium was saying. But unlike SeaWorld, it was also telling people specifically what they could do to reverse the trend.

The whales were not a lost cause whose only salvation was public display. The aquarium urged people to join the British Columbia Wild Killer Whale Adoption Program to "help support important research on wild killer whales and their conservation!" People could assist salmon habitat by supporting one of the Streamkeeper groups; they could reduce marine debris through the Great Canadian Shoreline Cleanup, or help marine conservation groups "teach others about protecting the oceans and the whales."

The message was clear: The answer to problems such as pollution and orca-habitat spoilage was not to build a Noah's ark of "human care." The answer was to heal the sea.

Species can and do come back, Naomi knew, and perhaps the Vancouver Aquarium's efforts were having an impact: In the previous five years, the Northern Resident community had experienced record population growth, despite all its challenges.

Back at SeaWorld, killer whales had over the decades generated billions in profits for the company and its investors. If SeaWorld had returned a good portion of that accumulated wealth into fighting pollution in Puget Sound, or helping salmon rebound by tearing down dams or eliminating fish farms, Naomi knew her mission would have been undermined. Opponents might accept captivity if the profit from orca display went toward saving orcas in the ocean. But that wasn't the case.

It was time for the beginning of the end, Naomi and so many of her allies felt. In show business, most performance animals were allowed to retire, to live out their lives in quiet dignity, freed from the daily routine of working for food, posing for cameras, or doing backflips on command. They had worked hard, performed well, and kept people employed. They had earned their retirement.

But at SeaWorld, whales were expected to perform virtually until they dropped dead. It was grossly unfair, Naomi thought. It was inhumane.

And *look* at how these noble animals died—so many of them prematurely, so often with gastric diseases, immune deficiency, shattered teeth, or deformed fins, despite their world-class veterinary care. Some were lost in brutal acts of aggression or self-injury, some in agonizing childbirth, and some from the unlucky bite of a disease-laden mosquito. One orca died when a gate crushed his skull. Was it any wonder they perished at two and a half times the annual rate of their free-ranging cousins? For that matter, was it a wonder that some of them harmed their trainers?

To Naomi, her allied scientists and activists, and ex-trainers such as Jeff Ventre and John Jett, the best solution was to end breeding programs and gradually retire captive orcas to sea pens—where they could still receive food, care, and human attention, but also live a more normal life in a more natural environment. Some might even become semi-independent, like Keiko. SeaWorld itself could run the sites and charge people to watch the whales swimming, foraging, breaching, and resting, though not performing triple bows and rocket launches.

Sea pens would be good for the whales, good for the public, and good for SeaWorld.

Meanwhile, technological advances in animatronics were now at the stage where robotic orcas might replace living ones. Robots required no food or veterinary care. They could perform all day long. They could take guests on thrilling rides at a cash premium for SeaWorld. And they would be highly entertaining to watch—even in a society saturated in entertainment options. Perhaps most important, robot Shamus would never hurt each other, or any people.

As lawyers, spectators, and OSHA expert witness Dave Duffus filed into the courtroom, Naomi reflected on what had brought everyone here today: a whale named Tilikum and his fateful encounter with a trainer named Dawn. Dawn's husband, Scott, walked in, the toll of personal pain carved deeply on his face. His world had been shattered.

Now, lawyers would battle over SeaWorld's culpability in what was perhaps the most notorious killing by a captive animal in human history.

Would there be justice in the Brancheau case? Naomi was not one to predict the outcome of a trial. But, she wondered, what about justice for Tilikum? What was the verdict for him?

Colleen Gorman had been sending e-mails to the orca group lately about animal advocates in Canada and Iceland who were actively pursuing a plan to get Tilikum out of Florida. According to Colleen's sources, the prime minister of Iceland had taken a personal interest in restoring the whale to his native waters. Some $50,000 had reportedly been secured to build an enclosure for Tilikum.

Naomi was skeptical. Talk of releasing Tilikum, especially at this late date, seemed far-fetched. Even if Iceland did want him back, he was still the property of SeaWorld.

She didn't think the PETA case had much chance, either. The likelihood of Tilikum's returning to Iceland seemed as remote as the fjord from which he was taken.

But what if the capture of the young calf had never occurred? Tilikum might still be swimming free in the frigid waters of the North Atlantic, chasing his cherished herring, perhaps alongside his mother. He might be surrounded by siblings, nieces, and nephews, and his grandmother might still be leading the pod.

An oceanic Tilikum would be gliding through his boundless home with fearless power and majestic grace, his fin erect, his teeth intact, his interactions with humans minimal and nonlethal. There would be no need for gelatin or Tagamet, antibiotics or isolation.

And of course, if Tilikum had never been wrenched away from his family and friends, entirely for the amusement of humans, the family and friends of Keltie Byrne, Daniel Dukes, and Dawn Brancheau might not be grieving to this day.

Tilikum was trying to tell us something. It was time to listen.

Epilogue

As federal administrative law hearings go, *Secretary of Labor v. Sea-World of Florida, LLC,* turned out to be a fairly engrossing tribunal. The mahogany-paneled courtroom was full, but not packed, as Judge Kenneth S. Welsch called the proceedings to order. Welsch, a sixty-something man with a soft, slightly gravelly voice, maintained a respectful and pleasant demeanor that could best be described as avuncular.

The conventions of an OSHA hearing are curious: Often, when Judge Welsch directed a procedural question to one of the government attorneys, he would refer to "the secretary," as if US labor secretary Hilda Solis were physically present in the room. However unintentionally, the formality served as a not-so-subtle reminder that SeaWorld was not just taking on some faceless agency lost amid the federal bureaucracy—it was challenging a member of Barack Obama's cabinet.

Secretary Solis's lead counsel was John Black, a seasoned litigator from the Labor Department's Atlanta office. Intense and combative, but adept at laying on charm at strategic moments to great effect, the tall, reedy lawyer with dark hair proved to be a nimble adversary. Black was assisted by a younger partner, Tremelle Howard-Fishburne, whose more reserved demeanor and generally softer lines of questioning complemented the fiery Black. Also sitting on the secretary's side were Lara Padgett and OSHA's expert witness, Dr. David Duffus.

SeaWorld's counsel, Carla Gunnin, was once a Labor Department attorney herself and appeared sociable with John Black—they knew each other from earlier days spent on the same side of the fence. Tall and attractive, with a gentle southern accent, Gunnin could also switch gears in court, from aggressive to magnetic, depending on her strategy that day. Usually her long blond hair was pulled back tightly, but some days she let it tumble lightly over her shoulders, softening her appearance remarkably. Also at the SeaWorld table were Kelly Flaherty Clark, curator of animal training at Orlando, and the company's chosen expert witness, Jeff Andrews, associate curator of mammals for the San Diego Zoo.

Dawn's family had counsel representing them at the trial as well, one or two attorneys whose sole purpose was to intervene against the showing of any video or still images from the attack and death scene, should the government try to enter them into evidence.

Everyone was in court to argue two basic questions: Was OSHA justified in issuing a willful violation; and were the proposed abatements feasible? Could they be employed without harming SeaWorld's ability to conduct business? SeaWorld would contend that the citation was unjustified, and the abatements—especially the end of physical contact with orcas without some sort of barrier—unfeasible in the extreme. What follows is a summary of highlights from the trial, which was scheduled to last five days, from the nineteenth to the twenty-third of September:[1]

Day 1: Monday, September 19, 2011. "Let me just say this," Judge Welsch began. "There is no issue before me regarding whether or not the whales should be held in captivity. Also, there is no issue before me as to whether or not SeaWorld was responsible for the death. The issues before me solely involve the OSHA citations that were issued."

Black launched into his opening statement: "Killer whales are large, powerful, and non-domesticated animals. They have the potential to cause serious physical harm or death to people who get near them. SeaWorld's killer-whale training program doesn't change the essential facts that harm or death to people is possible. Their program doesn't eliminate what SeaWorld itself recognizes as a calculated risk." Despite such a risk, SeaWorld was still allowing its trainers "during show performances to work in close and unprotected contact with its killer whales. This type of close, unprotected contact has been made off-limits with only one killer whale, and that's Tilikum."

The evidence was going to demonstrate that "SeaWorld's primary method for trying to keep trainers safe was to train their trainers, and SeaWorld trains its trainers how to recognize and how to avoid potential risks, and then, in effect, tells them, 'Be careful.' " But relying "primarily on training the trainers to be careful leaves gaps," Black said with barely concealed disdain, adding that new employees were obliged to sign a statement confirming that "their own skills are key to being safe." New hires were being obliged to sign what essentially amounted to a company liability waiver, in which they acknowledged the "inherent risk" of working with captive orcas. It was a wholly inadequate line of defense, Black charged.

"Training the trainer is not sufficient for providing protection against that risk." First, the trainer might not even *see* the precursor "and thus miss the opportunity to avoid the dangerous behavior," Black contended. Or the trainer might notice the precursor, but make a wrong decision and react in an incorrect and ineffective manner. Perhaps most critically of all—and most pertinent to the case at hand—a whale such as Tilikum might act out aggressively without "any previously identified precursor" at all.

But attorney Carla Gunnin countered that Dawn's death had been an isolated tragedy and certainly not the result of willful indifference. She argued that Tilikum had never shown any proclivity toward trying to pull people into the water. She said that SeaWorld had "lots of safety protocols and procedures in place. Training a killer whale trainer is a lengthy process, a lot of on-the-job training. You don't start day one at Shamu Stadium and go train a killer whale the next day [and] all of those protocols ultimately equate to the safety of the trainers."

Gunnin claimed that an across-the-board ban on water work with *all* orcas at SeaWorld would impede proper medical care and other husbandry procedures with the animals. "The burden is on the secretary to prove a feasible means of abatement." OSHA's abatement was to no longer have close contact with orcas, but "because of the nature of caring for the killer whales, it would be impossible to not have close contact with the killer whales. So, that actually does open up the case to a bigger consideration."

Much of the day was spent arguing over the way trainers were trained, especially in terms of recognizing "precursors" to aggression. A tense-looking Kelly Flaherty Clark, subjected to more than four hours of intensive questioning, supported Gunnin's claim by stating that Tilikum had "never given us any indication that he would pull somebody into the water with him."

But, Black countered, sometimes there were no precursors at all, and even where they had been present, the trainers were still at risk. "Harm could happen even if the trainer doesn't make a behavioral judgment error, right?" he asked Flaherty Clark. She answered yes. It was a substantial victory for Black, Naomi thought: He had just forced the witness to admit that, sometimes, all the training in the world didn't matter; unlike machines, the orcas had minds of their own. Naomi thought the only "feasible means of abatement" was to require only protected contact with the orcas, which, among other things, meant no water work.

Flaherty Clark tried to gain the upper hand by insisting that all trainers at Shamu Stadium had to complete eighteen to twenty-four months of intensive training before being granted proximity (within five feet) to any killer whale. Yes, the risks were "inherent," but errors were *extremely* rare. "The frequency of a trainer making a bad call or missing a behavioral cue is minimal," she said. "In twenty-five years, I've reviewed one behavioral incident that did not show something that I would have done differently, that there weren't behavioral cues." That incident was Tilikum's attack on Dawn.

Flaherty Clark did testify, however, that trainers had to be perpetually cognizant of each whale's mood, the proximity of other orcas, and even such things as weather conditions, which might affect animal behavior. She conceded that, sometimes, trainers failed in that task.

In a blow to some people who had been publicly defending SeaWorld, supported by some SeaWorld staff, Flaherty Clark flatly rejected any suggestions that Dawn Brancheau was responsible for the incident. "Dawn did not break protocol on the day of the accident," Flaherty Clark testified, in direct contradiction to what Thad Lacinak and some others were still saying about Dawn's culpability.

Another important detail covered on day one was whether the government's abatements to protect trainers with physical barriers would be applied only during shows, or in all interactions with whales. SeaWorld argued that trainers had to be in the water with the orcas, at some point, to offer them proper care. This "hands-on" approach to the whales was the only way to accurately assess their health. Kelly Flaherty Clark even suggested that Kalina would not have died if trainers had been allowed into the pool. "We would have picked up on it if we'd been in the water with her," she said, trembling with emotion as she spoke. "I think we might not have lost Kalina if we were able to be as close with her today as we were on February twenty-fourth."

John Black noted that SeaWorld, not OSHA, had removed trainers from the water. "It sounds like you want to pin the blame of this whale's death on SeaWorld's decision to take their trainers out of the water. Is that what I'm hearing?" he asked.

"No," Flaherty Clark replied icily.

"Who are you blaming it on?"

"I'm not blaming anybody. . . . I was answering the question somebody asked me."

Naomi looked over at Courtney Vail in frustration. Flaherty Clark was, essentially, blaming Kalina's death on dry work. It was an utterly illogical argument, Naomi thought. What about all those dozens of whales who'd died young at SeaWorld, even when trainers had all the close contact they wanted? It did nothing to save them.

As for Kalina, even if trainers had detected illness by swimming with her in the past, this time it might not have saved her. Kalina was eating normally and seemed fine the day before she died. The next afternoon she became sluggish, and early that evening she quickly died. Her septicemia set in rapidly. To even suggest that water work would have prevented Kalina's death, especially in court, aggravated Naomi to no end. "It's so irrational, she has absolutely no basis for it except her feelings," Naomi whispered to the person next to her. "It's just *not* a science-based argument."

Later, under questioning by the judge, Black clarified that OSHA's citation applied only to work during shows, even though SeaWorld insisted it was "impossible to draw a line between show behaviors and other behaviors, because so much of the work overlaps," as Jason Garcia reported in the *Orlando Sentinel*. "Flaherty Clark noted that trainers will often use performances to continue training whales on specific behaviors that are necessary for certain husbandry procedures, such as obtaining gastric samples."

It was a "crucial point," Garcia reported, because SeaWorld could "better argue that OSHA's recommendations are untenable if it can show that they would interfere with medical procedures or other husbandry work."

During the morning break, Naomi was interviewed by Orlando's NBC affiliate, WESH-TV. "Dawn Brancheau was put into a position where she could easily be pulled into the water," she noted. Even though SeaWorld had barred water work with Tilikum, the company had turned a blind eye toward trainers' lying down in a shallow ledge next to him. "He chose to grab part of her and pull her in, and she couldn't stop him," Naomi added.

That night, Carol Ray appeared on CNN's *AC-360* with Anderson Cooper (SeaWorld declined to send a representative). Carol insisted that Dawn was behaving in a way that many trainers had done with Tilikum over the years. "She was not in a position on that day that she hadn't been in several times before," Carol told Cooper. "If it had been brought to her attention that she shouldn't be doing that, I'm sure she wouldn't be doing that."

"Killer whales are so big in these tanks, which basically they just go around in circles," Cooper said. "Should they be used in these kinds of shows?"

Carol responded unequivocally, "One of the most important things we've learned from having these animals in captivity is that they're not suitable to have in captivity." She added, "I don't think there's a safe way to have humans interacting in close proximity with these whales."

Day 2: Tuesday, September 20. Much of this second day focused on the actual "takedown" of Dawn by Tilikum. For the past nineteen months SeaWorld had been saying—and the media had reported—that Dawn allowed her long ponytail to drift into the mouth of Tilikum, who simply viewed it as a novel plaything and did not act out of aggression. But Jeff Ventre, John Jett, members of The Orca Project, and others had been insisting that Tilikum had grabbed Dawn by her arm. They said the Connell video showed trainer and whale moving in unison in the water in the final seconds of footage, and that Tilikum already had Dawn's arm in his mouth.

Debunking the "ponytail theory" was important to the activists and ex-trainers—though not so to Naomi, who thought that Tilikum had acted deliberately no matter what body part he first grabbed. But many of the others thought the ponytail theory cloaked the purposeful nature of Tilikum's actions. The main witness who first described the ponytail grab—trainer Jay Topoleski—was not looking at Dawn when she went under, at least according to eyewitness Suzanne Connell.

Topoleski's statement to the sheriff's office was contradicted by another eyewitness at G Pool that day: security guard Fredy Herrera. The retired NYPD officer testified, "From my angle, I saw her left arm go underwater as the whale started descending." He said Dawn's arm was outstretched "like she was making a left-hand turn signal." Gunnin got Herrera to concede, however, he was not 100 percent certain, though it certainly looked that way from his observation.

The government then played the Connell video, which ended with Dawn on the shallow ledge next to Tilikum. Black asked if Herrera had seen trainers in the same position with Tilikum before. "I believe so," Herrera said.

Day 3: Wednesday, September 21. Early that morning, the *Today* show's Kerry Sanders delivered a live report from Orlando to cohost Matt Lauer at NBC studios in New York. Despite SeaWorld's long insistence that Tilikum had been "attracted to the trainer's bouncing ponytail," testimony had now been heard in court that "seriously challenges that claim," Sanders reported. "The answer to what happened may lie in the videotapes that Dawn's family and SeaWorld argue should not be played in court." The tapes "may be the only way to explain if he grabbed Dawn by the ponytail, or her arm." NBC also interviewed TOP's Colleen Gorman, in the courthouse foyer. "As you can see now with the evidence and the eyewitness coming forward, he already had her arm in his mouth," she said of Tilikum, "and he took her down from there." Sanders, in closing, said that because of "legal issues" SeaWorld would no longer comment "whether he grabbed her ponytail or her arm."

Day three saw the presentation of another video, never viewed before in public, of the 2006 attack by Kasatka on Ken Peters in San Diego. Without any audio, the gripping images of a wet-suited man being dragged underwater by his foot were even more surreal.

But most of the day was spent on testimony from Chuck Tompkins, corporate curator of zoological operations, and SeaWorld's record keeping of aggressive incidents among its killer whales. John Black alleged that the company had repeatedly failed to document such incidents and questioned the reliability of a corporate log that SeaWorld said included every single aggressive episode that had happened since it began keeping records in 1988. The log report contained around one hundred incidents, including two at Loro Parque. But Black cited several incidents listed in individual Animal Profiles that had never made it into the log, including an orca's lunging into a trainer. Another whale, Kayla, had five incidents in her profile that failed to appear in the log.

"What you're calling a complete incident report log is not complete and does not contain every incident between 1988 and 2009, correct?" Black demanded of Tompkins in one terse exchange. "We missed a few," Tompkins replied tetchily. He informed Black that only twelve of the one hundred incidents in the log resulted in significant injuries, and that was out of an estimated 1.6 million interactions between orcas and trainers since 1988.

Naomi loathed this overused statistical manipulation. If a percentage of cars of a certain model had faulty brakes, it didn't matter how many times those cars were driven safely before their brakes inevitably failed and people died. The same was true with orcas: It wasn't the number of *interactions* that mattered; it was the number of *whales* that had injured a trainer. But SeaWorld's incident log listed several whales who had injured people. The company claimed a "failure rate" of less than 0.001 percent, but by Naomi's calculations it was actually more than 10 percent.

Black next blasted SeaWorld for still not having entered the Brancheau killing into the incident log, but Tompkins pointed out that staff had reviewed that incident extensively at meetings. Everyone was well aware what happened that day, he said, but Black still took him to task anyway. "Nonetheless, you've made a decision not to record it now and not to provide a written description so that people will understand what SeaWorld's analysis was without [a trainer] having to come to somebody and say, 'Jeez, can you explain?' " Black growled.

Tompkins fired right back that at SeaWorld it was "not an option to know facts. It's mandatory."

Gunnin argued that SeaWorld trainers were required to read incident reports and the Animal Profiles alike, as well as other records, some of them kept daily. "There are a lot of redundancies," she told Judge Welsch. She was backed up by Tompkins, who testified that SeaWorld treated all safety issues with the utmost seriousness. He called the incidents "learning opportunities" that yielded increased knowledge of each whale and, by extension, greatly enhanced security. Mistakes were made, he conceded, "but we have a process by which we learn from those mistakes." As evidence for that, he said the number and frequency of incidents had dropped in recent years. "We have gotten a whole lot better."

"Maybe you were lucky?" Black shot back.

"We don't think so," Tompkins said dourly. "We think it was getting better."

But Black wasn't having it. "We have already established," he said, perhaps a bit haughtily, "that your incident reports are incomplete."

Naomi smiled as she scribbled in her notebook. Score one for the feds, she thought.

Day 4: Thursday, September 22. Two witnesses filled much of the day. The first was Jan "Jay" Topoleski, who had transferred from Shamu Stadium to Sea Lion and Otter Stadium after Dawn's death, then left SeaWorld entirely to work as an FBI forensic canine specialist. Topoleski, who first reported the ponytail grab, now offered a somewhat different account from what he originally told deputies. He testified in court that Dawn was doing a "mimic behavior" in which she had Tilikum turn on his back as she mimicked the motion on the shallow ledge.

"I saw her get up to her knees and put her hands on her ponytail, and I saw that she couldn't break free," he said. Dawn was struggling with both hands to extract her ponytail from Tilikum's mouth. But under cross-examination, Topoleski was obliged to acknowledge he did not actually see Dawn's ponytail in the animal's mouth.

Next up was Shana Groves, one of the first trainers to arrive on the scene when the alarm sounded. Still visibly shaken by the incident, she began sobbing when shown a picture of herself with Tilikum. Just a few weeks before the trial, Groves had requested a transfer out of Shamu Stadium, also to Sea Lion and Otter. Under government questioning, Groves testified about all the different cautions given to orca trainers, such as never turning one's back on an animal. Black pounced. Such warnings were vivid proof, he said, of orca unpredictability. But Groves countered that trainers were extremely cautious. "I was at SeaWorld for seven years before I could even open a Tilikum gate," she said. "That's how precautionary we were."

The late afternoon was spent arguing over the acceptability of Dr. Dave Duffus as OSHA's expert witness. Gunnin sought to have him disqualified because he had no expertise with captive killer whales, only wild ones. She fought hard, and for a moment Naomi thought the judge would side with her. In the end, he agreed to allow Duffus to testify, but warned that he would take the professor's lack of captivity experience into account when deciding how much weight he would lend to his testimony.

During his direct examination, the Canadian scientist calmly and methodically ripped into SeaWorld's complaint against the US government: "There's always going to be an uncertainty. I think it would be complete folly to put yourself in close proximity to wild killer whales where they could seize you or pull you into the water." He also politely mocked SeaWorld's claims that aggressive acts against trainers were exceedingly rare. "If it only happens once in a million, but if in that millionth time, the outcome is a catastrophe, I think it goes beyond the probability issue."

Day 5: Friday, September 23. In what was supposed to be the final day of testimony, Judge Welsch announced that the hearing would have to reconvene and set the date for November 15. Meanwhile, Duffus returned for Gunnin's harsh cross-examination trying to impeach his credibility, but he held his ground. Duffus was "alarmed" by Dawn's proximity to Tilikum, he said. "I don't want to second-guess an experienced trainer, but I would not, given my experience with killer whales . . . be that close to Tilikum. No way on earth."

Asked by OSHA attorney Howard-Fishburne why he felt that way, Duffus said simply, "I have a great deal of respect for the fundamental nature of large predators." As for the safety and predictability of interactions with trained whales, they could not be counted on "under every condition. If the interactions with Tilikum were sufficient under all conditions, we would not be here today."

Meanwhile, SeaWorld's safety training program was "influential," Duffus conceded to Gunnin, "it does work. My point is it does not work all the time." He added that he was struck by the similarities between the attacks on Dawn and on Keltie Byrne, back in 1991. But, he added somewhat ruefully, "Twenty years later, a lot has been done, yet I'm reading the same outcome."

The final witness was Les Grove, head of OSHA's Tampa office and Lara Padgett's boss. Gunnin grilled him on the inconsistency of calling for abatements during shows, but not for other activities at SeaWorld. "You're not saying, as long as it's not a show, they can do water work, are you?" she asked Groves. "What's the difference between any type of close contact?"

The citation applied only to shows, Groves replied. But then again, any "responsible employer" who was aware of "other interactions where [workers] are exposed to the hazard, you should take action." In other words, the citation was for the shows, but SeaWorld should apply the abatements across the board.

The hearing reconvened in mid-November in the same Sanford, Florida, courtroom. Naomi returned for part two, as did Colleen Gorman of TOP and Jared Goodman of PETA, which had since filed its controversial lawsuit against SeaWorld seeking the release of five wild-caught orcas under provisions of the antislavery Thirteenth Amendment. Courtney Vail of WDCS did not make the trip, though Tim Zimmermann and the documentary film team of Gabriela Cowperthwaite and Manny Oteyza flew to Florida for part of the second half.

This time, Naomi kept a running commentary on each day's proceedings and e-mailed them to her husband, Chris, and allies such as Nancy Yates, Courtney Vail, and Bill Rossiter at CSI.

Day 6: Tuesday, November 15, 2011. "Overall the day went to OSHA, but it was not a smooth victory." One point that OSHA did manage to score was that the incident reports "aren't really that useful as tools, because the trainers often do not re-review them, don't seem to remember what was in them, don't even seem to remember incidents from as recently as three years ago in their own parks," Naomi noted.

Scheduled that day were witnesses whose testimony SeaWorld had fought hard to suppress, but failed. First up was Ken Peters, involved in three aggressive incidents with Kasatka. He testified that trainers can predict whale behavior 99.9 percent of the time. But, as Naomi noted, he also admitted to not having looked at any of the pre-1994 incident reports in "a long, long time." Such inattention "hardly makes these reports useful as training tools if the trainers don't remember what's in 'em and never review them," she said.

Kasatka had tried to grab Ken Peters in 1999, yet SeaWorld continued water work with her until the 2006 incident. "So how many strikes does a killer whale get before SeaWorld gets trainers out of the water with them?" Naomi asked. According to Naomi, the day's "most shocking statement" had come from Peters himself. When John Black asked if Peters felt he might die that day, "he replied, 'I never felt my life was in danger.' I just found that absolutely bizarre," Naomi wrote. "I don't think he felt he was lying either. I just think he literally has amnesia about how scared shitless he was. Because I'm sorry, the close-ups of Ken Peters' face from that video DEFINITELY show a man in fear of his life."

Next up was Michael Scarpuzzi, San Diego's vice president of zoological operations, who Naomi said had delivered the "strongest testimony (for OSHA)" of the day. Scarpuzzi was a "very reluctant witness," Naomi noted, perhaps because "he has disagreed with his superiors on trainer safety. I think Scarpuzzi didn't want to be there and didn't want to answer any questions at all. He didn't want to lie and he didn't want to tell the truth."

Howard-Fishburne had clearly grown exasperated with Scarpuzzi's evasive tactics. "Mr. Scarpuzzi, we can play this game all day," the lawyer told him. "I'm asking you simple, basic questions, and you're giving me the runaround." Scarpuzzi denied he was playing a game, and Howard-Fishburne asked the judge if Scarpuzzi could be considered a hostile witness. "The judge said yes without hesitation," Naomi observed. "I think that shook Scarpuzzi a bit: he began answering more directly after that."

Scarpuzzi was the SeaWorld executive who had traveled to Tenerife when Keto killed Alexis. He testified that there had been a "series of incremental, minor behaviors by Keto, which alone were not significant," Naomi said. "I think he wanted to suggest that this therefore meant it would have been difficult to impossible to prevent Martinez's death, since there were no obvious precursors," and this was just an accident; no fault of SeaWorld's. "I think OSHA was able to suggest this actually meant once again that SW's training protocols were useless."

Day 7: Wednesday, November 16. OSHA rested its case, and Gunnin immediately moved to dismiss the citations against SeaWorld. "That's when Carla Gunnin took off!" Naomi wrote in her review. "We were wondering if the 'vibe' or tone of the hearing would change once SeaWorld started presenting its case and we were right! She was a different Carla—more aggressive, more articulate, more certain of her arguments. She actually had me going for a while there, but in the end, I think she went too far."

Gunnin argued forcefully that OSHA had failed to make a convincing case that SeaWorld was willful in its behavior. SeaWorld staff had repeatedly testified that company protocols were "in a constant state of being improved, precisely because SeaWorld recognizes problems via the incident reports and both minor and major changes have been made in response," Naomi wrote. "This is hardly 'indifferent' management."

Gunnin asserted that the law does not allow the government to simply say "you cannot do this," Naomi noted. The law demands that the *government* come up with feasible abatements, not the employer, so the company can continue doing business. "And the business of SeaWorld, boys and girls, is not to educate people or conserve nature—it's not even to entertain people," Naomi wrote. "It's to entertain people with WATERWORK SHOWS. (Gunnin said, 'The nature of SeaWorld's show IS waterwork.') Which means the abatement of 'no waterwork' is not feasible—it actually will prohibit the company from conducting its business."

At that point Gunnin "lost me, and I think that's where she lost the judge," Naomi said. Gunnin was equating SeaWorld's business with water work: remove it and the company would go bankrupt. "Now, we know this might actually be true, but they have shot themselves in the foot with this argument by voluntarily keeping their trainers out of the water for all these months," Naomi wryly noted. "The show has (as the saying goes) gone on, even with drywork only, so it's kind of unconvincing now to argue that waterwork is an essential element of SW's business." Even the judge noted there had been no water work for twenty months, and SeaWorld was still operating as normal. Gunnin was hardly going to admit that the new, "dry" Shamu show was not pulling in the crowds. It was a bit of a legal pickle for her side.

The judge rejected SeaWorld's motion to dismiss. Even so, Gunnin had carved some deep chinks into OSHA's armor, and Welsch agreed to take her arguments under consideration.

Later that day, Flaherty Clark returned to the stand. Much of her testimony related to incident reports and how they enhanced trainer safety. She seemed defiant. Gunnin asked whom the incident reports were written for. "They certainly are not written for lawyers or a courtroom," Flaherty Clark sniffed. Naomi didn't think the witness had done herself any favors with the flippant remark. "They were written for the trainers. They were written for the people that we are developing, and they're written in the name of enhancing our craft." Now Naomi had to suppress a chuckle. "Craft is right," she wrote in her notes, "cuz it sure ain't a science."

Flaherty Clark adamantly denied that she or SeaWorld showed plain indifference toward employees. Asked why, she said, "Because of my twenty-five-year career and watching the changes and watching the protocols and watching the science grow. We established protocols, we developed SOPs [standard operating procedures], we added levels of management, we added oversight, spotters, emergency protocols, training on how to recognize—you know, around this hearing, I would say, we trained—I mean, we train a lot." Those protocols "made me able to sleep at night as their director. I know that I did not disregard the safety of my animal trainers."

"Basically SeaWorld is trying to have it both ways," Naomi wrote to the gang that night. "They are saying that nothing went wrong on Feb. 24, 2010, that their SOPs work. . . . BUT they also are reviewing all their SOPs and have the trainers out of the water until they have completed that review. But if it wasn't broke, why are they trying to fix it?"

Day 8: Thursday, November 17. Much of the afternoon was spent wrangling over the qualifications—and alleged conflict of interest—of SeaWorld's expert witness, Jeff Andrews. Now employed by the San Diego Zoo, he was a former SeaWorld trainer and still consulted for Busch Gardens Tampa Bay, owned by SeaWorld Parks & Entertainment. Ultimately Judge Welsch approved him.

But the real drama came earlier, when John Black throttled Kelly Flaherty Clark, who'd returned to complete her testimony. "John Black pushed KFC pretty hard and rattled her cage enough to have her lose her cool at least once," Naomi commented.

It was a searing exchange. Black pressed the witness about whether Tilikum had been "desensed" to ponytails or not.

"I would assume that Tilikum, after nineteen years of not reacting to ponytails, was desensitized to ponytails," Flaherty Clark opined.

But whales "sometimes pull on things," Black reminded her, "as you have testified earlier."

The witness glowered. "Tilikum in nineteen years has never shown us that he pulled on anything."

"You testified . . . they *do* sometimes put things in their mouth and pull it, right?"

"Yes."

"So you put Tilikum in his own little box, if you will, and were not willing to generalize that he might engage in pulling behavior?"

"Tilikum never exhibited any behavior that would make me think he would pull."

"But you *knew* he was capable of pulling?" Black asked again.

"I know you are capable of rape," Flaherty Clark muttered. (Earlier, when asked if all orcas were physically capable of injuring trainers, she had disagreed with the generalization and snapped, "I could say to you that all men have the potential to be rapists.")

Welsch halted the deteriorating exchange. "We'll strike that comment," he said quietly.

Day 9: Friday, November 18. Naomi, and probably everyone else in court, was relieved that the last day had finally arrived. First up was Chris Dold, vice president of veterinary services for SeaWorld Parks & Entertainment, who corroborated Flaherty Clark's testimony about the benefits of physical contact to animal husbandry: "The long-term impact of no contact I would imagine would be negative. It certainly would not be a policy that I would endorse." Veterinary work was "multisensory," he said. "You touch, you feel. . . . This is a human and animal engagement. No contact would be to the detriment of certainly our responsibilities as stewards of these whales."

But, Black pointed out, there had been no contact with Tilikum ever since the attack. Was Dold suggesting that Tilikum was receiving suboptimum care? The answer, of course, was no.

Naomi found Dr. Dold arrogant and supercilious. At one point, John Black had objected to a question that Carla Gunnin put to Dold because it had already been answered by Kelly Flaherty Clark. "He's adding some veneer of expertise," Black complained. He was overruled. The witness was indignant. "Just briefly, *John,*" Dold said, turning to the government lawyer. "It was a one-hundred-thousand-dollar degree, my veterinary degree, and so the 'veneer of expertise' thing kind of stings, man." Black apologized, but Dold pressed on. "I know you've got to strike that," he said, this time to the judge. "Go ahead and strike it, but for crying out loud."

Given Dold's six-figure degree, Naomi couldn't help but gleefully note that when he was asked by the court reporter to spell *cetacean,* "he spelled it (very carefully) 'c-e-a-t-a-c-e-a-n.' "

Jeff Andrews, SeaWorld's expert witness, testified last. John Black opened by ripping into Andrews's credibility. As Naomi described the ambush: "He went after Andrews' expert report, which SeaWorld had not even submitted for the record (because this report contradicted the narrative that Kelly Flaherty Clark began and Chuck Tompkins and the others continued, that Dawn did not make a mistake that day, that she was following protocol, that nothing was out of the ordinary that day until Tilikum, completely out of the blue, killed her—Andrews had clearly written in his expert report that 'Dawn made a mistake'). His expert report was so bad that his side did not submit it for the record, but OSHA DID!"

Andrews had "confidently stated" his disagreement with Dave Duffus that captive orcas can exhibit predatory behavior, Naomi noted. "He said they are fed and satiated and therefore there is no need to show predatory behavior." When John Black pointed out that captive orcas kill seabirds, which even Chuck Tompkins described as "predatory" behavior, Andrews countered that it was more "fun" or play behavior. "But Black drove home that SeaWorld itself considered them predatory events," Naomi observed.

According to Naomi, Andrews's "biggest blooper" had been admitting that his expert report relied exclusively on what SeaWorld employees told him about Dawn's death and Tilikum's behavior that day. "None of the people he admitted speaking with had actually witnessed the attack—he never spoke with any eyewitnesses," Naomi marveled. But Andrews nonetheless offered these interviews as "facts" on which his "expert opinion" on the attack was based. "This was probably the most powerful example of his bias and lack of credibility and I think the judge took note," Naomi wrote.

When it was all over, Judge Welsch looked tired and ready to go home. He gave the two parties forty-five days to submit written briefs, once the hearing transcripts were completed. "I've been doing this for fifteen years, and this probably is one of the most unusual OSHA hearings I've ever had," he told the courtroom before adjourning. "I'm going to spend a lot of time dealing with this issue. I think it's a very complicated issue."

Naomi walked out into the pleasant evening air (November in Orlando is sublime) and headed for the airport. She had no idea how the judge would rule. OSHA had poked a lot of holes in SeaWorld's façade, but Carla Gunnin had scrambled hard to plug them, not always with the greatest assistance of her own witnesses.

Now, all there was left to do was go home and wait.

What did the future hold for SeaWorld and its killer whales? In early May of 2012 it was still anyone's guess. Many observers predicted that Judge Welsch would, in effect, split his decision, reducing the "willful" violation to "serious" (SeaWorld *had* argued quite forcefully that it was continuously reviewing and updating safety protocols), but still upholding most or all of the proposed "feasible abatements."

If that happened, thorny issues would linger. SeaWorld would almost surely appeal the ruling, and the case would be transferred to Washington, DC, where the Occupational Safety and Health Review Commission would hear the next stage of litigation. In the event that Welsch ruled completely against OSHA, Secretary of Labor Solis had the option of appealing the government's case as well. It could mean an ultimate showdown between the Obama administration and Stephen Schwarzman of the Blackstone Group, who had once likened the president to Adolf Hitler. Complicating political matters, Obama was now running for reelection and his probable opponent, Republican Mitt Romney, was the beneficiary of a big-donor fund-raising dinner at Schwarzman's Park Avenue home in January.[2] Most people familiar with the case doubted that a Romney administration labor secretary would pursue any appeals.

Assuming that Welsch maintained most or all of the abatements, it remained unclear if SeaWorld's adoption of such technologies as false bottoms in pools and spare air would put it in compliance with the feds. But as long as SeaWorld officials could keep the appeals process alive, they would not have to comply with any of the OSHA abatements.

Meanwhile, the Tilikum scandal, and the removal of orca trainers from the water, had done nothing to hurt SeaWorld's bottom line. The company claimed record earnings in 2011 and also announced its largest expansion in the park's history, none of which had anything to do with Shamu.[3] That might have signaled a desire by SeaWorld to move beyond orca shows as a central driver of corporate revenue. On the other hand, SeaWorld's killer whale breeding program appeared to be progressing at full tilt, and the company was reportedly scouting sites for overseas SeaWorld venues, particularly in China.

No matter what happened, killer whales were very much in the news. The iconic black-and-white animals were *everywhere* on TV, it seemed. The public had been enraptured recently by footage of wild orcas hunting sharks along the coast of New Zealand. Another killer whale had just been freed from entanglement in a crab-trap line, also in New Zealand. It made worldwide headlines and was covered on ABC's *Good Morning America*.

Killer whales had been sighted close to shore in Northern California, off Monterey, and in Southern California, including a large pod off Dana Point, not fifty miles from SeaWorld San Diego. Boatloads of people were sailing out to view the orcas, something that can be disruptive if the federal two-hundred-yard viewing limit is violated (another reason why sea pens are attractive alternatives). Still, the irony of spotting wild whales so close to captive ones was not lost on some people. "Why go to SeaWorld when you can see the real thing right there in the ocean?" one commenter said online.

Why indeed? Naomi thought.

Killer whales kept making news. On February 9, 2012, a Canadian appeals court issued a major ruling that the government of Canada was legally bound to protect the habitats of Northern and Southern Resident communities, including salmon stocks and marine environment quality.[4] The ruling might reduce fishing and vessel traffic in the Strait of Georgia and Juan de Fuca Strait in the south and Queen Charlotte Strait and Naomi's beloved Johnstone Strait in the north. It was not clear what impact it would have on all the Norwegian salmon farms in British Columbia, but their removal was important if wild salmon populations were to rebound, Naomi believed.

February did not bring equally good news to killer whales in captivity. On the eighth, a federal judge in San Diego granted SeaWorld's motion to dismiss PETA's lawsuit, with prejudice, to free five wild-caught whales—Katina, Kasatka, Ulises, Corky, and Tilikum.[5] The judge seemed sympathetic to the cause, calling the goal of protecting orcas "laudable." But he said the antislavery Thirteenth Amendment "affords no relief." That same day, the prime minister of Iceland's office denied any knowledge of plans to return Tilikum to his native waters.[6] It looked as if the mighty whale would live out his days in a tank at SeaWorld after all.

Meanwhile, the "SeaWorld Four," Jeff Ventre and the other former trainers, had just launched Voice of the Orcas,[7] a website dedicated to the whales, and to their own "change of heart" about captivity. Jeff and others were also arranging the next reunion—Superpod II—on San Juan Island for July 2012. Everyone was planning on going. Even Naomi hoped to attend. It was time, finally, to meet these ex-trainers face-to-face. She had come to respect Jeff, John Jett, Carol Ray, and Sam Berg, as well as John Kielty and Colleen Gorman of TOP, and she looked forward to seeing old friends such as Howie Garrett, Ken Balcomb, and Lori Marino. It would also be wonderful to see killer whales in their magnificent natural environment once again. Almost unbelievably, Naomi had never encountered a Southern Resident whale in the wild before.

There was nothing like witnessing a traveling family of killer whales cruising together up the strait, foraging, breaching, splashing, and playing, their cries of communication audible from your boat. The close-up sight of a mature male rising above the water's surface, his towering dorsal fin so lofty one sometimes had to look *up* to see its tip, was one of the most heart-pounding encounters the animal world had to offer.

Those experiences were not included in the price of admission at any marine mammal "entertainment" park. As Sam Berg so pithily declared one evening at the Superpod gathering, "SeaWorld whales are not real whales; they're *facsimiles* of whales."

Notes

Introduction

1. Some scientists disagree and note that male sperm whales are larger and more powerful than killer whales.
2. The Northern Resident killer whale population in 2011 was estimated at about 265 animals, a significant increase since 2004, when the total was about 205: Ellis, Towers, and Ford, Technical Report 2942, Department of Fisheries and Oceans (Canada). The Southern Resident population in 2011 was estimated at "about 88 whales, a decline from its estimated historical level of about 200 during the mid- to late 1800s. Beginning in about 1967, the live-capture fishery for oceanarium display removed an estimated 47 whales and caused an immediate decline in Southern Resident numbers. The population fell an estimated 30% to about 67 whales by 1971": www.nmfs.noaa.gov/pr/species/mammals/cetaceans/killerwhale.htm#note11.
3. According to the NMFS website: "One estimate put the worldwide population of killer whales at over 100,000 animals [R. R. Reeves and S. Leatherwood, "Dolphins, porpoises, and whales: 1994–1998 action plan for the conservation of cetaceans" (Gland, Switzerland: IUCN, 1994)]. However, the most recent estimate revised this figure to a minimum of about 50,000 animals [K. A. Forney and P. Wade, "Worldwide distribution and abundance of killer whales," 2006, in *Whales, Whaling and Ocean Ecosystems* (Berkeley: University of California Press, in press)]: www.nmfs.noaa.gov/pr/species/mammals/cetaceans/killerwhale.htm#note9.
4. Captive killer whale census data come from Orca Homepage, www.orcahome.de.

Prologue

1. Most accounts of this incident said that Keltie Byrne fell in the pool entirely, and that all three whales took part in her drowning. But eyewitnesses told a different story, as written in the prologue. The account was based on an interview by the author with witness Nadine Kallen, and a corroborating interview with Kallen's friend Corrine, conducted by filmmaker Gabriela Cowperthwaite in Seattle, Washington, in 2011.
2. Luke Rendell and Hal Whitehead, "Culture in whales and dolphins," *Behavioral and Brain Sciences* 24 (2001): 309–82.

2. Trainer

1. Erich Hoyt, *Orca: The Whale Called Killer* (Camden House, 1990 ed.).
2. John K. B. Ford, Graeme M. Ellis, and Kenneth C. Balcomb III, *Killer Whales: The Natural History and Genealogy of* Orcinus orca *in British Columbia and Washington State* (University of Washington Press, 2000).
3. "Whales caught," United Press International, August 10, 1970.
4. "Flying fat whale eyed as sire," Associated Press, October 19, 1976.
5. Jason Garcia, "SeaWorld Orlando killer whale death is third in four months, 24th in 25 years," *Orlando Sentinel,* October 6, 2010, http://articles.orlandosentinel.com/2010-10-06/business/os-seaworld-killer-whale-dies-20101005_1_dawn-brancheau-seaworld-orlando-killer-whale-seaworld-marine.
6. "Whale dies," Associated Press, April 29, 1986.
7. "Killer Whale (*Orcinus orca*)," NOAA Fisheries, Office of Protected Resources, www.nmfs.noaa.gov/pr/species/mammals/cetaceans/killerwhale.htm.
8. "SeaWorld moves Baby Shamu whale," *Fort Lauderdale Sun Sentinel* and wire services, February 14, 1990.
9. Patricia V. Barrios, "New whale makes splashdown in S.A.," *San Antonio Express-News,* May 31, 1991.
10. "Whales caught," *Saskatoon Star Phoenix* (SK, Canada), August 20, 1975.
11. Tim Zimmermann, "Do Orcas at Marine Parks Injure One Another?" September 14, 2010, http://timzimmermann.com/2010/09/14/do-orcas-at-marine-parks-injure-one-another.
12. Christopher Anderson, "Kalina celebrates SeaWorld birth to 'grandbaby Shamu' killer whale," *San Antonio Express-News,* February 3, 1993.
13. Ford, Ellis, and Balcomb, *Killer Whales.*
14. Ines Davis Parrish, "SeaWorld's latest birth provides whale of a time," *Orlando Sentinel,* June 19, 1995.
15. "Performing whale gives birth," Associated Press, November 5, 1988.
16. Sandra Mathers, "Orlando native Baby Shamu moves to SeaWorld in Ohio," *Orlando Sentinel,* April 22, 1991.
17. Tom Bower, "Whale's death stirs debate," *San Antonio Express-News,* May 7, 1999.
18. "Killer Whale (*Orcinus orca*)," NOAA Fisheries.
19. "A new Shamu, SeaWorld's Namu bears a bouncing baby killer whale," Associated Press, November 27, 1988.
20. "SeaWorld loses a Shamu," *Greensboro (NC) News-Record,* August 7, 1991.
21. "Female killer whale flown in for breeding," Associated Press, November 19, 1987.
22. "Baby Namu born," *Miami Herald,* July 13, 1989.
23. "Killer whale born at Orlando SeaWorld," *Business Wire,* January 1, 1994.

3. Capture

1. The account of Tilikum's life as a young calf in Iceland, his capture, his transfer to Canada, and his early years at SeaLand of the Pacific was put together from various published sources. The descriptions were based on informed speculation, consistent with other accounts of whale calf captures in Iceland and the Pacific Northwest.
2. Eric L. Walters, "Considerations for Keeping Marine Mammals in Captivity," document submitted to the Canadian Federation of Humane Societies, April 20, 1990.

7. Residents vs. Transients

1. John K. B. Ford and Graeme M. Ellis, "Selective foraging by fish-eating killer whales (*Orcinus orca*) in British Columbia," *Marine Ecology Progress Series* 316 (July 3, 2006): 185–99.

2. Pam J. Stacey and Robin W. Baird, "Birth of a 'Resident' killer whale off Victoria, British Columbia, Canada," *Marine Mammal Science* 13 (1998): 504–8.
3. J. K. B. Ford, G. M. Ellis, and K. C. Balcomb, *Killer Whales: The Natural History and Genealogy of* Orcinus orca *in British Columbia and Washington State* (University of Washington Press, 2000).

8. OrcaLab

1. Ted Griffin, *Namu—Quest for the Killer Whale* (Gryphon West Publishers, 1982).
2. Information on early captures from various sources, including PBS's *Frontline* and Rex Weyler, *Song of the Whale* (Anchor Press/Doubleday, 1986), and Erich Hoyt, *The Performing Orca—Why the Show Must Stop* (Bath, England: Whale and Dolphin Conservation Society, 1992).
3. Weyler, *Song of the Whale.*
4. M. A. Bigg, I. B. Macaskie, and G. Ellis, "Abundance and movements of killer whales off eastern and southern Vancouver Island, with comments on management" (Ste. Anne de Bellevue, Quebec: Arctic Biological Station, 1976).
5. M. A. Bigg, P. F. Olesiuk, G. M. Ellis, J. K. B. Ford, and K. C. Balcomb, "Social organization and genealogy of resident killer whales (*Orcinus orca*) in the coastal waters of British Columbia and Washington State," special issue, *Report of the International Whaling Commission* 12 (1990): 383–405; P. Olesiuk, M. A. Bigg, and G. M. Ellis, "Life history and population dynamics of resident killer whales (*Orcinus orca*) in the coastal waters of British Columbia and Washington State," special issue, *Report of the International Whaling Commission* 12 (1990): 209–43.
6. Alexandra Morton, *Listening to Whales—What the Orcas Have Taught Us* (Ballantine Books, 2002).

9. Happy Talk

1. The booklet of approved SeaWorld phrases was reproduced on the website of PBS's *Frontline,* "A Whale of a Business," www.pbs.org/wgbh/pages/frontline/shows/whales/seaworld/buzz.html.
2. A photocopy of "Difficult and Unusual Questions and Answers" was provided to HSUS by a former SeaWorld employee in 1991.
3. "Ask Shamu," SeaWorld Parks & Entertainment website, www.seaworld.org/ask-shamu/faq.htm.

11. Breakfast at SeaWorld

1. Bruce Hecker, "Ask the Experts: Environment—How do whales and dolphins sleep without drowning?," *Scientific American,* February 2, 1998.
2. "A Whale of a Business: Danger to Trainers," *Frontline,* www.pbs.org/wgbh/pages/frontline/shows/whales/debate/trainers.html.
3. Cindy Schreuder, "Report blames infection for killer whale's death," *Orlando Sentinel,* November 21, 1990.

12. Whale for Sale

1. Cindy Schreuder, "Yule baby may be a new Shamu," *Orlando Sentinel,* December 28, 1991.
2. "Verdict of Coroner's Inquest in the Death of Keltie Lee Byrne," Coroner's Court of British Columbia, Colwood, BC, June 5, 1991.
3. Eric L. Walters, personal letter to Canadian Federation of Humane Societies.
4. Eric L. Walters, memo to Norm Ellison, Coroner's Office, Victoria, BC, April 1, 1991.
5. Cindy Schreuder, "Orlando SeaWorld may import 2 whales," *Orlando Sentinel,* November 8, 1991.

6. "Killer whale's death probed," Associated Press, May 16, 1991.
7. "Experts try to establish death cause/SeaWorld whale seemed to be OK," Associated Press, May 16, 1991.
8. "SeaWorld loses a Shamu," Associated Press, August 7, 1991.
9. Cindy Schreuder, "Killer whale dies at Texas' SeaWorld," *Orlando Sentinel,* May 16, 1991.
10. "A Whale of a Business—Inside SeaWorld: The Tilikum Transaction," *Frontline,* www.pbs.org/wgbh/pages/frontline/shows/whales/seaworld/tilikum/.
11. Letter from Ann Terbush, NMFS Office of Protected Resources, to Brad Andrews, vice president for zoological operations, SeaWorld, December 17, 1991.
12. Letter to Brad Andrews, vice president for zoological operations, SeaWorld, from Nancy Foster, director, Office of Protected Resources, US Department of Commerce, National Oceanic and Atmospheric Administration, National Marine Fisheries Service, Silver Spring, MD, January 8, 1992.
13. Cindy Schreuder, "SeaWorld gets OK to import killer whale," *Orlando Sentinel,* January 9, 1992.
14. Letter to Andrews from Foster.
15. Letter from SeaWorld vice president for zoological operations Brad Andrews to Minister of Fisheries Thorsteinn Palsson, Reykjavík, Iceland, April 27, 1992.
16. Letter from Minister of Fisheries Thorsteinn Palsson to SeaWorld vice president for zoological operations Brad Andrews, May 7, 1992.
17. Interview transcript from PBS "A Whale of a Business," www.pbs.org/wgbh/pages/frontline/shows/whales/interviews/mcbain.html. The interview was with Brad Andrews and SeaWorld director of veterinarian services Jim McBain; it is not clear if Andrews, rather than McBain, made the statement, though it is probable.

14. Arrival

1. Letter from Erich Hoyt, Whale and Dolphin Conservation Society, to William Fox, assistant administrator for fisheries, NMFS, January 31, 1992.
2. "Infection kills 36-day-old killer whale," *Canadian Press,* March 11, 1992.
3. Letter from Nancy Foster, director, Office of Protected Species, NMFS, to Brad Andrews, vice president of operations, SeaWorld, Orlando, October 7, 1992; and NMFS Public Display Permit No. 774, issued to SeaWorld, Orlando, October 7, 1992.
4. "B.C. whales airlifted to new U.S. homes—Orca trio fly south in water-filled crates," *Hamilton Spectator* (ON, Canada), January 8, 1993

15. Humane Society

1. Information on early captures of killer whales was also derived from PBS's *Frontline*, "A Whale of a Business," www.pbs.org/wgbh/pages/frontline/shows/whales/etc/orcas1.html; and other sources, including Erich Hoyt, *Orca: The Whale Called Killer* (Camden House, 1990 ed.), and Tim Zimmermann, "The Killer in the pool," *Outside*, June 30, 2010.
2. www.imdb.com/title/tt0060737/.
3. Zimmermann, "Killer in the pool."
4. "Judge sets 'killer' whales free," Associated Press, March 13, 1976.

16. Backstage Doubts

1. Letter from F. B. den Herder, director, Dolfinarium Harderwijk, Netherlands, to Lanny Cornell, zoological director, SeaWorld, San Diego, January 1987. Available at www.pbs.org/wgbh/pages/frontline/shows/whales/seaworld/gudrun/amstltr.html.
2. "Killer whale born at Orlando SeaWorld," *Gainesville Sun,* January 1, 1994.
3. "Killer whale born," *Tuscaloosa News,* September 13, 1993.

4. "Killer whale born at Orlando SeaWorld," *Gainesville Sun*.
5. "SeaWorld loses second killer whale this year," Associated Press, April 2, 1996.
6. Information about Kalina's fighting for dominance with her mother, Katina, comes from the personal observations of Jeffrey Ventre and John Jett.

17. Blood in the Water

1. More information on deaths of captive killer whales is available from the Marine Mammal Inventory Report (MMIR), as well as several websites, including www.orcahome.de and www.cetacousin.bplaced.net.
2. Death of Kandu: Marine Mammal Inventory Report, www.databases.sun-sentinel.com.
3. "Shamu dies of infection," *Eugene (OR) Register-Guard,* August 31, 1971.
4. Death of Frankie: Marine Mammal Inventory Report, www.databases.sun-sentinel.com.
5. Death of Canuck: Erich Hoyt, *Orca: The Whale Called Killer,* "Appendix Six: Killer Whales Kept Captive" (Camden House, 1990 ed.).
6. Death of Kandu III: Ibid.
7. Death of Kona: Ibid.
8. Death of Kandu II: Ibid.
9. Rescue of Sandy: "No-Name finds unlikely friend," Associated Press, June 1, 1973.
10. Death of Kilroy: "Killer whale dies," Associated Press, September 24, 1978.
11. Death of Kanuck II: Hoyt, *Orca.*
12. Death of Ramu: Ibid.
13. Death of Baby Shamu II: "Baby killer whale dies," *Modesto (CA) Bee,* January 17, 1986.
14. Death of Winston: "Whale dies," Associated Press, April 29, 1986.
15. Death of Kona II: "Performing killer whale dies at SeaWorld," *Palm Beach Post,* October 29, 1987.
16. Death of Nootka (Knootka): "Killer whale dies of infection at SeaWorld," *Los Angeles Times,* March 14, 1990.
17. Death of Kahana: "Killer whale's death probed," Associated Press, May 16, 1991.
18. Vanessa Williams, "Captive Orcas 'Dying to Entertain You'—the Full Story," report for the Whale and Dolphin Conservation Society, Chippenham, UK, April 30, 2001, 44, www.wdcs.org/submissions_bin/orcareport.pdf.
19. Death of Samoa and calf: "Killer whale, calf die during birth," *Gainesville Sun*, March 15, 1992.
20. Death of Orky II: Linda Roach Monroe, "Orky, killer whale that fathered calf at SeaWorld, dies," *Los Angeles Times,* September 27, 1988.
21. Death of Kandu V: "Killer whales collide, SeaWorld attraction dies in accident," Associated Press, August 21, 1989.
22. "Performing whale dies in collision with another," *New York Times,* August 23, 1989.
23. Attack on Annette Eckis: "Killer Whale/Show Girl," *CBS Evening News,* April 20, 1971, footage available at www.youtube.com/watch?v=l9HwRvmfz6o.
24. Attack on Doug Cartlidge: Orca Homepage, "Violent Incidents Between Humans and Killer Whales in Captivity," www.orcahome.de/incidents.htm.
25. Attack on Jill Stratton: Cathleen Decker, "Trainer leaves hospital, isn't angry with whale," *Los Angeles Times,* May 1978.
26. Robert Reinhold, "At SeaWorld, stress tests whale and man," *New York Times,* April 4, 1988.
27. Attack on Joanne Hay: Mike Lee, "SeaWorld San Diego suspends Shamu show," *San Diego Union-Tribune,* February 24, 2010.
28. Attack on Georgia Jones: Ibid.
29. Attack on Bud Krames: Ibid.

30. Attacks on trainers at Marineland Ontario: Enzo di Matto, "Mayhem in Marineland," *NOW,* October 10, 1996.
31. Incident with Mark Beeler: Dayna Lynn Fried and John Wilkens, "Kandu bled to death," *San Diego Union-Tribune*, August 23, 1989.
32. Attack on Jonathan Smith: Erich Hoyt, *The Performing Orca—Why the Show Must Stop* (Bath, England: Whale and Dolphin Conservation Society, 1992).
33. Attack on Joanne Webber: "Ex-trainer suing SeaWorld for neck injury," *San Diego Union-Tribune,* June 15, 1987.
34. Attack on Chris Barlow: "Killer whale injures trainer," *Torrance (CA) Daily Breeze,* October 1, 1987.
35. Biting of Mark McHugh: Ibid.
36. Rash of attacks in San Diego: Newspapers reported on a "white paper" with at least fourteen orca-trainer injuries within five months at SeaWorld, only a few of which were reported to the media. Janny Scott, "Waves of trouble at SeaWorld," *Los Angeles Times,* December 20, 1987.
37. Injury of John Sillick: Hoyt, *Performing Orca;* and Reinhold, "At SeaWorld, stress tests."
38. "SeaWorld Investigation: Secrets Below the Surface Part 2," *10News San Diego* (ABC affiliate), May 17, 2007.
39. Chris Kraul and Greg Johnson, "SeaWorld president fired in aftermath of whale accident," *Los Angeles Times,* December 2, 1987.
40. Smith lawsuit against SeaWorld: Jenifer Warren and Greg Johnson, "Ex-trainer sues SeaWorld over whale injuries," *Los Angeles Times,* March 4, 1988.

18. The Case Against

1. The most notorious primate social-deprivation study was Harry F. Harlow et al., "Total Social Isolation in Monkeys," Department of Psychology Primate Laboratory and Regional Primate Research Center, University of Wisconsin, presented to the National Academy of Sciences USA on April 28, 1965.
2. "Recovery Plan for Southern Resident Killer Whales (*Orcinus orca*)," National Marine Fisheries Service, Northwest Region, Seattle, Washington, 2008.
3. "Response to Erich Hoyt Report Commissioned by the Whale and Dolphin Conservation Society," SeaWorld, October 1992.
4. Press release, American Association of Zoological Parks and Aquariums, Bethesda, MD, July 16, 1993.

19. Free Willy

1. Much of the biographical information on Keiko comes from PBS's *Frontline,* "A Whale of a Business," www.pbs.org/wgbh/pages/frontline/shows/whales/keiko; as well as Kenneth Brower, *Freeing Keiko* (Gotham Books, 2005).
2. The account of the meeting in Mexico City with Oscar Porter was confirmed by Naomi Rose and Kenneth Balcomb. The meeting was also captured on videotape.

21. Kiss of Death

1. Bill Hutchinson, "Dawn Brancheau lived her dream at Orlando SeaWorld before losing her life in Shamu whale attack," *New York Daily News,* February 25, 2010.
2. "SeaWorld fights to up whale population," Associated Press, September 22, 1984.
3. Tim Tucker, "This is one whale of an outdoorsman—SeaWorld trainer likes working with killer whales and hooking largemouth bass," *Palm Beach Post,* July 13, 1989.
4. "Death of SeaWorld whale may be linked to delivery," *Ocala Star-Banner,* September 15, 1994.
5. "SeaWorld welcomes new whale," Associated Press, November 22, 1994.

6. "Killer whale born in Texas," United Press International, November 20, 1994.
7. "Killer whale calf at SeaWorld dies," Associated Press, December 29, 1994.
8. Tim Zimmermann, "Do Orcas at marine parks injure one another?" posted at *Current TV* Web site, September 14, 2010, http://current.com/1b7n74c.

22. Catch and Release

1. Maria Ricapito, "Warner and the Whale—'Free Willy's' whale finds a new home," *Entertainment Weekly,* February 17, 1995.
2. "Keiko's Arrival Date Set for Newport, Ore.," news release, Oregon Coast Aquarium, October 17, 1995.
3. MARMAM archives may be accessed by joining the MARMAM list at https://lists.uvic.ca/mailman/listinfo/marmam.

23. Tilly's Willy

1. "Killer whale at SeaWorld gives birth to a dead calf," *Orlando Sentinel,* February 22, 1996.
2. Paul Spong, "New Orca Deaths at SeaWorld—Father Tilikum's Baby Dies Stillborn, Mother Gudrun's Death Follows Quickly," press release, OrcaLab, Alert Bay, BC, Canada, February 29, 1996.

24. Better Days

1. R. Small and D. DeMaster, "Survival of five species of captive marine mammals," *Marine Mammal Science* 11, no. 2 (1995): 209–26.
2. Douglas S. DeMaster and Jeannie K. Drevenak, "Survivorship patterns in three species of captive cetaceans," *Marine Mammal Science* 4, no. 4 (1988): 297–311.
3. Howard Garrett, "Keiko's Life," Orca Network, www.orcanetwork.org/captivity/keikostory.html.
4. "Keiko's Timeline," Jean-Michel Cousteau's Ocean Futures Society, www.oceanfutures.org/action/marine-mammals/keiko.
5. "Dispute over whale's care stalls plan to 'Free Willy,'" Associated Press, October 7, 1997.
6. More information on PBS's *Frontline* program "A Whale of a Business" is available at www.pbs.org/wgbh/pages/frontline/shows/whales.

25. The Salish Sea

1. Candace Calloway Whiting, "40th anniversary of Whidbey's Penn Cove orca capture," *Seattle Post-Intelligencer* blog, August 7, 2010, http://blog.seattlepi.com/whidbey/2010/08/07/40th-anniversary-of-whidbeys-penn-cove-orca-capture/.

26. The Strange Case of Daniel Dukes

1. Kasatka "Animal Profile," SeaWorld San Diego, revised June 2010.
2. "Killer whales: Other SeaWorld attacks," *Orlando Sentinel,* February 27, 2010.
3. "Man drowns in orca pool at SeaWorld Orlando," *Environmental News Service,* July 7, 1999.
4. Paul Wilborn, "Killer whale image questioned: One expert says SeaWorld's marketing efforts have 'turned them into Mickey Mouse,'" *St. Petersburg Times,* July 9, 1999.
5. "Report of Autopsy: Daniel Patrick Dukes, Case No. MEA 728-99," Office of the Medical Examiner, District Nine, Orlando, FL, July 7, 1999.

27. Transatlantic Ties

1. Howard Garrett, "Keiko's Life," Orca Network, www.orcanetwork.org/captivity/keikostory.html.
2. "Keiko the whale moves one step closer to home," Associated Press, June 10, 1998.

3. Garrett, "Keiko's Life."
4. "Keiko, Cousteau groups merge for education," Newhouse News Service, March 18, 1999.
5. E. C. M. Parsons, Sarah J. Dolman, Andrew J. Wright, Naomi A. Rose, and W. C. G. Burns, "Navy sonar and cetaceans: Just how much does the gun need to smoke before we act?" *Marine Pollution Bulletin* 56 (2008): 1248–57.
6. L. S. Weilgart, "Review: The impacts of anthropogenic pollution noise on cetaceans and implications for management," *Canadian Journal of Zoology* 85 : (2007) 1091–1116.
7. Donald McNeil Jr., "Keiko makes it clear: His 'Free Willy' was just a role," *New York Times,* November 6, 2001.
8. Patrick McMahon, "Killer whale project loses big source of funding," *USA Today,* March 28, 2002.
9. "Whale Expert: 'Best to destroy Keiko'—Human-loving whale from *Free Willy* in trouble," *Aftenposten* of Norway Multimedia, September 3, 2002.

28. Abnormal Activities

1. Shanna McCord, "Killer-whale trainer suffers broken arm in SeaWorld incident," *San Diego Union-Tribune,* August 8, 2002.
2. Linda Roach Monroe, "Orky, killer whale that fathered calf at SeaWorld, dies," *Los Angeles Times,* September 27, 1988.
3. Terry Rodgers, "Deaths of marine mammals hit staffers especially hard," *San Diego Union-Tribune,* April 24, 2005.
4. "News from Keiko's Veterinarian," Free Willy-Keiko Foundation website, November 2002, www.keiko.com/news2.html.
5. Kyuquot "Animal Profile," SeaWorld San Antonio, revised 2010.
6. James Aldridge, "Killer whale dies unexpectedly at SeaWorld San Antonio," *San Antonio Business Journal,* April 12, 2002.
7. "Killer Whale Attacks SeaWorld Trainer During Performance," *WFTV News,* San Antonio, July 27, 2004, www.wftv.com/news/news/killer-whale-attacks-seaworld-trainer-during-perfo/nFCcT/.
8. Christopher Sherman, "Killer whale jolts trainer," *Orlando Sentinel,* April 4, 2005.
9. Rodgers, "Deaths of marine mammals."
10. Patricia Dibsie, "SeaWorld gets a new pair of killer whales," *San Diego Union-Tribune,* March 9, 1999.
11. "SeaWorld San Antonio staff has a whale of a nursing job," *San Antonio Express-News,* October 23, 2005.

29. Ken and Kasatka

1. Terry Rodgers, "Marine park cited after whale attack," *San Diego Union-Tribune,* March 4, 2007.
2. "SeaWorld Trainer in Fair Condition After Killer Whale Attack," CNN, November 30, 2006.
3. Details from the 2006 Kasatka attack on Ken Peters were presented in the original Narrative Summary provided by Cal/OSHA in 2007, which can be viewed at www.orcanetwork.org/captivity/070302seaworldreport.pdf.
4. Terry Rodgers, "State withdraws report critical of SeaWorld—Marine park cited after whale attack," *San Diego Union-Tribune*, March 3, 2007.

30. Tenerife

1. "Killer whale calf is born at SeaWorld," *San Diego Union-Tribune,* May 4, 2002
2. "Killer whale at SeaWorld has new 350-pound bundle of joy," Associated Press, February 12, 2004.

3. "Breathing lessons: Killer whale Taima nudges her newborn calf towards the water's surface," caption, Associated Press wire photo, appearing in the *Lodi (CA) News-Sentinel,* November 9, 2000.
4. Tekoa page, Cetacean Cousins website, http://cetacousin.bplaced.net/orca/profile/tekoa.html.
5. Ines Davis Parrish, "SeaWorld's latest birth provides whale of a time," *Orlando Sentinel,* June 19, 1995.
6. "There's a new whale in town," Associated Press photo caption appearing in *Victoria (TX) Advocate,* February 21, 2001.
7. Tim Zimmermann, "Blood in the water," *Outside Online,* July 15, 2011, www.outsideonline.com/outdoor-adventure/nature/Blood-in-the-water-Keto.html?page=all.
8. Ibid.
9. "Woman survives killer whale ordeal," Associated Press, October 9, 2007.
10. Marine Mammal Inventory Report, www.databases.sun-sentinel.com.
11. "Killer whale born," Associated Press photo caption, *Tuscaloosa News,* September 13, 1993.
12. "SeaWorld announces killer whale Katina gives birth to seven foot long calf," *Environmental News,* September 22, 1996.
13. "Orca at SeaWorld San Antonio dies," Associated Press, June 17, 2008.
14. Wendy Cooke, "Orca Trainer Alexis Martinez, Death at Loro Parque Caused by Known Aggression," www.withoutmethereisnoyou.com, January 24, 2011.
15. Zimmermann, "Blood in the water."
16. Jason Garcia, "SeaWorld Orlando will flip to new owner—After months of talks, Busch Entertainment will be sold to Blackstone Group for up to $2.7 billion," *Orlando Sentinel,* October 7, 2009.
17. "Policy on Care of Animals," Merlin Entertainments, www.merlinentertainments.biz/en/environment/marinemammals.aspx.

32. Wake

1. Annie Hartnett, "SeaWorld Revives 'Dine with Shamu' Program After a Year of Deaths," www.change.org, January 25, 2011.
2. Eric Montenegro, "Ocean embassy abandons plans for dolphin park in San Carlos," *Panama America,* July 18, 2008, www.panama-guide.com/article.php/20080718123656879.
3. Carl Franzen, "SeaWorld's Killer Whale Shows Will Go On," AOL news, February 26, 2010.
4. "SeaWorld killer whale trainer: 'We treated Tilikum differently,'" *Guardian,* February 25, 2010.

33. Battle Stations

1. Anika Myers Palm and Eloísa Ruano González, "SeaWorld trainer tribute: Shamu Believe show resumes with standing ovation," *Orlando Sentinel,* February 27, 2010.
2. Ros Clubb and Georgia Mason, "Captivity effects on wide-ranging carnivores—Animals that roam over a large territory in the wild do not take kindly to being confined," *Nature* 425 (October 2, 2003).
3. Ann Oldenburg, "Bob Barker urges SeaWorld to release sea animals," *USA Weekend,* February 25, 2010.
4. From transcript of Tippi Hedren on CNN's *Larry King Live,* "SeaWorld Trainer Laid to Rest," March 1, 2010.
5. "Matt Damon Speaks Out on SeaWorld Tragedy," *Entertainment Tonight* and *ET Online,* February 25, 2010.
6. Paul Thompson, "Whale that killed its trainer at SeaWorld 'must be stoned to death,' says American religious group," *Mail Online* (UK), March 6, 2010, www.dailymail.co.uk/news/article-1255453/SeaWorld-whale-killed-Dawn-Brancheau-stoned-death.html#ixzz1leroEmzg.

7. James Quinn, "Blackstone chief Schwarzman likens Obama to Hitler over tax rises," *Telegraph* (UK), August 16, 2010.
8. Allegations of SeaWorld's withholding documents came from two former employees-turned-whistle-blowers and other contacts familiar with the investigation. SeaWorld denies all allegations of misconduct or foot-dragging in the Brancheau/OSHA probe.
9. "Autopsy Report for Dawn Brancheau, Case No. ME-2010-000248," Office of the Medical Examiner, District Nine, Orlando, FL, February 25, 2010. Final report released to the public on March 31, 2010, www.autopsyfiles.org/reports/Other/brancheau,%20dawn_report.pdf.

34. Oversight

1. Marine Mammal Poll, released by the Alliance of Marine Mammal Parks and Aquariums, conducted by Harris Interactive, March 2005, www.ammpa.org/_docs/HarrisPollResults.pdf.
2. Wording from sign at SeaWorld taken from photograph by Naomi Rose, Orlando, September 2011.
3. P. F. Olesiuk, G. M. Ellis, and J. K. B. Ford, "Life history and population dynamics of northern resident killer whales (*Orcinus orca*) in British Columbia" (abstract submitted to the 16th Biennial Conference on the Biology of Marine Mammals, San Diego, California, December 2005).
4. J. K. B. Ford, G. M. Ellis, and P. F. Olesiuk, "Linking prey and population dynamics: did food limitation cause recent declines of 'resident' killer whales (*Orcinus orca*) in British Columbia?" Canadian Science Advisory Secretariat Research Document 2005/042 (2005), http://www.dfo-mpo.gc.ca/CSAS/Csas/DocREC/2005/RES2005_042_e.pdf.
5. "What It's Like to Be Banned from SeaWorld's Facebook Page," The Orca Project, September 9, 2010, http://theorcaproject.wordpress.com/2010/09/09/what-its-like-to-be-banned-from-sea worlds-facebook-page/.
6. Stephen Jacob's Orca Homepage is at www.orcahome.de.
7. Diana Reiss and Lori Marino, "Mirror self-recognition in the bottlenose dolphin: A case of cognitive convergence," *Proceedings of the National Academy of Sciences of the United States of America,* 2001, 98, 5937-3942.
8. Coverage of the April 27, 2010, House subcommittee oversight hearing provided by CSPAN, www.c-spanvideo.org/program/293204-1.

35. Citation

1. Tim Zimmermann, "The killer in the pool," *Outside* and *Outside Online,* June 27, 2010, www.outsideonline.com/outdoor-adventure/nature/The-Killer-in-the-Pool.html?page=all.
2. "Killer Whales: Tanks and Tensions," *On Point with Tom Ashbrook,* WBUR-FM, Boston, July 21, 2010, http://onpoint.wbur.org/2010/07/21/killer-whales-tanks-and-tensions.
3. Comment from Mark Simmons posted on "More on Killer Whales—Tim Zimmermann's On Point Post" blog, *On Point with Tom Ashbrook,* WBUR-FM, Boston, July 23, 2010, http://onpoint.wbur.org/2010/07/23/more-on-killer-whales-tim-zimmermanns-on-point-post.
4. Ibid.
5. Howard Garrett, "Alan Grayson bows down to SeaWorld," *Daily Kos,* August 19, 2010, www.dailykos.com/story/2010/08/19/894701/-Alan-Grayson-bows-down-to-SeaWorld.
6. "US Labor Department's OSHA Cites SeaWorld of Florida Following Animal Trainer's Death," News Release Number 10-1124-ATL, US Department of Labor, Washington, DC, August 23, 2010, www.osha.gov/pls/oshaweb/owadisp.show_document?p_table=NEWS _RELEASES&p_id=18207.
7. SeaWorld staff, "SeaWorld Parks & Entertainment Will Contest OSHA Citation," "Sea Inside" blog, SeaWorld Parks & Entertainment, Orlando, FL, August 23, 2010, www.sea worldparksblog.com/seaworld-parks-entertainment-will-contest-osha-citation-0.

8. "SeaWorld Whistle Blower: Trainer Death Caused by Park's Negligence," ABC's *Good Morning America,* August 23, 2010, http://abcnews.go.com/GMA/sea-world-whistleblower-alleges-dangerous-practices-park-trainer/story?id=11459193#.TzBT09Sxu84.
9. Transcript, CNN's *Larry King Live,* "Did SeaWorld Help Cause Death of a Killer Whale Trainer?," August 23, 2010, http://transcripts.cnn.com/TRANSCRIPTS/1008/23/lkl.01.html.

36. Superpod

1. "Killer Whale Dies at SeaWorld, Sumar Was 12 Years Old; Cause of Death to Be Determined," *10News,* ABC San Diego, KGTV, September 7, 2010.
2. "Necropsy Performed on SeaWorld Killer Whale," "AM760 KFMB Talk Radio," San Diego, CA, September 8, 2010, www.760kfmb.com/Global/story.asp?S=13122291; and Michael Stetz, "SeaWorld hopes to learn what caused Sumar's death," *San Diego Union-Tribune*, September 8, 2010, www.utsandiego.com/news/2010/sep/08/seaworlds-hopes-cause-of-sumars-death-will-be-de/?print&page=all.
3. "Sumar, the Killer Whale, Dead at the Age of 12," The Orca Project, September 8, 2010.
4. "SeaWorld's killer whale Kalina dies unexpectedly," *Orlando Sentinel,* October 5, 2010.
5. "Tommy Lee explodes over whale sperm," *TMZ,* December 12, 2010.
6. "How Does SeaWorld Masturbate Their Stud Killer Whales? Rocker Tommy Lee Says 'Cow Vaginas' and He Is Almost Right," The Orca Project, December 9, 2010.
7. Ventre and Jett's paper is available at http://theorcaproject.wordpress.com/2011/01/20/keto-tilikum-express-stress-of-orca-captivity/.
8. Tim Zimmermann, "Blood in the water," *Outside*, July 15, 2011, www.outsideonline.com/outdoor-adventure/nature/Blood-in-the-Water-Keto.html?page=all.
9. Charles Buck, Grace P. Paulino, Daniel J. Medina, C. D. Hsiung, Terry W. Campbell, and Michael T. Walsh, "Isolation of St. Louis encephalitis virus from a killer whale," *Clinical and Diagnostic Virology* 1 (1993): 109–12, www.wdcs-na.org/story_details.php?select=275.
10. Naomi Rose's HSUS white paper on killer whales in captivity is available at www.hsi.org/assets/pdfs/orca_white_paper.pdf.
11. James McWilliams, "How PETA's lawsuit against SeaWorld could end factory farming," *Atlantic,* November 11, 2011.

Epilogue

1. The author attended each day of the OSHA hearing and also relied on media reports, court transcripts, and summaries from Naomi Rose for this section.
2. John Melloy, "Stephen Schwarzman Hosting Private Party for Romney, amid Criticism," *Fast Money Halftime Report,* CNBC, January 17, 2012.
3. Jason Garcia, "SeaWorld raises employees' pay after stellar 2011," *Orlando Sentinel,* January 12, 2012.
4. Judith Lavoie, "Government must protect orca habitat: court," *Victoria* (BC) *Times Colonist,* February 10, 2012.
5. *Tilikum, Katina, Corky, Kasatka, and Ulises, five orcas, by their Next Friends, People for the Ethical Treatment of Animals, Inc., Richard "Ric" O'Barry, Ingrid N. Visser, Ph.D., Howard Garrett, Samantha Berg, and Carol Ray, Case No. 11cv2476 JM(WMC), vs. SeaWorld Parks & Entertainment and SeaWorld, LLC,* Order Granting Motion to Dismiss Defendants, United States District Court, Southern District of California, February 8, 2012, http://pugetsoundblogs.com/waterways/files/2012/02/PETA.pdf.
6. E-mail to the author from Johann Hauksson, spokesman for Icelandic prime minister Jóhanna Sigurðardóttir, February 8, 2012.
7. Voice of the Orcas is at http://sites.google.com/site/voiceoftheorcas/home.

Acknowledgments

It is difficult to write about someone's life, and it is difficult to have one's life written about. I am exceedingly grateful to Naomi Rose of the Humane Society of the United States and to the four former trainers from SeaWorld for sharing their stories with me. Naomi gave her time, knowledge, documents, and sense of humor, and her husband, Chris Parsons, was nothing but kind and supportive. Jeff Ventre also was generous of time and effort, despite his hectic MD's schedule, and his girlfriend, Chica, was hospitable and filled with helpful observations. John Jett, Carol Ray, and Sam Berg were steadfast in their dedication to this project and tremendously helpful in pulling it all together. Carol's husband, Eric Peterson, and Sam's husband, Kevin Meddleton, were also wonderfully supportive.

Special thanks are heartily offered to journalist Tim Zimmermann, and also to Colleen Gorman and John Kielty at The Orca Project; to Kim Ventre and Marty Neese; and in the Salish Sea area, to Howard Garrett and Susan Berta of the Orca Network; Kenneth Balcomb of the Center for Whale Research; Candace Calloway Whiting, blogger for the *Seattle Post Intelligencer;* Suzanne Chisholm and Michael Parfit, producers of *The Whale;* Dave Duffus of the University of Victoria; NMFS researcher Dr. John Durban; and naturalist and boat captain Spencer Damico. Up in the Johnstone Strait/Telegraph Cove area, I offer deepest thanks to Dan and Sandra Kirby of Hidden Cove Lodge, Bill and Donna MacKay of MacKay Whale Watching, Jim and Mary Borrowman of Orcella Expeditions, and Paul Spong and Helena Symonds of OrcaLab.

In California, Wendy Cooke, ably assisted by Vicky Bailey, was essential in helping me track and fact-check the births, transfers, and deaths of various captive orcas. Many fond thanks also go to Lori Marino, Michael Mountain, Courtney Vail, Jared Goodman, Astrid van Ginneken, Ingrid Visser, Suzanne

Allee, Nancy Yates, Susan Millward, Bill Rossiter, Mark Berman, and Tim Ragen and Mike Gosliner of the Marine Mammal Commission.

Other members of the Superpod who kept me updated daily on all things killer whales were Pedro Bicchieri, Todd Bricker, Margaux Dodds of Marine Connection UK, Jeff Friedman, Trina Gadsen, Dean Gomersall of the Pacific Marine Mammal Center, Stefan Jacobs, dolphin activist and author Leah Lemieux, Ric O'Barry, Ester Quintana, Liz Sandeman of Marine Connection UK, Cara Sands, Ella Van Cleave of "Save the Dolphin Smile," Ann-Marie Van Dijk, Jordan Waltz, Gwen Williams, Steve Wise of the Non-Human Rights Project, and Jon Zern.

Nadine Kallen provided a firsthand account of the death of Keltie Byrne at SeaLand; former SeaWorld trainer Kim Ashdown sent me her recollections of working at SeaWorld Orlando; Michelle Dillard and her attorney Maurice Arcadier provided me statements on allegations of improprieties within SeaWorld; Todd Connell and Suzanne Connell provided eyewitness accounts of the last moments of Dawn Brancheau's life; and I am grateful to them all. I also thank author Erich Hoyt and filmmakers Gabriela Cowperthwaite and Manny Oteyza for moral and material support. Special thanks also to everyone at HSUS for providing me with workspace, access to documents, and photocopy services while I was in Gaithersburg, Maryland.

Of course, my amazing agent, Todd Shuster of Zachary Shuster Harmsworth, and my longtime and highly loyal editor, George Witte, of St. Martin's Press, deserve bundles of praise. Special thanks also go to Rachel Ekstrom, Terra Layton, copy editor Steven Boldt, production editor Geraldine Van Dusen, jacket designer Rob Grom, indexer Peter Rooney, and attorney Heather Florence, who has now expertly vetted two of my manuscripts. Nancy Hokkanen provided topflight transcription service, as always, backed up this time by my friend and neighbor Gwen O'Connor, who entered corrections.

I wish to offer extra-special thanks to my dear friend Jay Blotcher, who helped me edit, tighten, and enliven my first draft and is one of the most talented freelance editors that I have had the pleasure to know.

I first began the arduous journey of writing this book at the vacation home of my dear friends Sasha Silverstein and Len Heisler, and on the night I completed the manuscript, I celebrated with them at a delightful Indian restaurant in Brooklyn. I thank them both for their good humor and endless support.

Finally, special thanks to other friends who were especially supportive, including Shimon Attie, Jane Bayer, Scott and Laura Bono, Emily and Chris Collins, David France, Doug Fredman, Steve Goodman, Maria Angelina Jimeno, Bob Lenartz and Patty Glynn Lenartz, David and Margaret Leveson, Lou Pansulla, Laura Perry, Thomas Piel, John Ramos, Gabriel Rotello, Matthew Singer, and Abby Cassell. And of course, thanks always to those closest to me, who helped me the most: my mother and father, Barbara and Leo Kirby, plus Nancy, Michael, and Jenifer Bue, Carlos Arturo Jimeno Gallad, and though it may be frowned upon in certain circles, I offer thanks to my Irish terrier, Wilson, because he was most patient of all with me during this challenging endeavor.

Index